Topics in Applied Physics Volume 16

Topics in Applied Physics Founded by Helmut K. V. Lotsch

Volume 1 **Dye Lasers** Editor: F. P. Schäfer

Volume 2 **Laser Spectroscopy** of Atoms and Molecules
Editor: H. Walther

Volume 3 **Numerical and Asymptotic Techniques in Electromagnetics**
Editor: R. Mittra

Volume 4 **Interactions on Metal Surfaces** Editor: R. Gomer

Volume 5 **Mössbauer Spectroscopy** Editor: U. Gonser

Volume 6 **Picture Processing and Digital Filtering** Editor: T. S. Huang

Volume 7 **Integrated Optics** Editor: T. Tamir

Volume 8 **Light Scattering in Solids** Editor: M. Cardona

Volume 9 **Laser Speckle** and Related Phenomena Editor: J. C. Dainty

Volume 10 **Transient Electromagnetic Fields** Editor: L. B. Felsen

Volume 11 **Digital Picture Analysis** Editor: A. Rosenfeld

Volume 12 **Turbulence** Editor: P. Bradshaw

Volume 13 **High-Resolution Laser Spectroscopy** Editor: K. Shimoda

Volume 14 **Laser Monitoring of the Atmosphere** Editor: E. D. Hinkley

Volume 15 **Radiationless Processes** in Molecules and Condensed Phases
Editor: F. K. Fong

Volume 16 **Nonlinear Infrared Generation** Editor: Y.-R. Shen

Volume 17 **Electroluminescence** Editor: J. I. Pankove

Volume 18 **Ultrashort Light Pulses,** Picosecond Techniques and Applications
Editor: S. L. Shapiro

Nonlinear Infrared Generation

Edited by Y.-R. Shen

With Contributions by

R. L. Aggarwal T. J. Bridges R. L. Byer
T. Y. Chang R. L. Herbst B. Lax V. T. Nguyen
Y.-R. Shen P. P. Sorokin J. J. Wynne

With 134 Figures

Springer-Verlag Berlin Heidelberg New York 1977

Professor *Yuen-Ron Shen*

Department of Physics, University of California, Berkeley
Berkeley, CA 94720, USA

ISBN 3-540-07945-9 Springer-Verlag Berlin Heidelberg New York
ISBN 0-387-07945-9 Springer-Verlag New York Heidelberg Berlin

Library of Congress Cataloging in Publication Data. Main entry under title: Nonlinear infrared generation. (Topics in applied physics; v. 16). Includes bibliographies and index. 1. Infra-red sources. 2. Nonlinear optics. I. Shen, Yuen-Ron, 1935–. II. Aggarwal, Roshan Lal, 1937–. QC457.N68. 535'.012. 76-54647

Monophoto typesetting, offset printing, and bookbinding: Brühlsche Universitätsdruckerei, Giessen.
2153/3130–543210

Preface

The development of infrared radiation sources is of prime importance to the advancement of infrared science. This book is concerned with the progress of the field along a new direction, namely, generation of coherent infrared radiation by nonlinear optical methods. In the past, blackbody radiation had first been the only traditional infrared source. Its incoherent property, its rather weak spectral intensity, and its limited spectral purity left much to be asked for. Then, infrared lasers were invented. The laser beams have all the desired properties except that their frequencies are fixed and hardly tunable. With the advent of high-power lasers, however, one can now utilize various nonlinear optical effects to generate intense coherent infrared beams at new frequencies. The output in many cases is continuously tunable. This is then clearly a milestone in the advance of infrared science.

In recent years, we have witnessed a rapidly growing activity in the development of infrared sources. For instance, as many as six post-deadline papers in the recent *IXth International Quantum Electronics Conference (1976)* appeared to dwell on the subject. We, therefore, believe it is timely for us to present a book devoted solely to the nonlinear optical methods of generating coherent infrared radiation. We have divided the book into six chapters.

Chapter 1 is an introduction. It gives a brief historical survey and some background knowledge in nonlinear optics related to the subject of the book. It also presents a fairly detailed description on infrared generation by stimulated Raman and polariton scattering. A table at the end summarizes the present status of the field.

Chapter 2 describes generation of discretely tunable far-infrared radiation by difference frequency mixing of two CO_2 laser beams in crystals. It consists of a theoretical background for difference-frequency mixing, a discussion on various phase-matching techniques, choice of materials, and different mixing geometries, and a description of the present status of the experiments. It ends up with a section on the possible applications of the generated far-infrared radiation.

Chapter 3 is a survey on infrared generation by parametric oscillation and optical mixing. It begins with an introduction on the theory of nonlinear wave interaction in a medium, and proceeds on a discussion of continuously tunable infrared generation by optical mixing of different laser sources, by stimulated Raman scattering, by four-wave Raman mixing, and by parametric oscillation. It provides an overall view on the possibility of constructing a tunable coherent source covering the entire infrared spectral range down to 25 μm. Several tables in the chapter summarize the present experimental state of art.

Chapter 4 treats the problem of generation of continuously tunable far-infrared radiation by difference-frequency mixing via spin-flip transitions. Theoretical background and experimental achievement are given in great detail.

Chapter 5 discusses generation of tunable infrared radiation by stimulated Raman scattering and four-wave mixing in metal vapors. It shows how metal vapor can become an efficient optical mixer via resonant enhancement, and it gives a detailed account of many interesting experimental findings. Included in the chapter is also a section on tunable UV generation by four-wave mixing in metal vapors.

Chapter 6 is on infrared generation by optical pumping in gases. The technique has been proved to be most effective for creation of population inversion and laser emission. This chapter gives an up-to-date survey on optically pumped laser emissions in gases via electronic transitions, vibration-rotational transitions, and pure-rotational or inversion transitions, respectively.

The editor wants to express his deep appreciation to all contributors for their enthusiastic support. Without their patience and understanding, the completion of the book would be impossible.

Berkeley, Calif.
October, 1976 *Y.-R. Shen*

Contents

Contributors

Aggarwal, Roshan L.
Francis Bitter National Magnet Laboratory and Department of Physics, Massachusetts Institute of Technology, Cambridge, MA 02139, USA

Bridges, Thomas J.
Bell Telephone Laboratories, Inc., Holmdel, NJ 07733, USA

Byer, Robert L.
Microwave Laboratory, W.W. Hansen Laboratories of Physics, Stanford University, Stanford, CA 94305, USA

Chang, Tao Yuan
Bell Telephone Laboratories, Inc., Holmdel, NJ 07733, USA

Herbst, Richard L.
Microwave Laboratory, W.W. Hansen Laboratories of Physics, Stanford University, Stanford, CA 94305, USA

Lax, Benjamin
Francis Bitter National Magnet Laboratory and Department of Physics, Massachusetts Institute of Technology, Cambridge, MA 02139, USA

Nguyen, Van Tran
Bell Telephone Laboratories, Inc., Holmdel, NJ 07733, USA

Shen, Yuen-Ron
Physics Department, University of California, Berkeley, CA 94720, USA

Sorokin, Peter P.
IBM Thomas J. Watson Research Center, P.O. Box 218, Yorktown Heights, NY 10598, USA

Wynne, James J.
IBM Thomas J. Watson Research Center, P.O. Box 218, Yorktown Heights, NY 10598, USA

1. Introduction

Y.-R. Shen

With 2 Figures

This book is devoted to coherent infrared generation by nonlinear optical means, particularly tunable infrared generation in the medium and far-infrared range. From the scientific point of view, the need for novel infrared sources is quite obvious. Although an important field of science, infrared spectroscopy has always lagged behind optical spectroscopy in the visible. The reasons are twofold: a) infrared detectors are not as sensitive, and b) infrared sources are not as intense. Traditionally, black-body radiation has been the only practical infrared source. Yet governed by the Planck distribution, it has rather weak radiative power in the medium- and far-infrared regions. A $1\ \mathrm{cm}^2$ 5000 K black-body radiates a total power of about 3500 W over the 4π solid angle, but its far-infrared content in the spectral range of $50 \pm 1\ \mathrm{cm}^{-1}$ is only 3×10^{-6} $\mathrm{W/cm^2}$-sterad. Weak source intensity coupled with low detector sensitivity has been the major impedance in the development of infrared science in general. Recently, infrared lasers became available. Their intensities are of course more than sufficient for ordinary spectroscopic use, yet their discrete frequencies with essentially zero tunability have left much to be desired. With the help of lasers, however, it has become possible to generate coherent tunable infrared radiation over a wide range via nonlinear processes. These include difference-frequency mixing, parametric oscillation, stimulated Raman and polariton scattering, optically pumped laser emission, and others. Coherent infrared output at an intensity level far exceeding that of any available black-body source has already been achieved with either the cw or pulsed mode of operation. The high peak intensity of the pulsed output now opens up new areas of infrared spectroscopy, namely, transient coherent spectroscopy and nonlinear optical spectroscopy. One can expect that when these sources become readily available, the field of infrared sciences will definitely undergo a revolutionary change.

1.1 Historical Remarks

Infrared generation by nonlinear means is actually not limited to lasers as the pump sources. For many years, electron beam tubes, such as klystron, magnetron, and carcinotron, have been used as submillimeter and far-infrared sources up to $25\ \mathrm{cm}^{-1}$. Nonlinear frequency multiplication can then be used to obtain even higher frequencies. Unfortunately, the intensities of these devices drop off very rapidly with increasing frequency. With the advent of high-intensity lasers, nonlinear wave mixing can now be easily realized over a broad spectrum of electromagnetic radiation. In 1961, *Franken* et al. [1.1] first reported the detection of second-harmonic generation in a quartz crystal. Shortly after, *Bass* et al. [1.2] made the first sum-frequency mixing experiment with two different lasers. *Smith* and *Braslau* [1.3] first observed the difference-frequency generation and then *Bass* et al. [1.4] demonstrated the optical rectification effect. Immediately after the first second-harmonic generation experiment, the importance of phase (or momentum) matching for improvement of conversion efficiency was recognized by *Giordmaine* [1.5] and *Maker* et al. [1.6]. Other factors important for efficient nonlinear mixing such as focusing [1.7, 8] and double refraction [1.9] were also studied. Many other nonlinear optical effects were discovered in the early years. Among them, parametric oscillation was first demonstrated by *Giordmaine* and *Miller* [1.10] and stimulated Raman scattering by *Woodbury* and *Ng* [1.11].

In nonlinear infrared generation, *Zernike* and *Berman* [1.12] first reported detection of far-infrared output from mixing of a large number of modes of a Nd:glass laser pulse in a quartz crystal. Later, *Zernike* [1.13] also observed far-infrared generation by mixing of two CO_2 laser frequencies in InSb. Subsequently, *Yajima* and *Inoue* [1.14] and *Faries* et al. [1.15] used two simultaneously *Q*-switched ruby lasers with different frequencies as the pump sources in optical mixing to generate far-infrared radiation. The Berkeley group [1.15] emphasized the continuous tuning aspect and studied the spectral content of the output. The Bell Lab group [1.16–20] also reported a number of difference-frequency mixing experiments in semiconductors with CO_2 lasers. They studied various phase-matching schemes in far-infrared generations. Meanwhile, parametric oscillators have been developed to cover an extended frequency range from the visible to the near infrared [1.21]. Stimulated electronic Raman scattering in metal vapor was observed by *Sorokin* et al. [1.22] and by *Rokni* and *Yastiv* [1.23]. Stimulated spin-flip Raman lasers tunable in the mid-infrared region were invented by *Patel* and *Shaw* [1.24]. Optically pumped far-infrared molecular lasers were developed by *Chang* and *Bridges* [1.25]. Tunable far-infrared output from stimulated polariton scattering was generated by *Yarborough* et al. [1.26]. Tunable infrared generation by difference-frequency mixing in $LiNbO_3$ was achieved by *Dewey* and *Hocker* using dye lasers [1.27].

More recently, *Nguyen* and *Bridges* [1.28] demonstrated the possibility of using a tunable spin-flip transition to resonantly enhance the far-infrared output in optical mixing. *Matsumoto* and *Yajima* [1.29] and *Yang* et al. [1.30] succeeded in mixing two dye laser beams in nonlinear crystals to generate tunable far-infrared radiation over a wide range. *Sorokin* et al. [1.31] showed that tunable infrared radiation can also be generated in vapor via third-order four-wave mixing. *Lax*, *Aggarwal*, and coworkers [1.32, 33] studied extensively far-infrared generation by difference-frequency mixing of two CO_2 laser beams in semiconductors. They emphasized the importance of discrete but fine tunability of the output. They, as well as *Yang* et al. [1.30], demonstrated the improved efficiency of far-infrared generation by optical mixing with noncollinear phase matching. Using the multiple total-reflection scheme [1.34], *Aggarwal* et al. [1.33] were able to improve the conversion efficiency even further and observe for the first time a detectable cw far-infrared output from difference-frequency mixing. Recently, *Bridges* et al. [1.35] also obtained tunable cw far-infrared output by mixing of a spin-flip Raman laser beam with a CO_2 laser beam in Te. *Thompson* and *Coleman* [1.36] succeeded in using a GaAs far-infrared waveguide as the nonlinear medium for far-infrared generation. Aside from optical mixing, stimulated Raman scattering in atomic vapor [1.37] and in molecular gas [1.38] with a tunable dye laser as the pump source has also been used to generate tunable infrared radiation. Optically pumped high-pressure molecular gas lasers which could be tuned over limited ranges have been demonstrated [1.39]. Using a parametric oscillator for optical mixing, *Byer* and coworkers [1.40] are presently developing a system whose output can be continuously tuned all the way from near UV to infrared around 20 μm.

There exist some other schemes of nonlinear infrared generation. *Auston* et al. [1.41] studied far-infrared generation by optical mixing resulting from optical excitation of absorbing defects or impurities. *Yang* et al. [1.42] and also *Yajima* and *Takeuchi* [1.43] reported the generation of far-infrared short pulses through optical rectification of picosecond optical pulses. *Byer* and *Herbst* [1.44] succeeded in generating tunable infrared radiation by four-wave Raman mixing process in molecular gases. *Granastein* et al. [1.45] observed intense tunable submillimeter radiation from relativistic electrons moving in a spatially varying magnetic field.

The field of nonlinear infrared generation is relatively new. While there exist numerous review articles on special topics such as parametric oscillators (see Chapt. 3), stimulated spin-flip Raman lasers [1.46, 47] and optically pumped gas lasers (see Chapt. 6), the only one on far infrared generation is by *Shen* [1.48].

1.2 Infrared Generation by Optical Mixing

Optical mixing is by far the most frequently employed method for nonlinear infrared generation. The basic theory for optical mixing is well known [1.49]. The field $\boldsymbol{E}(\omega)$ generated by optical mixing is governed by

$$\begin{aligned} &\nabla\cdot[\overleftrightarrow{\varepsilon}(\omega)\cdot\boldsymbol{E}(\omega)+4\pi\boldsymbol{P}^{\mathrm{NL}}(\omega)]=0 \\ &[\nabla x(\nabla x)-\omega^2\varepsilon(\omega)/c^2]\boldsymbol{E}(\omega)=(4\pi\omega^2/c^2)\boldsymbol{P}^{\mathrm{NL}}(\omega) \end{aligned} \tag{1.1}$$

where $\overleftrightarrow{\varepsilon}$ is the linear dielectric constant of the medium and $\boldsymbol{P}^{\mathrm{NL}}(\omega)$ is the nonlinear polarization induced by beating of the pump fields. Consider for example difference-frequency mixing in a crystal. We have in the second-order electric-dipole approximation

$$\boldsymbol{P}^{\mathrm{NL}}(\omega)=\overset{\leftrightarrow}{\chi}^{(2)}(\omega=\omega_1-\omega_2):\boldsymbol{E}(\omega_1)\boldsymbol{E}^*(\omega_2) \tag{1.2}$$

where $\overset{\leftrightarrow}{\chi}^{(2)}$ is the second-order nonlinear susceptibility and $\boldsymbol{E}(\omega_1)$ and $\boldsymbol{E}(\omega_2)$ are the pump fields.

The solution of (1.1) and (1.2) in the diffractionless limit is fairly simple and is to be discussed in some detail in Chapters 2 and 3. For difference-frequency generation of infinite plane waves in a semi-infinite medium along $\hat{z}$, we find the well-known result

$$|\boldsymbol{E}(\omega, z=l)|^2=\left|\frac{2\pi\omega^2}{c^2k_z}P^{\mathrm{NL}}(\omega)\right|^2\left[\frac{\sin(\Delta kl/2)^2}{(\Delta kl/2)}\right]l^2 \tag{1.3}$$

where $\Delta k=(\boldsymbol{k}_1-\boldsymbol{k}_2-\boldsymbol{k})\cdot\hat{z}$ is the phase mismatch in the $\hat{z}$ direction. It shows that the efficiency of difference-frequency generation is a maximum at $\Delta k=0$. In practice, one should also consider the effects of finite beam cross sections, focusing, absorption, double refraction, etc. These will be discussed in Chapters 2 and 3 by adopting the formalism developed in the literature for sum-frequency and second-harmonic generation.

However, if the difference frequency is small, then there are actually important differences between sum- and difference-frequency generation. First, when the wavelength becomes comparable to the beam dimensions, diffraction is no longer negligible. Second, for a crystal slab of thickness l, the approximate phase matching condition $\Delta kl \ll 1$ can be satisfied for difference-frequency wavevectors extending over a broad cone angle. Third, the refractive index of a condensed matter at the infrared is often large (~ 5), and therefore, the boundary effects on far-infrared difference-frequency generation can be very important. This is manifested by the fact that part of the generated radiation may suffer total reflections in the medium and never get out of the medium. Finally, the collection angle of the detector may even be smaller than the cone angle of the emitted radiation. These effects are of course more appreciable for smaller difference frequencies.

We can use a simple model to take into account these long-wavelength effects approximately. Physically, the nonlinear polarization $\boldsymbol{P}^{\mathrm{NL}}(\omega)$ is simply a set of oscillating dipoles in the medium so that (1.1) essentially describes radiation from a dipole antenna array. The general solution of such a problem with appropriate boundary conditions is difficult, but in special cases, approximate solutions can be obtained. For example, assume there is a thin slab of nonlinear medium and the normally incident pump beams induce a $\boldsymbol{P}^{\mathrm{NL}}(\omega)$ which can be approximated by

$$\begin{aligned}\boldsymbol{P}^{\mathrm{NL}}(\boldsymbol{r},\omega)&=\hat{x}P^{\mathrm{NL}}\exp(\mathrm{i}k_s z-\mathrm{i}\omega t) \quad \text{for} \quad (x^2+y^2)\leqq a^2\\ &=0 \quad \text{for} \quad (x^2+y^2)>a^2\end{aligned}$$

where P^{NL} is independent of $\boldsymbol{r}$. If we neglect the boundary effects of the slab, then the usual dipole radiation theory gives us immediately the far-field solution of (1.1)

$$\boldsymbol{E}(\boldsymbol{r},\omega)=(\omega/c)^2 \int_v d^3r'(1-\hat{r}\hat{r})\cdot \boldsymbol{P}^{NL}(\boldsymbol{r}',\omega)e^{ik|\boldsymbol{r}-\boldsymbol{r}'|}/|\boldsymbol{r}-\boldsymbol{r}'|\,. \tag{1.4}$$

By integrating the intensity over the area of a circular detector sitting on the $\hat{z}$-axis, we find the difference-frequency power collected by the detector to be [1.15]

$$W=\frac{\omega^4\varepsilon^{\frac{1}{2}}(\omega)}{c^3}|P^{NL}(\omega)|^2 l^2(\pi a^2)^2 \int_0^{\theta_m} d(\sin\theta)\sin\theta\left(\frac{\sin\alpha}{\alpha}\right)^2\left(\frac{2J_1(\beta)}{\beta}\right)^2 \tag{1.5}$$

where l is the thickness of the slab, θ is the angle between $\hat{r}$ and $\hat{z}$, θ_m is the maximum collection angle of the detector, $\alpha=kl(1-\cos\theta+\Delta k/k)/2$, $\Delta k=k_s-k$, and $\beta=ka\sin\theta$. The Bessel function term $[2J_1(\beta)/\beta]^2$ is a description of diffraction from a circular aperture, while $(\sin\alpha/\alpha)^2$ takes into account the effect of phase mismatch. In the diffractionless limit $ka\ll 1$, the term $[2J_1(\beta)/\beta]^2$ is significant only when $\theta<1/ka$, and the integral asymptotically reduces to $(2/k^2a^2)\sin^2(\Delta kl/2)/(\Delta kl/2)^2$. Then (1.5) reduces to the same form as the one obtained from (1.3) in the plane-wave approximation.

We can of course use a more realistic spatial distribution for $\boldsymbol{P}^{NL}(\boldsymbol{r},\omega)$, e.g., a Gaussian profile as induced by Gaussian pumped beams [1.50]. The integration in (1.4) may be more complicated, but can always be done numerically. The difficulty of the above approach is, however, in the boundary effects. Because of the large refractive indices at long wavelengths, reflection and refraction of the difference-frequency waves at the boundaries can be very important. As a crude approximation for the thin slab case, we can use an average Fabry-Perot transmission factor

$$F=\mathrm{T}/|1-\mathrm{R}e^{i2kl}|^2 \tag{1.6}$$

to take into account the boundary effects, where T and R are the average transmission and reflection coefficients, respectively. The power collected by the detector then becomes WF. This approximation is good when the collection angle θ_m is sufficiently small. The above equations should provide a good order-of-magnitude estimate of the difference-frequency output power.

More rigorously, to take focusing, diffraction, double refraction, boundary effects, etc., all properly into account, we must use the method of Fourier analysis [1.50, 51]. We first decompose the fields and the nonlinear polarization into spatial Fourier components and then solve the wave equation for each Fourier component of the difference-frequency field with the proper boundary conditions. The formalism is quite similar to the one used by *Bjorkholm* [1.7] and *Kleinman* et al. [1.8] for second-harmonic generation, although it is somewhat more complicated in the long-wavelength limit.

Preliminary calculations [1.50] indicate that the results obtained from (1.4–6) are in fact a good approximation if the double-refraction effect can be neglected. The frequency spectrum and the angular distribution of the difference-frequency output and the effects of focusing and double refraction on the output have also been calculated [1.50].

The experimental status of difference-frequency generation in crystals is reviewed and summarized in Chapter 2 for output in the far-infrared region and in Chapter 3 for output in the mid- and near-infrared region. Roughly speaking, tunable output in the range between 400 cm^{-1} and 10000 cm^{-1} has been obtained by using either a dye laser or a parametric oscillator as the tunable pump source. Tunable output in the range between 1 cm^{-1} and 200 cm^{-1} has been obtained with ruby lasers, dye lasers, CO_2 lasers, or spin-flip Raman lasers (see Chapt. 4) as the pump sources. Report on difference-frequency generation between 200 cm^{-1} and 400 cm^{-1} has, however, been rare. As shown in (1.3) and (1.5), the difference-frequency output is proportional to ω^2 in the diffractionless limit and to ω^4 in the limit where diffraction dominates. Therefore, if the pump beam intensities and the nonlinear susceptibility remain unchanged, the efficiency of difference-frequency generation should drop appreciably as the output wavelength increases. The result is that for mid- and far-infrared generation, the power conversion efficiency is usually much less than 1%.

Nonlinear infrared generation can also be achieved with four-wave mixing in a medium with inversion symmetry [1.31, 44]. In this case, three pump beams beat against one another to produce a third-order nonlinear polarization $\boldsymbol{P}^{NL}(\omega)$ at the infrared frequency. The calculation of the output is then exactly the same as the one described earlier. Normally, as a third-order term, the nonlinear polarization is very small. However, if the frequencies are close to sharp resonances, $\boldsymbol{P}^{NL}(\omega)$ can be greatly enhanced through resonant enhancement. This happens for example in gas media. Tunable infrared generation by four-wave optical mixing in atomic vapor is reviewed in Chapter 5 and the same process in hydrogen gas is discussed in Chapter 3. In experiments, long gas cells are often used to increase the interaction length and hence improve the conversion efficiency. However, the improvement drops off quickly with decrease of output frequency as diffraction becomes more important. No far-infrared generation in a gas medium has yet been reported presumably because of the poor conversion efficiency.

1.3 Infrared Parametric Oscillators

The subject of parametric oscillators as coherent tunable infrared sources is reviewed in Chapter 3. Physically, parametric amplification can be considered as the inverse process of sum-frequency generation ($\omega_1+\omega_2=\omega_3$). Energy in the pump field at ω_3 is transferred to the signal and idler fields at ω_1 and ω_2. The amplification is maximum or the oscillation threshold is minimum when the collinear phase matching condition $k_1+k_2=k_3$ is approximately satisfied. This phase-matching condition together with the cavity condition of the parametric oscillator then selects the particular set of frequencies ω_1 and ω_2 appearing at the output. External perturbation such as temperature, crystal orientation, and dc electric field can change the wavevectors and vary the phase-matching condition, and therefore can be used as a means to tune the output frequencies. The theory of parametric oscillators is given in detail in Chapter 3.

Parametric amplification was first observed by *Wang* and *Racette* [1.52] in 1965. Subsequently, parametric oscillation was observed by *Giordmaine* and *Miller* [1.10]. While work in the early days was mainly in constructing a coherent tunable source in the visible and near-infrared, the more recent activity in the field is to extend the operating range of parametric oscillators farther into the infrared. At present, with a pulsed Nd:YAG laser as the pump source, parametric oscillation has been observed down to 10.4 μm in CdSe [1.53] and from 1.22 μm to 8.5 μm in proustite [1.54]. In most cases, pulsed Nd:YAG lasers were used to pump parametric oscillators because of their good mode quality and high peak intensity. The present state of the art on parametric oscillators is more thoroughly described in Chapter 3.

If the parametric amplification in a nonlinear crystal is sufficiently high, then the tunable output resulting from noise amplification in one traversal through the crystal can be very strong. This is known as parametric superfluorescence (see Chapt. 3). Clearly, for such a process to occur, a pulsed laser with a very high peak intensity is usually required to pump the nonlinear crystal. Recently, with a Nd:glass mode-locked laser as the pump source, parametric superfluorescence has been used to generate intense tunable picosecond pulses in the infrared [1.55].

1.4 Infrared Generation by Stimulated Raman and Polariton Scattering

Stimulated Raman scattering (SRS) is another method now being used to construct practical devices that generate coherent tunable infrared radiation. The effect was discovered accidentally in 1962 by *Woodbury* and *Ng* [1.11]. In the subsequent years, it has contributed a great deal to the advance of nonlinear optics (see the review articles by *Bloembergen* [1.56], by *Kaiser* and *Maier* [1.57], and by *Shen* [1.58]).

The theory of stimulated Raman process involving a dispersionless final excitation can be described simply as follows [1.56, 58]. As shown in Fig. 1.1. Raman scattering is a direct two-photon process. We shall consider here only the Stokes process. Following an intuitive physical argument, we can express the rate of Stokes amplification as

$$\frac{d\bar{n}_2}{dz}=\left(\frac{dW_{fi}}{d\omega_2}\varrho_i-\frac{dW_{if}}{d\omega_2}\varrho_f\right)\varepsilon_2^{\frac{1}{2}}/c-\alpha_2\bar{n}_2 \tag{1.7}$$

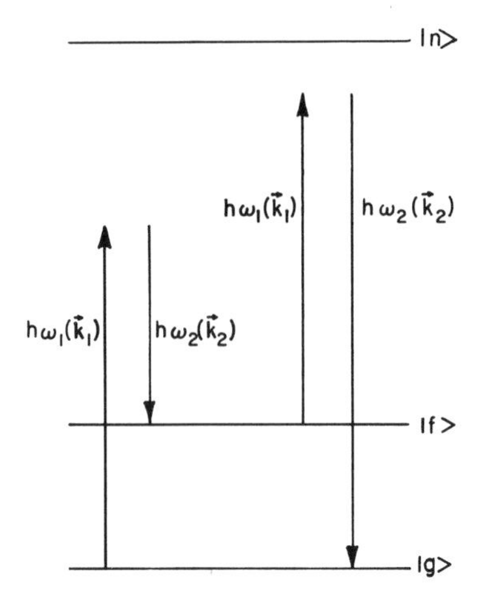

Fig. 1.1. Schematic drawing showing the Stokes ($\omega_1 > \omega_2$) and anti-Stokes ($\omega_1 < \omega_2$) Raman transition between the ground state $|g\rangle$ and the excited state $|f\rangle$

where $\bar{n}_1$ and $\bar{n}_2$ are the average densities of photons in the incoming laser and scattered Stokes modes, respectively, ϱ_i and ϱ_f are the populations in the initial and final states, respectively, and α_2 is the attenuation coefficient at ω_2. The differential transition probability W_{fi} from $|i\rangle$ to $|f\rangle$ can be easily derived from the second-order perturbation calculation. Physically, we expect to find $W_{fi} = (A/\varrho_i)(d\sigma/d\Omega)\bar{n}_1(\bar{n}_2+1)$ and $W_{if} = (A/\varrho_i)(d\sigma/d\Omega)\,\bar{n}_1+1)\bar{n}_2$ where A is a proportional constant and $(d\sigma/d\Omega)$ is the differential spontaneous Raman cross section. If $\bar{n}_1, \bar{n}_2 \gg 1$, we then have from (1.7),

$$\frac{d\bar{n}_2}{dz} = (G-\alpha_2)\bar{n}_2 \tag{1.8}$$

with $G = (A\varepsilon_i^{\frac{1}{2}}/c\varrho_i)(d^2\sigma/d\omega_2 d\Omega)\bar{n}_1(\varrho_i - \varrho_f)$. The solution of (1.8) is $\bar{n}_2(z) = \bar{n}_2(o)\exp[(G-\alpha)z]$ assuming constant $\bar{n}_1$. This shows that if $G > \alpha$, then the Stokes intensity will grow exponentially until depletion of $\bar{n}_1$ or saturation sets in. The gain factor G is directly proportional to the spontaneous Raman cross section and the pump laser intensity.

One can also obtain the above result from the usual semiclassical derivation [1.49, 59]. The wave equation for the field E_2 at ω_2 is

$$\nabla \times (\nabla \times \boldsymbol{E}_2) - \frac{\omega_2^2}{c^2}\varepsilon_2 \boldsymbol{E}_2 = \frac{4\pi\omega_2^2}{c^2}\boldsymbol{P}^{(3)}(\omega_2) \tag{1.9}$$

where the third-order nonlinear polarization $\boldsymbol{P}^{(3)}$ is given by

$$\boldsymbol{P}^{(3)}(\omega_2) = (\chi_{\mathrm{R}2}^{(3)}|E_1|^2 + \chi_2^{(3)}|E_2|^2)E_2\,.$$

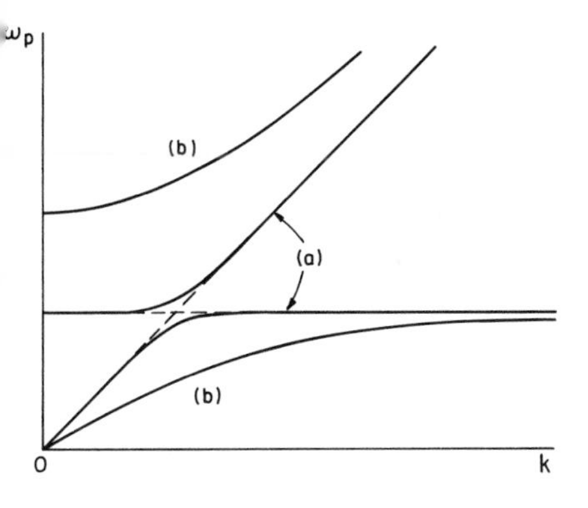

Fig. 1.2a and b. Polariton dispersion curves. (a) Weak coupling and (b) strong coupling between electromagnetic and material excitational waves

The solution of (1.9) is, assuming constant $|E_1|^2$,

$$|E_2|^2(z) = |E_2|^2(\mathrm{o}) \exp[(G_R - \alpha)z] \tag{1.10}$$

where $G_R = -(4\pi\omega_2^2/c^2k_2)(\operatorname{Im}\chi_{R2}^{(3)})|E_1|^2$. From semiclassical derivation for a dispersionless excitation, we can show that $\operatorname{Im}\chi_{R2}^{(3)} \propto (d^2\sigma/d\omega_2 d\Omega)(\varrho_i - \varrho_f)/\varrho_i$, and also $G_R = G$, so that (1.10) is in fact consistent with the solution of (1.8).

More generally, however, SRS should be considered as the result of coupling between light and material excitational waves [1.59, 60]. The pump field E_1 at ω_1 beats with the Stokes field E_2 at ω_2 to excite a material excitational wave ψ at $\omega_3 = \omega_1 - \omega_2$; the latter in turn beats with the pump field to amplify the Stokes field. In general, for an infrared-active excitation, ψ can also interact directly with an infrared field E_3 at ω_3. Direct coupling of ψ and E_3 leads to a new class of excitations known as polaritons [1.61]—mixed em and material excitations. Typical polariton dispersion curves are shown in Fig. 1.2. The Raman process now excites a particular polariton on the dispersion curve as governed by energy and momentum conservation. Stimulated Raman scattering by polariton is known as stimulated polariton scattering (SPS). The output of SPS can be tuned if the excitation is varied along the polariton dispersion curve.

The wave equations for the coupled waves can be written as [1.58, 62].

$$\begin{aligned}
&[\nabla^2 + \omega_2^2\varepsilon_2/c^2]E_2 = -(4\pi\omega_2^2/c^2)P^{\mathrm{NL}}(\omega_2) \\
&[\nabla^2 + \omega_3^2\varepsilon_2/c^2]E_3 = -(4\pi\omega_3^2/c^2)[P^{\mathrm{L}}(\omega_3) + P^{\mathrm{NL}}(\omega_3)] \\
&[\beta\nabla^2 + \hbar(\omega_3 - \omega_0 + i\Gamma)]\psi = f^{\mathrm{L}}(\omega) + f^{\mathrm{NL}}(\omega)
\end{aligned} \tag{1.11}$$

where β, ω_0, and Γ denote, respectively, dispersion, resonant frequency, and damping of the material excitation, $P^{\mathrm{NL}}(\omega_2) = -\partial F/\partial E_2^*$, $P^{\mathrm{L}}(\omega_3) + P^{\mathrm{NL}}(\omega_3) = -\partial F/\partial E_3^*$, and $f^{\mathrm{L}}(\omega) + f^{\mathrm{NL}}(\omega) = -\partial F/\partial[N(\varrho_i - \varrho_f)\psi^*]$, with $F = \{N[AE_3\psi^* + ME_1E_2^*\psi^*](\varrho_i - \varrho_f) + \chi^{(2)}E_1E_2^*E_3^* + \text{complex conj.}\}$ being the average coupling energy. N is the density of molecules or unit cells, and A, M, and $\chi^{(2)}$ are coupling constants. For simplicity, we have neglected, in the polarizations, terms which are independent of the excitation. We have also assumed $|E_1|^2$ to be a constant and neglected the pumping of population from the initial to the final state by SRS. In the language of magnetic resonance, pumping of population is known as longitudinal excitation, while excitation of ψ is known as transverse excitation.

Consider first the special case where E_3 vanishes because $A=0$ and $\chi^{(2)}=0$. Then, if the dispersion of ψ is negligible ($\beta=0$), we can immediately find $P^{\rm NL}(\omega_2)=\chi_{\rm R2}^{(3)}|E_1|^2E_2$ with $\chi_{\rm R2}^{(3)}=N|M|^2(\varrho_i-\varrho_f)/\hbar(\omega_3-\omega_0-{\rm i}\Gamma)$ which is exactly the same expression one would find for Raman susceptibility from a semi-classical derivation. Therefore, this is clearly the case of ordinary SRS.

Consider next SPS with $\beta=0$. We can eliminate ψ in (1.11) and reduce (1.11) to the form

$$\begin{aligned}[][\nabla^2+(\omega_2^2/c^2)(\varepsilon_2)_{\rm eff}]E_2&=-4\pi(\omega_2^2/c^2\chi_{\rm eff}^{(2)*}E_1E_3^*\\ [\nabla^2+(\omega_3^2/c^2)(\varepsilon_3)_{\rm eff}]E_3&=-4\pi(\omega_3^2/c^2)\chi_{\rm eff}^{(2)}E_1E_2^*\end{aligned}\tag{1.12}$$

where

$$\begin{aligned}(\varepsilon_2)_{\rm eff}&=\varepsilon_2+4\pi\chi_{\rm R2}^{(3)}|E_1|^2\\ (\varepsilon_3)_{\rm eff}&=\varepsilon_3-N|A|^2(\varrho_i-\varrho_f)/\hbar(\omega_3-\omega_0+{\rm i}\Gamma)\\ \chi_{\rm eff}^{(2)}&=\chi^{(2)}-NA^*M(\varrho_i-\varrho_f)/\hbar(\omega_3-\omega_0+{\rm i}\Gamma)\,.\end{aligned}\tag{1.13}$$

Note that $k_3=(\omega_3/c)(\varepsilon_3)_{\rm eff}$ governing the dispersion of E_3 is the dispersion relation for polaritons. Equation (1.12) is now in the form identical to those describing parametric amplification (see Chapt. 3) [1.49]. The solution is well known and can be written as

$$\begin{aligned}E_2^*&=[\mathscr{E}_{2+}^*\exp({\rm i}\Delta K_+z)+\mathscr{E}_{2-}^*\exp({\rm i}\Delta K_-z)]\exp(-{\rm i}\boldsymbol{k}_2\cdot\boldsymbol{r})\\ E_3&=[\mathscr{E}_{3+}\exp({\rm i}\Delta K_+z)+\mathscr{E}_{3-}\exp({\rm i}\Delta K_-z)]\exp({\rm i}\boldsymbol{k}_3\cdot\boldsymbol{r}+{\rm i}\Delta kz)\end{aligned}\tag{1.14}$$

where

$$\begin{aligned}k&=(\omega/c)(\varepsilon'_{\rm eff})^{\frac{1}{2}}\\ \Delta k&=k_{1z}-k_{2z}-k_{3z},\qquad k_z=\boldsymbol{k}\cdot\hat{z}\,,\\ \Delta K_\pm&=\tfrac{1}{2}(\gamma_2-\gamma_3)\pm\tfrac{1}{2}[(\gamma_2+\gamma_3)^2-4\Lambda]^{\frac{1}{2}}\\ \gamma_2&=(k_2/2k_{2z})({\rm i}\alpha_2+2k_{\rm R})\\ k_{\rm R}&=(\omega_2^2/2k_{2z}c^2)4\pi\chi_{\rm R2}^{(3)}|E_1|^2\\ \gamma_3&=-\Delta k-{\rm i}(k_3/2k_{3z})\alpha_3\\ \Lambda&=(4\pi^2\omega_2^2\omega_3^2/c^2k_{2z}k_{3z})(\chi_{\rm eff}^{(2)})^2|E_1|^2\\ |\mathscr{E}_3/\mathscr{E}_2|_\pm&=(\omega_3^2k_{2z}/\omega_2^2k_{3z})^{\frac{1}{2}}|\Lambda^{\frac{1}{2}}(\Delta K_\pm+\gamma_3)|\,.\end{aligned}\tag{1.15}$$

Either ΔK_+ or ΔK_- has a negative imaginary part indicating an exponential gain for E_2 and E_3. The gain is maximum at $\Delta K=0$ if the ω_0 resonance is sufficiently narrow.

From what we have discussed, we realize that the output of SRS and SPS can be tuned by 1) using a tunable pump laser, 2) varying the excitation frequency ω_0 with an external perturbation, or 3) in the case of SPS, varying the excited polariton frequency by adjusting the relative directions of beam propagation. All these methods have been used with success to generate tunable infrared radiation. Thus, SRS in vapor with a tunable pulsed dye laser as the pump source can produce tunable output from the visible down to $\sim 670\,\mathrm{cm}^{-1}$ [1.37]. The peak output power can be as high as 60 MW [1.38]. The process is strongly enhanced when the pump laser frequency approaches a resonance. The experimental status of SRS in vapor is briefly reviewed in Chapters 3 and 5.

For a magnetic excitation, ω_0 can be easily adjusted by an external magnetic field. This is the case of spin-flip transitions in solids where $\omega_0 = 2g\mu_B B$, with μ_B being the magneton and B the applied magnetic field. While the spin g factors for most solids are small, the one for InSb is as large as 50 so that if B is varied from 0 to 100 kG, ω_0 can be tuned from 0 to $250\,\mathrm{cm}^{-1}$. It happens that the spin-flip Raman cross section $d\sigma/d\Omega$ in InSb is also unusually large. *Yafet* [1.63] has shown that

$$d\sigma/d\Omega \cong (e^2 g/2m)(\omega_2/\omega_1)\left[E_g\hbar\omega_1/(E_g^2-\hbar^2\omega_1^2)\right]^2 \tag{1.16}$$

where m is the electron mass and E_g is the gap energy. For InSb, $E_g=1900\,\mathrm{cm}^{-1}$, and then at $\omega_1 \sim 1000\,\mathrm{cm}^{-1}$, we find $d\sigma/d\Omega \cong 10^{-23}\,\mathrm{cm}^2/\mathrm{sterad}$. In addition to the large $d\sigma/d\Omega$, spin-flip Raman scattering in InSb also has a very narrow linewidth at low temperatures ($\sim 0.3\,\mathrm{cm}^{-1}$ in an n-type InSb with $n=10^{16}\,\mathrm{cm}^{-3}$). Consequently, InSb has the largest known Raman gain for all materials. With $\omega_1 = 940\,\mathrm{cm}^{-1}$ (10.6 μm), *Patel* and *Shaw* [1.64] found a Raman gain of $G = 1.7\times10^{-5}\,I\,\mathrm{cm}^{-1}$ in an n-type InSb with $n=3\times10^{16}\,\mathrm{cm}^{-3}$, where I is the laser intensity in $\mathrm{W/cm^2}$. One therefore expects than spin-flip SRS in InSb is readily observable. Tunable output from 10.9 to 13.0 μm was first detected by *Patel* and *Shaw* [1.24] using a Q-switched CO_2 laser at 10.6 μm. With an input power of 1 kW focused into an area of $10^{-3}\,\mathrm{cm}^2$ in a 5 mm long InSb crystal ($n\approx10^{16}\,\mathrm{cm}^{-3}$ at 18 K), they obtained a Stokes output of 10 W. The output linewidth was less than $0.03\,\mathrm{cm}^{-1}$. Equation (1.16) shows that if $\hbar\omega_1$ approaches E_g, the Raman cross section can be resonantly enhanced by several orders of magnitude. Then, the spin-flip SRS can even be operated on a cw basis, as has actually been demonstrated by *Brueck* and *Mooradian* [1.65] with a CO laser at 5.3 μm. They found a pump threshold of less than 50 mW, a power conversion efficiency of larger than 50%, and an output power in excess of 1 W. The output linewidth could be less than 1 kHz [1.66]. More recently, using an optically pumped NH_3 laser operating at 12.8 μm, *Patel* et al. [1.67] obtained a tuning range of SRS in InSb from 13.9 to 16.8 μm.

Spin-flip SRS has also been observed in InAs with an HF laser pump [1.68] and in $Hg_{0.77}Cd_{0.23}Te$ with a CO_2 TEA laser [1.69]. The former has a threshold power of 15 W and a conversion efficiency of 20%, and may become an important tunable source in the 3–5 μm range. Anti-Stokes radiation and Stokes radiation up to the 4th order have also been observed in spin-flip SRS [1.65, 66, 70]. They can be used to extend the spectral range of the output. Spin-flip SRS as a tunable infrared source has recently been reviewed by *Patel* [1.46] and by *Colles* and *Pidgeon* [1.47].

Strictly speaking, spin-flip SRS should be treated as a SPS process [1.62] since the spin-flip excitation can also be excited directly be an infrared wave via magnetic-dipole transition, or in other words, the spin-flip excitation is coupled directly with the infrared wave. The coupling is weak and therefore the corresponding dispersion curve shows very little splitting at the intersection as shown by curve (a) of Fig. 1.2. For this case, the polariton is a highly mixed em and material excitational wave only around the knee of the dispersion curve. If the spin-flip SRS operates with $\boldsymbol{k}_3(=\boldsymbol{k}_1-\boldsymbol{k}_2)$ around the knee, then by the theory of SPS we should expect to find output radiation at both ω_2 and $\omega_3(\sim\omega_0)$. Using a CO_2 TEA laser, *Shaw* [1.71] has recently observed simultaneous output at $\omega_3\sim\omega_0$ from spin-flip SRS in InSb. It is of course more efficient to generate the far-infrared radiation at $\omega_3\sim\omega_0$ by pumping the crystal with input beams at both ω_1 and ω_2. This has been done by *Nguyen* and *Bridges* [1.28] and is reviewed in Chapter 4.

In many polar crystals, the infrared phonon modes are strongly coupled to the infrared radiation. They form polaritons with dispersion curves that resemble curve (b) in Fig. 1.2. Wiht a fixed pump laser frequency ω_1, the output of SPS at ω_2 and ω_3 can now be tuned by adjusting the relative angle between $\boldsymbol{k}_1$ and $\boldsymbol{k}_2$ so as to move ω_3 along the polariton dispersion curve to satisfy the phase-matching condition $\boldsymbol{k}_1=\boldsymbol{k}_2+\boldsymbol{k}_3$. The tuning range depends on the shape of the polariton curve. SPS was first observed by *Kurtz* and *Giordmaine* [1.72] and by *Gelbwachs* et al. [1.73] in $LiNbO_3$ on the 248 cm^{-1} mode. In a resonator, 70% of the laser power can be converted into Stokes at ω_2. Tunable far-infrared output at ω_3 on the same polariton mode in $LiNbO_3$ was later observed by *Pantell* and coworkers [1.26, 74]. It was tuned from 50 μm to 700 μm. The peak power could be as high as 100 W.

Intense anti-Stokes and higher-order Raman radiation can also be generated in SRS. They can be used to extend the output tuning range. We will not discuss the higher-order Raman processes here, but simply refer the readers to the literature (see references in [1.58]).

1.5 Other Methods of Nonlinear Infrared Generation

Among other methods of nonlinear infrared generation, optically pumped stimulated emission (OPSE) is most attractive. OPSE can be considered as a special case of SRS. We mentioned earlier in Section 1.4 that the stimulated

Raman gain increases sharply as the pump laser frequency approaches resonance. The resonant enhancement is of course much more significant with narrow but strong resonances. This happens in the case of gas media. In metal vapor, for example, resonant SRS leads to strong tunable infrared output (see Chapt. 5). As the pump frequency gets even closer to the resonant line, however, direct absorption of the pump field becomes important and induces a net population change from the ground state $\langle g|$ to the resonant intermediate state $\langle n|$ (see Fig. 1.1). This is known as optical pumping. Following optical pumping, an inverted population can be established between the intermediate state $\langle n|$ and the final state $\langle f|$ so that laser emission may occur. Actually, in this case of a three-level system, both laser emission and SRS operate simultaneously [1.75].

The above picture describes an optically pumped three-level laser system. Such a scheme has been used in atomic and molecular gases to generate a large number of discrete infrared lines. More generally, optical pumping in a multi-level system with subsequent radiative and non-radiative transfer of population between different levels can establish inverted population for many pairs of states. It can then lead to many more laser lines than the three-level scheme. Thus, for example, in the far-infrared range, several hundreds of OPSE laser lines have already been discovered. Continuous tuning of the laser frequency may become possible with OPSE in high-pressure gas systems where the rotational fine structure is smeared out by pressure broadening [1.39]. While OPSE normally generates only discrete infrared laser frequencies, it has the advantages of being highly efficient and operable in a cw mode at the mW level or in a pulsed mode with peak power at the MW level. Infrared generation by OPSE in gases is discussed in great detail in Chapter 6.

Far-infrared radiation of very long wavelength can also be generated by picosecond laser pulses in a crystal [1.76]. This is because a 1 ps pulse has a spectral bandwidth of about 15 cm^{-1} and mixing of the various spectral components of the pulse in the crystal generates far-infrared radiation. Mixing can be achieved either through the second-order nonlinearity in the crystal [1.42, 43] or through impurity (or defect) absorption [1.41]. In the former case, the far-infrared output is tunable and can have a linewidth of $\sim 1\ \mathrm{cm}^{-1}$ as determined by the phase-matching condition. In the latter case, picosecond excitation of impurities in an acentric polar lattice induces a rapidly varying nonlinear polarization via the pyroelectric effect or/and a change of dipole moment on the impurities. In both cases, with picosecond input pulses, the far-infrared output is often limited to the range between 1 and 30 cm^{-1}. It is however in the form of a short pulse with a pulsewidth in the picosecond range.

Another novel method of generating coherent tunable infrared radiation is by letting relativistic electrons undergo cyclotron motion in a magnetic field. *Granastein* et al. [1.45] observed a narrow-band radiation with its peak at the relativistic cyclotron resonance frequency. The peak power was over 1 MW at 7.8 GHz and was greatly enhanced by periodic perturbation of

Table 1.1. Present status of various methods of nonlinear infrared generation. The output of a black-body radiation is shown for comparison

Technique	Mode of operation	Tuning range (cm^{-1})	Average or peak power obtained	Spectral width (cm^{-1})	Pumped lasers
Difference-frequency mixing in crystals	Pulsed	1–200	0.1 W	<0.1	Ruby/dye or
	Pulsed	600–10000	300 W	<0.1	2 dye lasers
	Pulsed	420–6000	10 W	<0.1	2 parametric oscillators
	Pulsed	5–140 in 3200 steps	10 W	<0.1	2 CO_2 lasers
	cw	5–140 in 3200 steps	0.1 μW	$<3\times10^{-3}$	2 CO_2 lasers
	cw	2380–4550	1 μW	5×10^{-4}	Ar^+/dye
Mixing via spin-flip transition	Pulsed	possibly 10–300	2 μW	<0.1	CO_2/spin-flip
Mixing in metal vapor	Pulsed	400–5000	0.1 W	<0.1	2 dye lasers
Parametric oscillator	Pulsed	900–>10000	~1 MW	0.03	Nd:YAG
Stimulated Raman scattering in vapor	Pulsed	670–10000	60 MW	0.2	Dye
Stimulated spin-flip Raman scattering	Pulsed	590–1100 1540–2000	200 W	0.03	CO_2 or NH_3
	cw	1540–2000	1 W	3×10^{-6}	CO
Stimulated polariton scattering	Pulsed	14–200	100 W	<1	Ruby
1-cm^2 black-body radiation at 1000 K With a 10 mrad angular spread	cw		0.06 $\mu W/cm^{-1}$ at 1000 cm^{-1} 1.2×10^{-3} $\mu W/cm^{-1}$ at 100 cm^{-1} 1.3×10^{-5} $\mu W/cm^{-1}$ at 10 cm^{-1}		

the magnetic field along the field axis. Relativistic electrons in a spatially periodic transverse magnetic field both emit and absorb radiation via bremsstrahlung and inverse bremsstrahlung, respectively. The wavelength for emission is however slightly longer than the wavelength for absorption. This then makes stimulated emission of bremsstrahlung possible [1.77]. Using the relativistic electrons from the superconducting linear accelerator at Stanford, *Elias* et al. [1.78] recently reported the observation of a stimulated emission gain of 7% per pass at 10.6 μm. The physics of stimulated emission of bremsstrahlung is quite similar to that of stimulated Compton scattering proposed earlier by *Pantell* et al. [1.79]. However, no experiment on stimulated Compton scattering has yet been reported. Intense microwave emission has also been observed in coherent Cherenkov radiation from a relativistic electron beam interacting with a slow-wave structure [1.80]. The peak power was as high as 500 MW with a 17% conversion efficiency. In principle, the scheme can also be extended to radiation at much shorter wavelengths.

1.6 Summary

We summarize in Table 1.1 the experimental state of art of the various different methods of nonlinear infrared generation. Radiation from a black body at 1000 K is also listed as a comparison. We note that even at this early stage of development, nonlinear infrared generation has already led to tunable infrared sources with both higher peak power and higher average power than the black-body radiation.

Acknowledgements. This work is supported by the U.S. Energy Research and Development Administration. The author is indebted to the Miller Institute of the University of California for a research professorship.

References

1.1 P.A.Franken, A.E.Hill, C.W.Peters, G.Weinreich: Phys. Rev. Lett. **7**, 118 (1961)
1.2 M.Bass, P.A.Franken, A.E.Hill, C.W.Peters, G.Weinreich: Phys. Rev. Lett. **8**, 18 (1962)
1.3 A.W.Smith, N.Braslau: IBM J. Res. **6**, 361 (1962)
1.4 M.Bass, P.A.Franken, J.F.Ward, G.Weinreich: Phys. Rev. Lett. **9**, 446 (1962)
1.5 J.A.Giordmaine: Phys. Rev. Lett. **8**, 19 (1962)
1.6 P.D.Maker, R.W.Terhune, N.Nisenoff, C.M.Savage: Phys. Rev. Lett. **8**, 21 (1962)
1.7 J.E.Bjorkholm: Phys. Rev. **142**, 126 (1966)
1.8 D.A.Kleinman, A.Ashkin, G.D.Boyd: Phys. Rev. **145**, 338 (1966)
1.9 G.D.Boyd, A.Ashkin, J.M.Dziedzic, D.A.Kleinman: Phys. Rev. A **137**, 1305 (1965)
1.10 J.A.Giordmaine, R.C.Miller: Phys. Rev. Lett. **14**, 973 (1965)
1.11 E.J.Woodbury, W.K.Ng: Proc. IRE **50**, 2347 (1962)
1.12 F.Zernike, P.R.Berman: Phys. Rev. Lett. **15**, 999 (1965)
1.13 F.Zernike: Bull. Am. Phys. Soc. **12**, 687 (1967); Phys. Rev. Lett. **22**, 931 (1969)
1.14 T.Yajima, K.Inoue: IEEE J. QE-**4**, 319 (1968); QE-**5**, 140 (1969); Phys. Lett. A **26**, 281 (1968)
1.15 D.W.Faries, K.A.Gehring, P.L.Richards, Y.R.Shen: Phys. Rev. **180**, 363 (1969); D.W.Faries, P.L.Richards, Y.R.Shen, K.H.Yang: Phys. Rev. A **3**, 2148 (1971)

1.16 T.Y.Chang, V.T.Nguyen, C.K.N.Patel: Appl. Phys. Lett. **13**, 357 (1968); T.J.Bridges, T.Y.Chang: Phys. Rev. Lett. **22**, 811 (1969)
1.17 C.K.N.Patel, V.T.Nguyen: Appl. Phys. Lett. **15**, 189 (1969); V.T.Nguyen, C.K.N.Patel: Phys. Rev. Lett. **22**, 463 (1969)
1.18 V.T.Nguyen, A.R.Strnad, A.M.Jean-Louis, G.Duroffoug: in *The Physics of Semimetals and Narrow Gap Semiconductors*, ed. by D.L.Carter, R.T.Bates (Pergamon Press, New York 1971) p. 231
1.19 T.J.Bridges, A.R.Strnad: Appl. Phys. Lett. **20**, 382 (1972)
1.20 G.D.Boyd, T.J.Bridges, C.K.N.Patel, E.Buchler: Appl. Phys. Lett. **21**, 553 (1972)
1.21 See the review articles by R.G.Smith: In "Laser Handbook", ed. by F.T.Arecchi, E.O. Schulz-Dubois (North Holland, Amsterdam 1972), p. 837; R.L.Byer: In "Treatise in Quantum Electronics", ed. by H.Rabin, C.L.Tang (Academic Press, New York 1975); see also Chapter 3
1.22 P.O.Sorokin, N.S.Shiren, J.R.Lankard, E.C.Hammond, T.G.Kazyaka: Appl. Phys. Lett. **10**, 44 (1967)
1.23 M.Rokni, S.Yatsiv: Phys. Lett. A **24**, 277 (1967); IEEE J. QE-**3**, 329 (1967)
1.24 C.K.N.Patel, E.D.Shaw: Phys. Rev. Lett. **24**, 451 (1970)
1.25 T.Y.Chang, T.J.Bridges: Opt. Commun. **1**, 423 (1970)
1.26 J.M.Yarborough, S.S.Sussman, H.E.Puthoff, R.H.Pantell, B.C.Johnson: Appl. Phys. Lett. **15**, 102 (1969)
1.27 C.F.Dewey, L.O.Hocker: Appl. Phys. Lett. **18**, 58 (1971)
1.28 V.T.Nguyen, T.J.Bridges: Phys. Rev. Lett. **29**, 359 (1972)
1.29 N.Matsumoto, T.Yajima: Japan J. Appl. Phys. **12**, 90 (1973)
1.30 K.H.Yang, J.R.Morris, P.L.Richards, Y.R.Shen: Appl. Phys. Lett. **23**, 669 (1973)
1.31 P.P.Sorokin, J.J.Wynne, J.R.Lankard: Appl. Phys. Lett. **22**, 342 (1973)
1.32 R.L.Aggarwal, B.Lax, G.Favrot: Appl. Phys. Lett. **22**, 329 (1973); B.Lax, R.L.Aggarwal, G.Favrot: Appl. Phys. Lett. **23**, 679 (1973)
1.33 R.L.Aggarwal, B.Lax, H.R.Fetterman, P.E.Tannenwald, B.J.Clifton: J. Appl. Phys. **45**, 3972 (1974); N.Lee, R.L.Aggarwal, B.Lax: Opt. Commun. **11**, 339 (1974)
1.34 J.A.Armstrong, N.Bloembergen, J.Ducuing, P.S.Pershan: Phys. Rev. **127**, 198 (1962)
1.35 T.J.Bridges, V.T.Nguyen, E.G.Burkhardt, C.K.N.Patel: Appl. Phys. Lett. **27**, 600 (1975)
1.36 D.E.Thompson, P.D.Coleman: IEEE Trans. MTT-**22**, 995 (1974)
1.37 J.L.Carlsten, T.J.McIlrath: J. Phys. B **6**, L 80 (1973); J.L.Carlsten, P.C.Dunn: Opt. Commun. **14**, 8 (1975); D.Cotter, D.C.Hanna, P.A.Karkkainen, R.Wyatt: Opt. Commun. **15**, 143 (1975); D.Cotter, D.C.Hanna, R.Wyatt: Opt. Commun. **16**, 256 (1976)
1.38 W.Schmidt, W.Appt: Naturforsch **27**a, 1373 (1972); Abstracts of 8th Intern. Quantum Electronic Conf., San Francisco (1974), post-deadline paper R 6; R.Frey, F.Pradere: Opt. Commun. **12**, 98 (1974)
1.39 T.Y.Chang, O.R.Wood: Appl. Phys. Lett. **23**, 370 (1973); **24**, 182 (1974)
1.40 R.L.Byer, R.L.Herbst, R.N.Fleming: In Proc. 2nd Intern. Conf. Laser Spectroscopy, ed. by S.Haroche, J.C.Pebay-Peyroula, T.W.Hansch, S.E.Harris (Springer, Berlin, Heidelberg, New York 1975) p. 207
1.41 D.H.Auston, A.M.Glass, A.A.Ballman: Phys. Rev. Lett. **28**, 897 (1972); D.H.Auston, A.M.Glass: Appl. Phys. Lett. **20**, 398 (1972); A.M.Glass, D.H.Auston: Opt. Commun. **5**, 45 (1972); D.H.Auston, A.M.Glass, P.LeFur: Appl. Phys. Lett. **23**, 47 (1973)
1.42 K.H.Yang, P.L.Richards, Y.R.Shen: Appl. Phys. Lett. **19**, 385 (1971)
1.43 T.Yajima, N.Takeuchi: Japan J. Appl. Phys. **10**, 907 (1971)
1.44 R.L.Byer, R.L.Herbst: Chapter 3 of this book
1.45 V.L.Granastein, M.Herndon, R.K.Parker, S.P.Schlesinger: IEEE J. QE-**10**, 651 (1974); IEEE Trans. MTT-**22**, 1000 (1974)

1.46 C.K.N.Patel: In *Fundamental and Applied Laser Physics*, ed. by M.S.Feld, A.Javan, N.A. Kurnit (J. Wiley, New York 1972) p. 689; In *Laser Spectroscopy*, ed .by R.G.Brewer, A.Mooradian (Plenum Press, New York 1974) p. 471
1.47 M.J.Colles, C.R.Pidgeon: Repts. Progr. in Phys. **38**, 387 (1975)
1.48 Y.R.Shen: Progr. Quantum Electron. **4**, 207 (1976)
1.49 See, for example, N.Bloembergen: *Nonlinear Optics* (Benjamin, New York 1965)
1.50 J.R.Morris, Y.R.Shen: (to be published)
1.51 D.W.Faries: Ph. D. Thesis, Univ. of California at Berkeley (1970)
1.52 C.C.Wang, G.W.Racette: Appl. Phys. Lett. **6**, 169 (1965)
1.53 R.L.Herbst, R.L.Byer: Appl. Phys. Lett. **21**, 189 (1972)
1.54 D.C.Hanna, B.Luther-Davis, R.C.Smith: Appl. Phys. Lett. **22**, 440 (1974)
1.55 A.Laubereau, L.Greiter, W.Kraiser: Appl. Phys. Lett. **25**, 87 (1974); T.Kushida, Y.Tanaka, M.Ojima, Y.Nakazaki: Japan J. Appl. Phys. **14**, 1097 (1975)
1.56 N.Bloembergen: Am. J. Phys. **35**, 989 (1967)
1.57 W.Kaiser, M.Maier: In *Laser Handbook* ed. by F.T.Arecchi, E.O.Schulz-Dubois (North-Holland, Amsterdam 1972), p. 1077
1.58 Y.R.Shen: In *Light Scattering in Solids*, Topics in Applied Physics, Vol. 8, ed. by M.Cardona (Springer, Berlin, Heidelberg, New York 1975) p. 275
1.59 Y.R.Shen, N.Bloembergen: Phys. Rev. **137**, A 1786 (1965)
1.60 Y.R.Shen: Phys. Rev. **138**, A 1741 (1965)
1.61 K.Huang: Nature **167**, 779 (1951); Proc. Roy. Soc. (London) A **208**, 352 (1951); J.J.Hopfield: Phys. Rev. **112**, 1555 (1958)
1.62 Y.R.Shen: Appl. Phys. Lett. **26**, 516 (1973)
1.63 Y.Yafet: Phys. Rev. **152**, 858 (1966)
1.64 C.K.N.Patel, E.D.Shaw: Phys. Rev. B **3**, 1279 (1971)
1.65 S.J.Brueck, A.Mooradian: Appl. Phys. Lett. **18**, 229 (1971)
1.66 C.K.N.Patel: Phys. Rev. Lett. **28**, 649 (1972)
1.67 C.K.N.Patel, T.Y.Chang, V.T.Nguyen: Appl. Phys. Lett. **28**, 603 (1976)
1.68 R.S.Eng, A.Mooradian, H.R.Fetterman: Appl. Phys. Lett. **25**, 453 (1974)
1.69 J.R.Sattler, B.A.Weber, J.Nemarich: Appl. Phys. Lett. **25**, 451 (1974)
1.70 A.Mooradian, S.R.J.Brueck, F.A.Blum: Appl. Phys. Lett. **17**, 481 (1970); E.D.Shaw, C.K.N.Patel: Appl. Phys. Lett. **18**, 215 (1971) C.S.DeSilets, C.K.N.Patel: Appl. Phys. Lett. **22**, 543 (1973)
1.71 E.Shaw: Bull. Am. Phys. Soc. **21**, 224 (1976)
1.72 S.K.Kurtz, J.A.Giordmaine: Phys. Rev. Lett. **22**, 192 (1969)
1.73 J.Gelbwachs, R.H.Pantell, H.E.Puthoff, J.M.Yarborough: Appl. Phys. Lett. **14**, 258 (1969)
1.74 M.A.Piestrup, R.N.Fleming, R.H.Pantell: Appl. Phys. Lett. **26**, 418 (1975)
1.75 A.Javan: Phys. Rev. **107**, 1579 (1975); K.Shimoda, T.Shimizu: Progr. Quantum Electron. **2**, 61 (1972); Y.R.Shen: Phys. Rev. B **9**, 622 (1973)
1.76 J.R.Morris, Y.R.Shen: Opt. Commun **3**, 81 (1971)
1.77 M.J.Madey: J. Appl. Phys. **42**, 1906 (1971); M.J.Madey, H.A.Schwettman, W.M.Fairbank, IEEE Trans. NS-**20**, 980 (1973)
1.78 L.R.Elias, W.M.Fairbank, J.M.J.Madey, H.A.Schwettman, T.I.Smith: Phys. Rev. Lett. **36**, 717 (1976)
1.79 R.H.Pantell, G.Soncini, H.E.Puthoff: IEEE J. QE-**4**, 905 (1968)
1.80 Y.Carmel, J.Ivers, R.E.Kribel, J.Nation: Phys. Rev. Lett. **33**, 1278 (1974)

2. Optical Mixing of CO_2 Lasers in the Far-Infrared

R. L. Aggarwal and B. Lax

With 36 Figures

2.1 Historical Background

Development of lasers in the visible and near infrared regions of the electromagnetic spectrum in the early nineteen sixties prompted several studies into the use of nonlinear optical effects in crystals for the generation of coherent radiation at new frequencies. *Franken* and coworkers [2.1] at the University of Michigan in 1961 were the first to have observed experimentally second-harmonic generation (SHG) in the ultraviolet at 347.15 nm by focussing a 3 kW pulse of 694.3 nm red ruby laser light onto a quartz crystal. This work spurred considerable experimental and theoretical interest in nonlinear optics [2.2].

The generation of far-infrared (FIR) radiation by difference-frequency mixing of laser radiation in a nonlinear crystal was first reported by *Zernike* and *Berman* [2.3] in 1965. Again quartz was used as the nonlinear mixer. FIR radiation around 100 μm ($100\,cm^{-1}$) was generated by irradiating the crystal with a high-power pulsed Nd:glass laser emitting several frequencies between 1.059 μm and 1.075 μm. In this three-wave mixing process phase matching between the FIR frequency and the two incident laser frequencies was achieved through the use of birefringence of the mixing crystal as first demonstrated by *Giordmaine* [2.4] and *Maker* et al. [2.5], independently in 1962 in their SHG experiments.

An important milestone for FIR generation by difference-frequency mixing was the development of CO_2 gas laser at the Bell Telephone Laboratories by *Patel* [2.6] in 1964. A CO_2 laser can be tuned by an intracavity grating or by some other device to more than 100 vibration-rotational transitions in the wavelength region from ~9.1 μm to 11.0 μm. When radiation from two such lasers is mixed, FIR radiation at more than 5000 difference frequencies can be generated. These difference frequencies cover the FIR region from ~70 μm to several mm with an average spacing of approximately $0.04\,cm^{-1}$ between neighboring lines. Therefore, difference-frequency mixing of two CO_2 lasers can be used to construct a step-tunable or quasi-tunable source of radiation to cover a broad spectral range in the FIR.

In 1967 *Zernike* [2.7] reported the mixing of two *Q*-switched CO_2 laser beams at 9.6 μm and 10.6 μm for FIR generation around 100 μm. Since quartz is opaque to the CO_2 laser radiation, mixing was done in a crystal of InSb cooled to 77 K. Being a cubic crystal, InSb lacks birefringence and therefore birefringent phase matching could not be employed. In 1969 *Zernike* [2.8] achieved phase matching in InSb through the use of temperature dependence of the anomalous dispersion, between FIR and the 10 μm region of the CO_2 laser, arising from the Restrahlen band around 55 μm [2.9].

In other experiments several new phase-matching schemes have been reported. In 1969 *Van Nguyen* and *Patel* [2.10], in FIR generation around 100 μm by CO_2 laser mixing in InSb, achieved phase matching by using the free carrier contribution to the refractive index in a magnetic field so as to counteract the anomalous dispersion of the crystal. *Boyd* and *Patel* [2.11] obtained phase matching by using the phase change upon total internal reflection between the FIR wave and the nonlinear polarization at the difference frequency.

A common feature of the phase-matching schemes cited above is that they use collinear mixing configuration in which the incident laser radiation and the generated FIR radiation propagate in the same direction. In 1969 *Zernike* [2.8] pointed out that it is possible to achieve phase-matched difference-frequency mixing in a crystal which possesses anomalous dispersion if one uses noncollinear mixing geometry. With *Q*-switched CO_2 laser beams propagating at an angle of approximately 2° inside an InSb crystal, ~1 μW peak power at 94 μm (106 cm^{-1}) was obtained. The use of the noncollinear mixing geometry is important since it provides a phase-matching mechanism applicable to any crystal which possesses anomalous dispersion.

Another important milestone for the FIR generation by difference-frequency mixing has been the development of transversely excited atmospheric (TEA) pressure pulsed CO_2 lasers in Canada by *Beaulieu* [2.12] in 1970. High peak intensities in the 1–10 MW/cm^2 range are readily available from relatively small and inexpensive TEA CO_2 lasers. In 1973 *Aggarwal, Lax,* and *Favrot* [2.13, 14] reported the use of grating tuned TEA CO_2 lasers and noncollinear phase-matched mixing in GaAs for generating a large number of FIR frequencies from ~70 μm to 2 mm (140 cm^{-1} to 5 cm^{-1}). Following this experiment, *Aggarwal, Lax* and coworkers [2.15] reported a similar experiment for continuous-wave (cw) FIR generation. Using a "folded" non-collinear mixing geometry proposed by *Lee* et al. [2.16], they reported cw FIR power output of 0.05 μW at 43.4 cm^{-1} with input power of ~25 W in each CO_2 laser beam.

Thompson and *Coleman* [2.17] have reported still another approach to phase matching for FIR generation in isotropic crystals. By making a GaAs crystal in the form of a planar dielectric waveguide they achieved collinear phase matching for the mixing of CO_2 lasers, with the waveguide dispersion cancelling the dispersion of bulk GaAs. Using these GaAs waveguides, they have reported obtaining pulsed FIR signals in the 100–1000 μm range with peak powers in the mW range.

There have been a number of other experiments reported on the generation of FIR by mixing radiation from a variety of other laser sources. *Matsumoto* and *Yajima* [2.18] in 1973 were the first to use difference-frequency mixing of dye lasers to generate FIR radiation in the wavelength region above 100 μm. With the self-beating of a dye laser in crystals of ZnTe and ZnSe, they reported the observation of pulsed FIR in the range 300 μm to 2 mm. *Yang* et al. [2.19] have reported the use of a dual frequency dye laser system for the generation of continuously tunable FIR radiation over the spectral range from 20 cm^{-1} (500 μm) to 190 cm^{-1} (52.5 μm) with maximum peak power in the 10–100 mW range and FIR bandwidth of $\sim 3\ cm^{-1}$. *Bridges* and *Nguyen* [2.20] have mixed the output of a magnetically tuned spin-flip Raman laser with the pumped CO_2 laser to obtain FIR continuously tunable from 90 cm^{-1} to 110 cm^{-1} with peak power levels in the μW range and FIR linewidth of $\sim 0.1\ cm^{-1}$.

In addition to FIR generation by difference-frequency mixing, there have been a number of other significant advances made in recent years toward the development of intense sources of tunable or quasi-tunable FIR radiation. *Piestrup* et al. [2.21] have reported the generation of continuously tunable pulsed FIR radiation in a $LiNbO_3$ crystal with peak power in the kilowatt range from 20 cm^{-1} to 60 cm^{-1}. FIR was produced by stimulated polariton scattering of Nd:YAG laser providing 1.7 MW peak power in a 20-ns pulse. More importantly, optically pumped FIR molecular gas lasers first developed by *Chang* [2.22] in 1970 now provide cw output power at the multimilliwatt level at more than 300 discrete frequencies in the FIR region.

Development of these highly monochromatic and high-power tunable FIR radiation sources is expected to attract many more spectroscopists to this heretofore "energy-starved" region of the electromagnetic spectrum. Perhaps one of the most important applications of the high-power FIR sources is in the field of thermonuclear fusion research. FIR sources appear to be best suited for plasma diagnostics such as the determination of ion temperature by Thomson scattering [2.23] and the investigation of a host of other parameters for plasmas in Tokamak-type fusion machines. *Lax* and *Cohn* [2.24] have suggested some very interesting experiments on the breakdown and heating of gases at low pressures under cyclotron resonance conditions. Also similar studies of impurity levels and interband effects in semiconductors were suggested by *Lax* and *Cohn* [2.25].

2.2 Nonlinear Difference-Frequency Mixing

2.2.1 Theoretical Background

When a dielectric medium is subjected to an applied electric field $\boldsymbol{E}$, it tends to become polarized due to the distortion of its internal charge distribution under the influence of $\boldsymbol{E}$. The resultant electric dipole moment per unit volume is called the electric polarization $\boldsymbol{P}$. In general, $\boldsymbol{P}$ is related to $\boldsymbol{E}$ through a field-dependent susceptibility tensor $\underset{\sim}{\chi}(\boldsymbol{E})$. In MKS units this relationship is

$$\boldsymbol{P} = \varepsilon_0 \underset{\sim}{\chi}(\boldsymbol{E}) \cdot \boldsymbol{E} \tag{2.1}$$

where $\varepsilon_0 = 8.85 \times 10^{-12}\ \mathrm{C^2\ s^2/kgm^3}$ is the permittivity of free space. Expanding $\underset{\sim}{\chi}(\boldsymbol{E})$ as a power series in $\boldsymbol{E}$, (2.1) may be rewritten as

$$\boldsymbol{P} = \varepsilon_0 [\underset{\sim}{\chi}^{(1)} \cdot \boldsymbol{E} + \underset{\approx}{\chi}^{(2)} : \boldsymbol{EE} + \underset{\approx}{\chi}^{(3)} \vdots\, \boldsymbol{EEE} + \ldots] \tag{2.2}$$

where $\underset{\sim}{\chi}^{(1)}$ is the usual linear susceptibility tensor and $\underset{\approx}{\chi}^{(2)}$, $\underset{\approx}{\chi}^{(3)}$, etc., are respectively the quadratic, cubic, etc., nonlinear susceptibility tensors. The latter are generally much smaller than $\underset{\sim}{\chi}^{(1)}$ and hence contribute noticeably only at high-amplitude fields.

Similarly, when an electromagnetic wave is incident upon a medium, $\boldsymbol{P}$ will vary with the electric field component of the wave in accordance with (2.2). The large variety of diverse optical phenomena is due to the first three terms in series expansion of (2.2). $\underset{\sim}{\chi}^{(1)}$ is responsible for the linear optical properties such as reflection and refraction. The quadratic nonlinear susceptibility $\underset{\approx}{\chi}^{(2)}$ gives rise to nonlinear phenomena of second-harmonic generation [2.1] as well as sum [2.26] and difference-frequency mixing, optical rectification [2.27], linear electro-optic effect [2.28], and parametric generation [2.29]. The cubic nonlinear susceptibility $\underset{\approx}{\chi}^{(3)}$ is responsible for third-harmonic generation [2.30], quadratic electro-optic effect [2.28], two photon absorption [2.31], and stimulated *Raman* [2.32], *Brillouin* [2.33], and *Rayleigh* scattering [2.34].

The electric fields associated with ordinary or conventional monochromatic light sources are far too small for the observation of nonlinear phenomena. That is why the experimental discovery of many nonlinear optical effects had to wait for the development of powerful lasers so that sufficient brute force could be brought to bear at optical frequencies. The peak amplitude E_0 of the electric field component associated with an optical wave of intensity I in a medium of refractive index n is given by

$$E_0 = \left(\frac{2}{\varepsilon_0 nc} I\right)^{1/2} = 27.4 \left(\frac{I}{n}\right)^{1/2} \tag{2.3}$$

where c is the velocity of light, and n is related to $\chi^{(1)}$ as

$$n^2 = \chi^{(1)} + 1\,. \tag{2.4}$$

Very large values of E_0 are readily attained even with modest lasers. For instance, one watt of laser power focussed into a 100 μm spot yields an intensity of about $10^8\ \mathrm{W/m^2}$ which corresponds to a peak amplitude of about 2×10^5 V/m in a medium with $n \approx 1.5$. With present technology, laser pulses with peak power in the 10 to 100 MW range can be easily obtained from Q-switched or pulsed lasers. When focussed, these high-power lasers can provide electric fields of about 10^9 V/m which is more than enough to cause the breakdown of air ($\sim 3 \times 10^8$ V/m). Still orders of magnitude higher electric fields can be obtained from large mode-locked lasers which can deliver pulses as short as 10^{-10} s in duration and energy as high as 100 J per pulse.

Since in this chapter we are interested in the difference-frequency of two laser sources for the generation of radiation in the far-infrared region, we shell consider the case of two light waves of angular frequencies ω_1 and ω_2 being incident upon a crystalline medium. The induced polarization at the sum or difference frequencies $\omega_3=\omega_1 \pm \omega_2$ arises from the second-order nonlinear susceptibility $\chi^{(2)}$. In Cartesian coordinates, the i-th component of this polarization is

$$P_i(\omega_3=\omega_1 \pm \omega_2)=g\varepsilon_0 d_{ijk}(\omega_3=\omega_1 \pm \omega_2)E_j(\omega_1)E_k(\pm \omega_2) \tag{2.5}$$

where summation over repeated indices is implied. Here d_{ijk} is the nonlinear optical coefficient, E_j and E_k are the j-th and k-th electric field Fourier components of the waves at the frequencies given in the parentheses, and g is the degeneracy factor equal to the number of distinguishable permutations of the frequencies. Thus $g=1$ for $\omega_1=\omega_2$ corresponding to second-harmonic generation and optical rectification, and $g=2$ for $\omega_1 \mp \omega_2$ corresponding to sum or difference-frequency mixing. It should be noted that for electric field components at negative frequencies in (2.5), we use the convention $E(-\omega)=E^*(\omega)$ where * denotes the complex conjugate.

The nonlinear coefficient d_{ijk} is a third-rank tensor which vanishes in crystals that possess a center-of-inversion symmetry [2.35]. Considering that the sum or difference-frequency mixing involves three frequencies ω_1, ω_2, and ω_3, there are 81 components required for this three-wave frequency mixing process. However, *Armstrong* et al. [2.36] have deduced certain permutation symmetry relations between these components of the type [2.2]

$$d_{ijk}(\omega_3=\omega_1 \pm \omega_2)=d_{jik}(\omega_1=\omega_3 \mp \omega_2)=d_{kij}(\omega_2=\pm \omega_3 \mp \omega_1)\,. \tag{2.6}$$

Furthermore, if $\omega_1=\omega_2$, one has additional permutation symmetry relations of the type

$$d_{ijk}=d_{ikj}\,. \tag{2.7}$$

In this special case corresponding to second-harmonic generation or optical rectification, the symmetry properties of the nonlinear optical tensor $\underset{\approx}{d}$ are the same as those of the piezoelectric tensor in that both are symmetric in the last two indices. These symmetry considerations reduce the independent components to 18. *Kleinman* [2.37] has shown that if the dispersion of the refractive index in the frequency range containing ω_1, ω_2, and ω_3 is negligible, the nonlinear tensor is symmetric in all indices which are freely interchangeable. In particular, it is symmetric in the last two indices.

Miller [2.38] has argued that the dispersion of the nonlinear susceptibility is related to the dispersion of the linear susceptibility. According to *Miller*, d_{ijk} can be expressed as

$$d_{ijk}(\omega_3=\omega_1\pm\omega_2)=\chi_{ii}^{(1)}(\omega_3)\chi_{jj}^{(1)}(\omega_1)\chi_{kk}^{(1)}(\omega_2)\delta_{ijk} \tag{2.8}$$

where $\chi_{ii}^{(1)}$, $\chi_{jj}^{(1)}$, and $\chi_{kk}^{(1)}$ represent the diagonal components of the linear susceptibility tensor at the frequencies indicated, and δ_{ijk}, known as "Miller's delta", is a frequency-independent parameter which is essentially the same for all materials within the same crystal symmetry class. An important consequence of (2.8) is that those crystals which have large linear susceptibilities at the frequencies considered also exhibit large nonlinear susceptibilities.

Even when $\omega_1 \neq \omega_2$, it is generally assumed that the components of the nonlinear tensor are symmetric in the last two indices. Therefore, these are most often expressed in the contracted form d_{il} where the single suffix l replaces the symmetric suffixes j and k. In this notation i takes the values 1 to 3 corresponding to $x\leftrightarrow 1$, $y\leftrightarrow 2$, $z\leftrightarrow 3$, and l takes the values 1 to 6 corresponding to $xx\leftrightarrow 1$, $yy\leftrightarrow 2$, $zz\leftrightarrow 3$, yz or $zy\leftrightarrow 4$, zx or $xz\leftrightarrow 5$, and xy or $yx\leftrightarrow 6$.

Requirements of crystal symmetry reduce the number of nonzero components of d_{il} even further. This depends upon the crystal class in question. For example, in tellurium, which has the point-group symmetry 32, there are only 5 nonzero components: d_{11}, $d_{12}=-d_{11}$, d_{14}, $d_{25}=-d_{14}$, and $d_{26}=-d_{11}$ so that there are only two independent constants. In GaAs, which belongs to the point group $\bar{4}3$ m, there are 3 nonzero components $d_{14}=d_{25}=d_{36}$ so that there is only one independent constant. Complete tensors for all the 20 crystal classes with nonzero components have been compiled by many authors [2.39].

From the microscopic point of view, the linear susceptibility arises from the response of the electrons as well as ions to the electromagnetic radiation. If a given frequency is higher than those of the ionic vibrations, i.e., phonons in a crystal, the ions cannot follow the time variations of the electric field of the radiation. Therefore, the susceptibility is essentially due to electrons. On the other hand, if the given frequency is below those of the phonons, both the ionic and electronic contributions must be considered. *Garrett* [2.40] has extended Miller's phenomenological model for the nonlinear susceptibility to include both the ionic and electronic contribution. In Garrett's model, $d_{ijk}(\omega_3=\omega_1\pm\omega_2)$ is expressed in terms of four Miller δ coefficients which are frequency independent. Application of this model to the component, say, $d_{123}(\omega_3=\omega_1\pm\omega_2)$ gives [2.41]

$$\begin{aligned} d_{123}(\omega_3=\omega_1\pm\omega_2)=\{&\chi_1^{\mathrm{i}}(\omega_3)\chi_2^{\mathrm{i}}(\omega_1)\chi_3^{\mathrm{i}}(\omega_2)\delta_{\mathrm{A}}+\chi_1^{\mathrm{i}}(\omega_3)\chi_2^{\mathrm{i}}(\omega_1)\chi_3^{\mathrm{e}}(\omega_2)\delta_{\mathrm{B}}\\ &+\chi_1^{\mathrm{i}}(\omega_3)\chi_2^{\mathrm{e}}(\omega_1)\chi_3^{\mathrm{e}}(\omega_2)\delta_{\mathrm{C}}+\chi_1^{\mathrm{e}}(\omega_3)\chi_2^{\mathrm{e}}(\omega_1)\chi_3^{\mathrm{e}}(\omega_2)\delta_{\mathrm{D}}\}\end{aligned} \tag{2.9}$$

where χ^{i}'s denote the ionic contribution to the linear susceptibility and χ^{e}'s denote the electronic contribution to the linear susceptibility; δ_A, δ_B, δ_C, and δ_D are the four frequency-independent parameters. If ε_0 and ε_∞ represent the dielectric constant (in units of ε_0) at low and high frequencies, respectively, then χ^e and χ^i can be expressed in terms of ε_0 and ε_∞ as

$$\chi^e = \varepsilon_\infty - 1 \tag{2.10}$$

and

$$\chi^i = \varepsilon_0 - \varepsilon_\infty \,. \tag{2.11}$$

When all the three frequencies ω_1, ω_2, and ω_3 are far above the phonon frequencies, the relevant nonlinear susceptibility is referred to as "optical" nonlinear susceptibility. In this case all χ^{i}'s are zero so that (2.9) gives

$$d^{o}_{123}(\omega_3 = \omega_1 \pm \omega_2) = (\chi^e)^3 \delta_D \tag{2.12}$$

where we have ignored the dispersion of χ^e. Another interesting case is that corresponding to far-infrared generation by difference-frequency mixing of high-frequency laser beams. In particular, we consider the case where ω_1 and ω_2 are above the phonon frequencies and $\omega_3 = \omega_1 - \omega_2$ is below the phonon frequencies. The relevant nonlinear susceptibility for far-infrared generation d^f is the same as obtained in the electro-optic effect. In the latter case, a high-frequency laser beam is modulated at a low frequency so that the frequency of the modulated laser beam is also high. In both these cases, only one of the three frequencies is below and the other two are above the phonon frequencies. Consequently, (2.9) yields

$$d^{f}_{123}(\omega_3 = \omega_1 - \omega_2) = \chi^i (\chi^e)^2 \delta_C + (\chi^e)^3 \delta_D = d^{eo}_{123} \,. \tag{2.13}$$

A comparison of (2.12) and (2.13) shows that the relative magnitudes of optical and far-infrared or electro-optic nonlinear susceptibilities will be determined by the relative signs of δ_C and δ_D. When δ_C and δ_D are of the same sign, $|d^f| = |d^{eo}| > |d^o|$, which is the case for a number of ferroelectrics such as $BaTiO_3$, $LiTaO_3$, $LiNbO_3$, $LiIO_3$, KDP, and ADP. On the other hand, if δ_C and δ_D are of opposite sign, $|d^f| = |d^{eo}| < |d^o|$ which is the case for a number of semiconductors such as GaAs, GaP, CdTe, and CdS.

In order to calculate the power generated at the FIR difference-frequency ω_3, we consider a nonlinear medium traversed by three plane waves of frequencies ω_1, ω_2 and $\omega_3 = \omega_1 - \omega_2$ so that the electric fields associated with these waves are described by

$$\boldsymbol{E}_1(\boldsymbol{r}, t) = \boldsymbol{E}_1(\boldsymbol{r}) \cos(\omega_1 t - \boldsymbol{k}_1 \cdot \boldsymbol{r}) \tag{2.14}$$

with similar equations for the other two waves. Here $\boldsymbol{k}_1$, $\boldsymbol{k}_2$, and $\boldsymbol{k}_3$ are wave vectors for the waves of frequency ω_1, ω_2, and ω_3, respectively. Thus the medium is subjected to a total electric field

$$\boldsymbol{E}(\boldsymbol{r},t)=\boldsymbol{E}_1(\boldsymbol{r},t)+\boldsymbol{E}_2(\boldsymbol{r},t)+\boldsymbol{E}_3(\boldsymbol{r},t)\,. \tag{2.15}$$

On substituting for $\boldsymbol{E}(\boldsymbol{r},t)$ from (2.15) into Maxwell's equations, one obtains the following equation (in MKs units) [2.42]:

$$\nabla^2\boldsymbol{E}=\mu_0\underset{\sim}{\boldsymbol{\sigma}}\cdot\frac{\partial\boldsymbol{E}}{\partial t}+\mu_0\underset{\sim}{\varepsilon}\cdot\frac{\partial^2\boldsymbol{E}}{\partial t^2}+\mu_0\frac{\partial^2\boldsymbol{P}_{\mathrm{NL}}}{\partial t^2} \tag{2.16}$$

where $\underset{\sim}{\boldsymbol{\sigma}}$ is the conductivity tensor of the medium, μ_0 is the permeability of free space, $\underset{\sim}{\varepsilon}=\varepsilon_0(\underset{\sim}{\chi}^{(1)}+1)$ is the dielectric constant tensor, and $\boldsymbol{P}_{\mathrm{NL}}$ is the nonlinear polarization. Many salient features of frequency mixing can best be illustrated by consideration of a medium with a scalar dielectric constant ε and a scalar conductivity σ, as for instance in a cubic crystal. We further simplify the problem by assuming that $\boldsymbol{P}_{\mathrm{NL}}$ is parallel to $\boldsymbol{E}$, and limit our discussion to the quadratic nonlinearity so that

$$P_{\mathrm{NL}}=\varepsilon_0\chi^{(2)}E^2=\varepsilon_0(2d)E^2 \tag{2.17}$$

where d is the lowest-order nonlinear coefficient. Let us assume, without further loss of generality, that all the three waves are propagating in the z-direction so that (2.15) for the total electric field gives

$$\begin{aligned}E(z,t)=&E_1(z)\cos(\omega_1t-k_1z)+E_2(z)\cos(\omega_2t-k_2z)\\&+E_3(z)\cos(\omega_3t-k_3z)\end{aligned} \tag{2.18}$$

where $k_1=n_1\omega_1/c$, $k_2=n_2\omega_2/c$, and $k_3=n_3\omega_3/c$ with n_1, n_2, and n_3 being the refractive indices at the frequencies ω_1, ω_2, and ω_3, respectively. Substituting for E from (2.18) into (2.17), one can show, after some algebraic manipulation, that in addition to the dc component equal to $d[E_1^2(z)+E_2^2(z)+E_3^2(z)]$, the nonlinear polarization has Fourier components at the second-harmonic frequencies $2\omega_1$, $2\omega_2$, $2\omega_3$, sum frequencies $\omega_1+\omega_2$, $\omega_2+\omega_3$, $\omega_3+\omega_1$, and difference frequencies $\omega_1-\omega_2$, $\omega_2-\omega_3$, and $\omega_1-\omega_3$. Only components of the above polarization at the frequencies $\omega_2+\omega_3=\omega_1$, $\omega_1-\omega_2=\omega_3$, and $\omega_1-\omega_3=\omega_2$ will be able to drive the oscillation at ω_1, ω_3, or ω_2. Others being in general, nonsynchronous, will not be able to do so. Consequently, we need not consider these other components. Physically, the component of polarization at the frequency $\omega_2+\omega_3=\omega_1$ can act as a source polarization for the wave at ω_1 with power flow from the fields at ω_2 and ω_3 into that at ω_1. Similarly, the polarization component at $\omega_1-\omega_2=\omega_3$ can act as a source of polarization for the difference-frequency ω_3 with power flow from the fields at ω_1 into those at ω_2 and ω_3.

When we substitute the component of nonlinear polarization at the difference frequency ω_3 as given by

$$P_{NL}(\omega_3=\omega_1-\omega_2,z,t)=\varepsilon_0(2d)E_1(z)E_2(z)\cos[(\omega_1-\omega_2)t-(k_1-k_2)z] \quad (2.19)$$

into the scalar form, we get the following equation for the spatial growth of E_3 [2.42]

$$\frac{\partial E_3(z)}{\partial z}=-\frac{\sigma_3}{2}\left(\frac{\mu_0}{\varepsilon_3}\right)^{1/2}E_3(z)-\frac{i}{2}\omega_3\left(\frac{\mu_0}{\varepsilon_3}\right)^{1/2} \cdot \varepsilon_0(2d)E_1(z)E_2(z)e^{i(k_3-k_1+k_2)z}\,. \quad (2.20)$$

Similarly, one gets

$$\frac{\partial E_2(z)}{\partial z}=-\frac{\sigma_2}{2}\left(\frac{\mu_0}{\varepsilon_2}\right)^{1/2}E_2(z)-\frac{i}{2}\omega_2\left(\frac{\mu_0}{\varepsilon_2}\right)^{1/2} \cdot \varepsilon_0(2d)E_1(z)E_3(z)e^{i(k_2-k_1+k_3)z} \quad (2.21)$$

$$\frac{\partial E_1(z)}{\partial z}=-\frac{\sigma_1}{2}\left(\frac{\mu_0}{\varepsilon_1}\right)^{1/2}E_1(z)-\frac{i}{2}\omega_1\left(\frac{\mu_0}{\varepsilon_1}\right)^{1/2} \cdot \varepsilon_0(2d)E_2(z)E_3(z)e^{i(k_1-k_2-k_3)z} \quad (2.22)$$

for the growth (or attenuation) of fields at the frequencies ω_2 and ω_1, respectively.

Equations (2.20–2.22) are the basic equations describing the phenomenon of parametric amplification which is based on the quadratic response of the medium in which the waves propagate. The complete solution of these three coupled equations is very complicated. Fortunately, however, in most experimental situations the decrease or increase in the amplitude of either one or two of the three frequencies can be considered negligible. For the particular case of difference-frequency generation at $\omega_3=\omega_1-\omega_2$, we consider $E_1(z)$ and $E_2(z)$ as constant over the mixing length of the crystal. Then we need consider only (2.20) which may be rewritten as

$$\frac{\partial E_3(z)}{\partial z}=-\frac{\sigma_3}{2}\left(\frac{\mu_0}{\varepsilon_3}\right)^{1/2}E_3(z)-\frac{i\omega_3}{2}\left(\frac{\mu_0}{\varepsilon_3}\right)^{1/2}\varepsilon_0(2d)E_1E_2e^{-i\Delta kz} \quad (2.23)$$

where

$$\Delta k=(k_1-k_2)-k_3 \quad (2.24)$$

is the momentum mismatch between the polarization wave at ω_3 and the electromagnetic wave at ω_3. If we set $\sigma_3=0$, i.e., neglect absorption at ω_3, and integrate (2.23) over the crystal length L, we obtain amplitude of the electric field just inside the exit face of the crystal

$$E_3(L) = \frac{\varepsilon_0\omega_3}{2}\left(\frac{\mu_0}{\varepsilon_3}\right)^{1/2}(2d)E_1E_2\,\frac{e^{-i\Delta kL}-1}{\Delta k}. \tag{2.25}$$

The output intensity I (power per unit area) is related to the field as

$$I_{\omega_3} = \frac{1}{2}\left(\frac{\varepsilon_3}{\mu_0}\right)^{1/2} E_3E_3^*. \tag{2.26}$$

Substituting for $E_3(L)$ from (2.25) into (2.26), we obtain the output intensity as

$$I_{\omega_3} = \frac{1}{2}\left(\frac{\mu_0}{\varepsilon_0}\right)^{1/2}\frac{\omega_3^2(2d)^2}{n_1n_2n_3c^2}I_{\omega_1}I_{\omega_2}\frac{\sin^2(\frac{1}{2}\Delta kL)}{(\frac{1}{2}\Delta kL)^2}\cdot L^2 \tag{2.27}$$

where I_1 and I_2 are the intensities at ω_1 and ω_2, respectively, and n_1, n_2, and n_3 are the corresponding refractive indices. Equation (2.27) can be written in terms of the external input and output powers as

$$P_{\omega_3} = \frac{1}{2}\left(\frac{\mu_0}{\varepsilon_0}\right)^{1/2}\frac{\omega_3^2(2d)^2L^2}{n_1n_2n_3c^2}\left(\frac{P_{\omega_1}P_{\omega_2}}{A}\right)T_1T_2T_3\frac{\sin^2(\frac{1}{2}\Delta kL)}{(\frac{1}{2}\Delta kL)^2} \tag{2.28}$$

where T_1, T_2, and T_3 are the single surface power transmission coefficients at ω_1, ω_2, and ω_3, respectively, and A is area of the beam cross section.

When the effect of absorption is included [2.43], (2.28) becomes

$$P_{\omega_3} = \frac{1}{2}\left(\frac{\mu_0}{\varepsilon_0}\right)^{1/2}\frac{\omega_3^2(2d)^2L^2}{n_1n_2n_3c^2}\left(\frac{P_{\omega_1}P_{\omega_2}}{A}\right)T_1T_2T_3\,e^{-\alpha_3L} \cdot\frac{1+e^{-\Delta\alpha L}-2e^{-\frac{1}{2}\Delta\alpha L}\cos(\Delta kL)}{(\Delta kL)^2+(\frac{1}{2}\Delta\alpha L)^2} \tag{2.29}$$

where Δk is the momentum mismatch as given by (2.24) and

$$\Delta\alpha = \alpha_1+\alpha_2-\alpha_3 \tag{2.30}$$

where α_1, α_2, and α_3 are the power absorption coefficients at ω_1, ω_2, and ω_3, respectively. If there is no reflective or antireflective coating on the input and output faces of the crystal, the transmission coefficient T is related to the refractive index as

$$T = \frac{4n}{(n+1)^2}. \tag{2.31}$$

Substituting the above expression for T into (2.29), it becomes

$$P_{\omega_3}=32\left(\frac{\mu_0}{\varepsilon_0}\right)^{1/2}\left(\frac{2\omega_3 dL}{(n_1+1)(n_2+1)(n_3+1)}\right)^2\left(\frac{1}{c^2}\right)\frac{P_{\omega_1}P_{\omega_2}}{A}$$
$$\cdot\, e^{-\alpha_3 L}\frac{1+e^{-\Delta\alpha L}-2e^{-\frac{1}{2}\Delta\alpha L}\cos(\Delta kL)}{(\Delta kL)^2+(\frac{1}{2}\Delta\alpha L)^2}. \qquad (2.32)$$

Assuming no absorption losses, i.e., $\alpha_1=\alpha_2=\alpha_3=0$, the last two factors on the rhs of (2.32)

$$e^{-\alpha_3 L}\frac{1+e^{-\Delta\alpha L}-2e^{-\frac{1}{2}\Delta\alpha L}\cos(\Delta kL)}{(\Delta kL)^2+(\frac{1}{2}\Delta\alpha L)^2}$$

are reduced to unity for the phase-matched condition $\Delta k=0$. In the absence of absorption losses, *Manley* and *Rowe* [2.44] relations for the power exchange between the three fields are

$$\frac{\Delta P_{\omega_3}}{\omega_3}=\frac{\Delta P_{\omega_2}}{\omega_2}=-\frac{\Delta P_{\omega_1}}{\omega_1} \qquad (2.33)$$

where ΔP denotes the change in the power at the frequency indicated. The above Manley-Rowe relations written for the difference-frequency generation at $\omega_3=\omega_1-\omega_2$ tell us that the power at ω_1 is lost to the generated signal wave at ω_3 and also to the so-called "idler" wave at ω_2. In the photon picture, a photon of frequency ω_1 splits into two photons of frequencies ω_2 and ω_3, respectively. In other words, generation of the difference-frequency wave also results in parametric amplification at the lower of the two input frequencies. It should be pointed out that the Manley-Rowe relations do not require any assumption of negligible amplitude variations, as was assumed in the derivation of (2.27–2.29) and (2.32). According to the Manley-Rowe relations, the maximum power conversion efficiency from ω_1 to ω_3 is given by

$$\eta_{\omega_3}^{\max}=\frac{\omega_3}{\omega_1} \qquad (2.34)$$

which corresponds to 100% quantum conversion efficiency. According to (2.34), if CO_2 lasers operating in the 10 μm region are used for FIR generation, the maximum power conversion efficiency cannot exceed 10% at 100 μm and 1% at 1000 μm or 1 mm. If dye lasers are used instead of CO_2 lasers for FIR generation, the maximum conversion efficiencies will even be lower by more than an order of magnitude.

2.2.2 Phase-Matching Techniques

When absorption losses are negligible, (2.27) shows that the intensity of the radiation at the difference frequency ω_3 varies as the coherence factor

$$F_c = L^2 \frac{\sin^2(\frac{1}{2}\Delta kL)}{(\frac{1}{2}\Delta kL)^2} \tag{2.35}$$

with the distance L traversed by the mixing optical beams in the crystal. For a given momentum mismatch $\Delta k \neq 0$, F_c will go through a series of maxima and minima with L. Maxima in F_c will occur when

$$\tfrac{1}{2}\Delta kL = \frac{\pi}{2}, \quad \frac{3\pi}{2}, \quad \frac{5\pi}{2}, \quad \frac{7\pi}{2}, \quad \text{etc.} \tag{2.36}$$

The characteristic thickness of the crystal corresponding to the first maximum

$$L_c = \frac{\pi}{\Delta k} \tag{2.37}$$

is usually referred to as a "coherence length". The successive maxima or minima in F_c are separated by $2L_c$, and all the maxima have the same value of $(4/\pi^2)L_c^2$. Consequently the maximum intensity or power at the difference frequency ω_3 is reached in a crystal whose thickness is equal to the coherence length, and there is no advantage in using a thicker crystal.

The validity of (2.35) was verified experimentally by *Maker* et al. [2.5] in their experiments on second-harmonic generation. They observed periodic variations in the intensity of the second harmonic by tilting a thin quartz plate so as to vary its effective thickness for the ruby laser radiation passing through it. Values of L_c deduced from the intensity maxima were found to be in excellent agreement with those calculated from the dispersion of the refractive index between the fundamental and the second-harmonic wavelengths. In physical terms, periodic variations in the intensity of the second-harmonic with crystal thickness can be visualized as follows: Consider a series of planes in the crystal perpendicular to the direction of propagation of the fundamental wave. At each of these planes, the second-order nonlinear susceptibility produces a second-harmonic polarization whose phase velocity is

$$v_p = \frac{c}{n_\omega}, \tag{2.38a}$$

n_ω being the refractive index at the fundamental frequency ω since the spatial variation of the second-order polarization is anchored to the spatial variation of the fundamental. However, the second-harmonic wave generated by this polarization wave propagates, in general, with a different velocity

$$v_2 = \frac{c}{n_{2\omega}} \tag{2.38b}$$

where $n_{2\omega}$ is the refractive index at the frequency of the second-harmonic 2ω. Consequently, the second-harmonic wave generated at one plane will not necessarily be in phase with the second-harmonic wave propagating from a previous plane unless $n_\omega = n_{2\omega}$. In particular, if the separation between the consecutive planes is L_c, the second-harmonic wave which was generated at the previous plane will be exactly 180° out of phase on arrival at the given plane with the second-harmonic generated at the given plane. Thus these two waves will tend to cancel each other producing a series of minima and maxima in the intensity of the second harmonic with the thickness of the crystal.

However, it is not so easy to visualize the effect of dispersion in the three-wave interaction involved in the difference-frequency mixing. But mathematically the situation is quite similar. The momentum mismatch Δk for difference-frequency mixing with collinear beams is related to the dispersion of the refractive index as

$$\Delta k = k_3 - k_1 + k_2 = \frac{1}{c}(n_3\omega_3 - n_1\omega_1 + n_2\omega_2) \tag{2.39}$$

where n_1, n_2, and n_3 are the refractive indices at the frequencies ω_1, ω_2, and ω_3, respectively. Assuming $n_1 \simeq n_2 = n$, and $n_3 = n + \Delta n$, (2.39) gives

$$\Delta k \simeq \frac{\omega_3}{c} \Delta n \tag{2.40}$$

so that the coherence length

$$L_c \simeq \frac{\pi c}{\omega_3 \Delta n} = \frac{\lambda_3}{2\Delta n} \tag{2.41}$$

where λ_3 is the vacuum wavelength of the difference-frequency radiation. $2\Delta n \lesssim 1$ for the typical crystals used for FIR generation by difference-frequency mixing and therefore L_c is of the order of FIR wavelength being generated.

If the momentum mismatch Δk is made zero, L_c would become infinitely large. This requirement of zero momentum mismatch

$$\Delta k = 0 \tag{2.42}$$

is usually referred to as the phase-matching condition. When this condition is satisfied, the coherence function

$$F_c = L^2 \tag{2.43}$$

so that the intensity at ω_3 increases quadratically with the thickness of the crystal traversed by the input laser beams. A number of techniques have been employed to achieve phase matching in order to increase the mixing efficiency by taking advantage of the quadratic dependence of the intensity on L under phase-matched conditions. When optical mixing is obtained with laser beams propagating in the same direction, we use the term "collinear" mixing in contrast to "noncollinear" mixing in which the input laser beams and the beam generated at the new frequency may propagate at a certain angle to one another in the crystal. Consequently the various phase-matching techniques may also be divided into two categories: i) collinear and ii) noncollinear.

Collinear Phase-Matching Techniques

a) Birefringent Phase Matching. In isotropic crystals such as those belonging to the cubic crystal class, the refractive index is, in general, a function of frequency. In an anisotropic crystal, e.g., a uniaxial crystal, the refractive index also depends upon the state of polarization and direction of propagation of the light beam relative to the optic axis, usually referred to as the z-axis or c-axis of the crystal. This property of a crystal known as birefringence which means double refraction was first discovered in 1669 in calcite which was then known as Iceland spar. It was discovered that the refractive index of light rays polarized normal to the optic axis ($\boldsymbol{E} \perp \boldsymbol{c}$), is independent of the direction of propagation just as in the optically isotropic crystals. For this reason such rays are called ordinary or o-rays. In comparison the refractive index of light rays with a component of polarization parallel to the c-axis ($\boldsymbol{E} \| \boldsymbol{c}$) varies with the direction of propagation relative to the optic axis. Therefore, the rays with this unusual behavior are known as the extraordinary or e-rays. Consequently uniaxial crystals have two principal indices of refraction n_o for o-rays with $\boldsymbol{E} \perp \boldsymbol{c}$ and n_e for e-rays with $\boldsymbol{E} \| \boldsymbol{c}$. The difference $\Delta n = n_e - n_o$ is a measure of the birefringence. There are two types of uniaxial crystals. If $n_e < n_o$, Δn is negative, and the crystal is said to be uniaxial negative, e.g., KDP. On the other hand, if $n_e > n_o$, the crystal is uniaxial positive. Quartz belongs to the latter class.

In certain birefringent crystals, it is possible to select a direction of propagation such that the effect of normal dispersion on phase mismatch Δk is completely cancelled by the presence of birefringence. Let us first consider the application of this idea to the generation of second-harmonic radiation. Assume that the nonlinear crystal possesses normal frequency dispersion, i.e., $dn/d\omega > 0$ in the frequency region between the fundamental and the second-harmonic. This implies that $n(2\omega) > n(\omega)$. Now if the nonlinear crystal is uniaxial negative, it is possible that $n_o(\omega)=n_e(2\omega)$ for a particular direction of propagation in the crystal. If θ denotes the angle between the direction of propagation and the optic axis, the refractive index of e-waves for this direction of propagation is given by

$$\frac{1}{n_e^2(\theta)}=\frac{\cos^2\theta}{n_o^2}+\frac{\sin^2\theta}{n_e^2} \tag{2.44}$$

where n_o and n_e are the principal indices of refraction. This idea of using birefringence to achieve phase matching, also known as index matching, was first demonstrated independently by *Giordmaine* [2.4] and *Maker* et al. [2.5] in their experiments on second-harmonic generation with the ruby laser. If light propagates within a KDP crystal at a specific angle θ_m, called the phase-matching angle, the refractive index $n_o(\omega)$ of the fundamental wave as the ordinary wave will be precisely equal to the refractive index $n_e(2\omega)$ of the second-harmonic propagating as an extraordinary wave. Using a KDP crystal cut at this particular angle, the intensity of the second-harmonic of the ruby laser was found to be orders of magnitude above that obtained in the non-phase-matched case. Ten to twenty percent conversion efficiencies from the fundamental to the second-harmonic can be obtained in this manner. By placing a phase matched nonlinear crystal of barium sodium niobate inside a Nd^{3+}: YAG laser with a cw output of 1 W at 1.06 μm, *Geusic* et al. [2.45] obtained 1-W output in the second-harmonic at 0.53 μm. In a sense, this corresponds to 100% conversion efficiency from the fundamental to the second-harmonic.

For the three-wave mixing involved in the difference-frequency generation at $\omega_3=\omega_1-\omega_2$, two types of phase matching in birefringent crystals are possible: Type I in which ω_1 and ω_2 are of the same polarization (both o-waves or e-waves), and Type II in which ω_1 and ω_2 are of orthogonal polarization (one is an o-wave and the other is an e-wave). In either case the phase-matching condition is

$$n_3\omega_3=n_1\omega_1-n_2\omega_2 \tag{2.45a}$$

or

$$n_3=\frac{\omega_1}{\omega_3}n_1-\frac{\omega_2}{\omega_3}n_2 \tag{2.45b}$$

where n_1, n_2, and n_3 are the appropriate refractive indices for the o-waves or the e-waves as the case may be at the frequencies ω_1, ω_2, and ω_3, respectively. Suppose that ω_1 and ω_2 are so close to each either that $n_1\simeq n_2=n_{12}$, the phase matching condition (2.45a) becomes

$$n_3\simeq n_{12}\,. \tag{2.46}$$

Usually the crystals employed for FIR generation have anomalous dispersion, i.e., $dn/d\omega<0$, i.e., $n_3>n_{12}$ in the frequency region between ω_3 on the one hand and ω_1, ω_2 on the other hand. If the nonlinear crystal is uniaxial positive ($n_o<n_e$), it follows that (2.45) could be satisfied only if ω_1 and ω_2 are the e-waves, and ω_3 propagates as the o-wave, corresponding to Type I phase matching. For FIR generation with CO_2 laser mixing in a uniaxial positive birefringent ternary semiconductor $ZnGeP_2$, *Boyd* et al. [2.46] employed Type II phase matching in which ω_3 and ω_2 were the o-waves, and ω_1 was the e-wave. In this manner, they generated ω_3 in the range $70\,cm^{-1}$ to $110\,cm^{-1}$ using θ_m in the range from 40° to 55°.

While birefringent phase matching is most commonly used for second-harmonic, sum and difference-frequency generation in the visible and near-infrared regions, there are relatively few birefringent nonlinear crystals suitable for FIR difference-frequency generation by mixing of laser beams in the visible and near-infrared frequencies, and in particular for mixing of CO_2 laser beams. However, there are some cubic nonlinear semiconductors such as InSb, GaAs, CdTe, etc., which can be used for FIR difference-frequency generation with CO_2 lasers. But these crystals being cubic lack birefringence. Therefore, other methods for phase matching in isotropic crystals must be considered.

b) Magneto-Optic Phase Matching. If the nonlinear crystal is a semiconductor such as InSb containing free carriers, the free-carrier plasma contribution to the dielectric constant at the FIR difference-frequency ω_3 is given by [2.47, 48]

$$\Delta\varepsilon_{FC}(\omega_3)=-\varepsilon_\infty\frac{\omega_p^2}{\omega_3^2} \tag{2.47}$$

where

$$\omega_p=\left(\frac{Ne^2}{m^*\varepsilon_\infty}\right)^{1/2} \tag{2.48}$$

is the plasma frequency. N is the density of the free carriers, e is the electronic charge, m^* is the effective mass of the free carriers, and ε_∞ is the background high-frequency dielectric constant. In (2.47) it has been assumed that the effective mass is isotropic and the plasma is lossless. The dispersion of the lattice contribution to the FIR refractive index is given by

$$\Delta\varepsilon_{TO}(\omega_3)=\frac{(\varepsilon_0-\varepsilon_\infty)}{\omega_{TO}^2-\omega_3^2} \tag{2.49}$$

where ε_0 is the low-frequency or static dielectric constant, and ω_{TO} is the frequency of the transverse optical phonon. We have assumed here, for the sake of simplicity, that there is only one infrared-active TO phonon. For $\varepsilon_0>\varepsilon_\infty$ which is the usual case, free-carrier plasma and lattice contributions to the dielectric constant are of opposite sign. Therefore, it should be possible to adjust the carrier density so that the two contributions cancel each other. In other words, $\Delta\varepsilon_{FC}$ can be used to achieve phase matching for FIR difference-frequency generation.

Unfortunately, a given value of N will provide phase matching only for a particular value of ω_3. A different value of N will be required to achieve phase matching at a different frequency. Clearly it is not easy to vary N at will. In fact, a new crystal will be required for each FIR frequency. However, if the semiconductor is subjected to an applied magnetic field, magnetoplasma contribution to the refractive index can be adjusted by varying the strength of the applied magnetic field. For example, consider light propagation along the magnetic field (Faraday configuration) [2.47, 48]

$$\Delta\varepsilon_{FC}(\omega_3) = -\varepsilon_\infty \frac{\omega_p^2}{\omega_3(\omega_3 \mp \omega_c)} \tag{2.50}$$

where $\omega_c = eH/m^*$ is the cyclotron frequency, and H is the strength of the applied magnetic field. Here $\pm$ signs refer to right and left circular polarization, respectively. Similarly, for light propagation perpendicular to the magnetic field (Voigt configuration) [2.47, 48]

$$\Delta\varepsilon_{FC}(\omega_3) = -\varepsilon_\infty \frac{\omega_p^2(\omega_3^2 - \omega_p^2)}{\omega_3^2(\omega_3^2 - \omega_p^2 - \omega_c^2)}\,. \tag{2.51}$$

The idea of using magnetoplasma effect for phase matching has been demonstrated by *Nguyen* and *Patel* [2.10, 49] for FIR generation in the region ~75–100 μm with CO_2 laser mixing in n-type InSb. The drawback of this technique for phase matching arises from the fact that the free carriers necessary for phase matching are strongly absorbing in the FIR.

c) Dielectric Waveguide Phase Matching. When light propagates in a dielectric waveguide, the relevant dielectric constant is no longer equal to the dielectric constant ε of the bulk material. For a waveguide of thickness $2d$, the waveguide dielectric constant ε_w is given by [2.17]

$$\varepsilon_w = \varepsilon - \frac{h^2}{\mu_0\omega^2} = \varepsilon_0 + \frac{p^2}{\mu_0\omega^2} \tag{2.52}$$

with

$$h\tan\left(hd - \frac{m\pi}{2}\right) = \begin{cases} (\varepsilon/\varepsilon_0)\,p\,, & \text{for TM}_m \\ p\,, & \text{for TE}_m \end{cases}. \tag{2.53}$$

Here m is a positive integer. TM and TE denote the transverse magnetic and transverse electric modes of propagation inside the waveguide. For a given m, p and h are obtained from the solution of (2.52) and (2.53). Equation (2.52) shows that the waveguide dielectric constant is smaller than the dielectric constant of the bulk material. This property of the waveguide can be used to achieve phase matching in FIR difference-frequency generation since the FIR refractive index is, in general, higher than that in the near infrared. It turns out that the waveguide thickness required for phase-matching low-order propagation modes is of the order of $(1/2)\,\lambda_3$ where λ_3 is the FIR wavelength. *Thompson* and *Coleman* [2.17] have successfully used GaAs waveguide for phase-matched FIR generation with CO_2 lasers.

d) Phase Matching with Periodic Structures. *Bloembergen* and coworkers [2.36, 50] were the first to suggest the possibility of using the periodic variation of the nonlinear coefficient or the index of refraction for phase matching. It was later considered by *Somekh* and *Yariv* [2.51] for phase matching in light waveguides. More recently, *Yacoby*, *Aggarwal* and *Lax* [2.52] have analyzed a lamella structure with a periodic nonlinear coefficient for phase-matched generation of FIR difference-frequency radiation.

Consider a lamella structure made up of a number of plates of thickness d and the same refractive index. The plates are so arranged that the nonlinear coefficient is of the same magnitude for all plates but alternates in sign from one plate to another. Such a structure provides a periodic variation of the nonlinear coefficient. The effective wave vector of such lamellae for the difference-frequency ω_3 is [2.52]

$$k_l = \frac{(2m+1)\,\pi}{d} \tag{2.54}$$

where $m=0$, 1, 2, 3, etc. For forward wave collinear mixing in the above lamellae, the phase matching is obtained when

$$\Delta k = k_3 - k_1 + k_2 = k_l\,. \tag{2.55}$$

In terms of the coherence length given by (2.37), the phase-matching condition reduces to

$$d = (2m+1)\,L_c\,. \tag{2.56}$$

Under phase-matched generation in lamellae, of total thickness L, the intensity at ω_3 turns out to be

$$I_{\omega_3}(\Delta k = k_l) = \left(\frac{4}{\pi^2}\right)\left(\frac{1}{2m+1}\right)^2 I_{\omega_3}(\Delta k = 0) \tag{2.57}$$

where $I_{\omega_3}(\Delta k=0)$ represents the intensity which would be obtained from bulk phase-matched crystal whose thickness is equal to the total thickness L of the lamellae. It is obvious from (2.57) that the maximum intensity will be obtained from lamellae with $m=0$ which corresponds to $d=L_c$, i.e., the thickness of each individual plate is equal to the coherence length. Since L_c is of the order of the FIR wavelength being generated, as discussed previously, it is not too difficult to fabricate such lamellae. But compared with bulk phase matching, the FIR intensity from a lamellae structure is at best 40%. However, the advantage of this scheme is that it can be used with crystals which are optically isotropic.

Noncollinear Phase Matching

Zernike [2.8] and *Aggarwal* et al. [2.13] have shown that it is possible to obtain phase-matched FIR difference-frequency generation by noncollinear mixing of two laser beams provided that the nonlinear crystal possesses anomalous dispersion, i.e., the index of refraction at the difference-frequency, is larger than those at the two input frequencies. This can be shown as follows: Consider two beams of frequencies ω_1 and ω_2 are incident on a nonlinear medium such that they are propagating at an angle θ with respect to each other as shown in Fig. 2.1. If $\boldsymbol{k}_3$ is the wave vector of the difference-frequency radiation at ω_3, and $\boldsymbol{k}_1$ and $\boldsymbol{k}_2$ are the corresponding wave vectors for the input frequencies ω_1 and ω_2, the phase-matching condition or momentum conservation requires

$$\Delta\boldsymbol{k}=\boldsymbol{k}_3-\boldsymbol{k}_1+\boldsymbol{k}_2=0 \tag{2.58}$$

which is the vector analog of the scalar (2.42). Simple geometrical considerations show that (2.58) is satisfied if the angles θ and φ are given by the following expressions

$$\sin\left(\tfrac{1}{2}\theta\right)=\left[\frac{(n_3\omega_3)^2-(n_1\omega_1-n_2\omega_2)^2}{4n_1n_2\omega_1\omega_2}\right]^{1/2} \tag{2.59}$$

and

$$\cos\varphi=\left[1+2(\omega_2/\omega_3)\sin^2\left(\tfrac{1}{2}\theta\right)\right]\left[1+4\left(\frac{\omega_1\omega_2}{\omega_3^2}\right)\sin^2\left(\tfrac{1}{2}\theta\right)\right]^{-1/2} \tag{2.60}$$

where n_1, n_2, and n_3 are the refractive indices for radiation of frequencies ω_1, ω_2, and ω_3, respectively. If we assume for the sake of simplicity that $n_1=n_2=n$, and $n_3=n+\Delta n$, (2.59) and (2.60) may be rewritten as

$$\sin\theta\simeq\theta\simeq\frac{\omega_3}{\sqrt{\omega_1\omega_2}}\sqrt{\frac{2\Delta n}{n}}, \tag{2.61}$$

and

$$\varphi\simeq\sqrt{\frac{2\Delta n}{n}}. \tag{2.62}$$

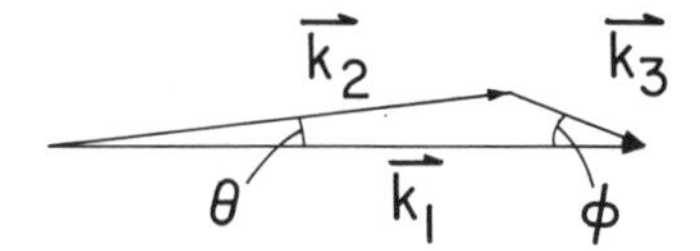

Fig. 2.1. Wave vector triangle showing the direction of propagation of the incident laser beams at frequencies ω_1 and ω_2 and that of the difference-frequency radiation at ω_3 inside a nonlinear crystal (after *Aggarwal* et al. [2.13])

It follows from (2.61) that $\sin\theta$ and therefore θ will be real only if $\Delta n \geqq 0$, i.e., noncollinear phase matching is possible only when the crystal possesses anomalous dispersion. Equation (2.62) indicates that φ is essentially independent of the laser frequencies ω_1 and ω_2 as well as the FIR frequency ω_3. It is this feature that facilitates broadband mixing for an FIR tunable system.

The use of noncollinear phase matching for FIR difference-frequency generation was first reported by *Zernike* [2.8] who mixed 9.6 μm and 10.6 μm CO_2 laser beams in a crystal of InSb to generate FIR at ~100 μm. *Aggarwal* et al. [2.13, 14] have applied this technique of noncollinear phase matching to CO_2 laser mixing in GaAs for the FIR difference-frequency generation from ~70 μm to several mm. *Yang* et al. [2.19] have reported the use of noncollinear phase matching even in a birefringent crystal of $LiNbO_3$ for FIR generation with mixing of tunable dye laser beams.

2.3 Choice of Nonlinear Crystal

There are a number of considerations for the optimum choice of nonlinear crystals to be employed in the generation of FIR radiation by difference-frequency mixing of higher-frequency infrared, visible, or UV lasers:

i) The crystals should have low absorption losses or high transparency at the two input laser frequencies as well as that at the FIR frequency to be generated. Neglecting multiphoton processes, the absorption losses at frequencies below the fundamental gap or absorption edge of the crystal are essentially due to free carriers and phonons. Research efforts towards the development of window materials for high-power laser applications have produced certain crystals of very high purity. When such crystals are compensated to obtain high-resistivity "semi-insulating" material, the free carrier absorption is reduced to negligible levels so that the residual absorption is essentially due to intrinsic phonon processes. The frequency dependence of this intrinsic absorption is such that the absorption falls off rapidly on either side of the fundamental lattice absorption bands. However, superimposed on this decreasing background are small absorption peaks which reflect the structure in the joint density-of-states of the phonons involved in the absorption process. Consequently, one should look for a nonlinear crystal in which the Restrahlen band or bands are well above the FIR frequency but well below the input laser frequencies which in turn should be well below the fundamental gap.

ii) Since the efficiency for FIR generation is proportional to the square of the nonlinear coefficient, crystals with large nonlinearity are desirable. The relevant nonlinear coefficient for FIR difference-frequency generation is the one obtained from the electro-optic coefficient when the FIR frequency is well below and the input laser frequencies are well above the Restrahlen bands.

iii) The FIR generation efficiency is also proportional to the square of the effective interaction length L_{eff} of the crystal. Therefore, one should look for crystals which are available in relatively large sizes.

iv) Also crystals with high damage threshold to the input laser intensity are desirable in view of the fact that the FIR intensity is proportional to the product of the input laser intensities.

As we are primarily concerned with the mixing of CO_2 laser radiation for FIR generation, we will consider a number of nonlinear crystals which have been used or could be used for this purpose.

GaAs

GaAs is one of the most popular materials currently used for electro-optic modulators and CO_2 laser windows because of its relatively large nonlinear coefficient and small absorption loss. Its fundamental electronic absorption edge at room temperature is $\sim$0.9 μm ($E_g = 1.35$ eV) in the near infrared. The FIR absorption of high-resistivity GaAs has been measured by *Johnson* et al. [2.53]. As shown in Fig. 2.2, there is a strong absorption band centered around 37 μm due to the long wavelength, i.e., $\boldsymbol{q}=0$ zone-center transverse optical (TO) phonons. On either side of this fundamental Restrahlen band, there is additional absorption due to multiphonon processes. This absorption is reduced significantly on lowering the temperature as can be seen from a comparison of the curves in Fig. 2.2 for 300 K and 8 K.

Stolen [2.54] has made rather careful measurements of the absorption on the long wavelength side of the Restrahlen band. Figure 2.3 shows his results replotted in terms of the absorption coefficient vs. frequency in cm^{-1} at room temperature and liquid nitrogen temperature. *Stolen* [2.55] has also studied the temperature dependence of this absorption, as shown in Fig. 2.4 for three frequencies: 110 cm^{-1} (91 μm), 80 cm^{-1} (125 μm), and 50 cm^{-1} (200 μm). From an analysis of these data, it has been shown that this long wavelength absorption is predominantly due to two-phonon difference processes involving the absorption of one phonon accompanied by the emission of a higher frequency phonon. In particular the absorption is dominated by the (O-LA) process in which a longitudinal acoustic phonon is absorbed and an optical phonon is emitted, along with the (LA-TA) process involving the emission of an LA phonon and absorption of a TA phonon. In addition, the three-phonon difference processes make a small contribution to the observed absorption.

Infrared absorption in high-resistivity GaAs has been studied by a number of investigators for the 10 μm CO_2 laser region. Figure 2.5 shows the

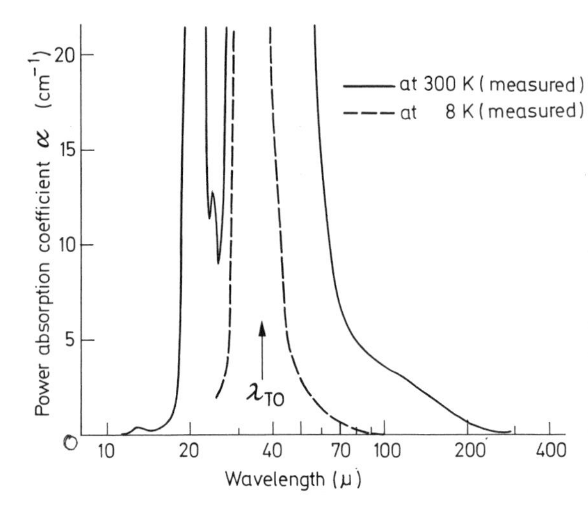

Fig. 2.2. Infrared absorption coefficient of high-sensitivity GaAs at 300 K and 8 K. λ_{TO} denotes the wavelength of the zone-center transverse optical phonon (after *Johnson* et al. [2.53])

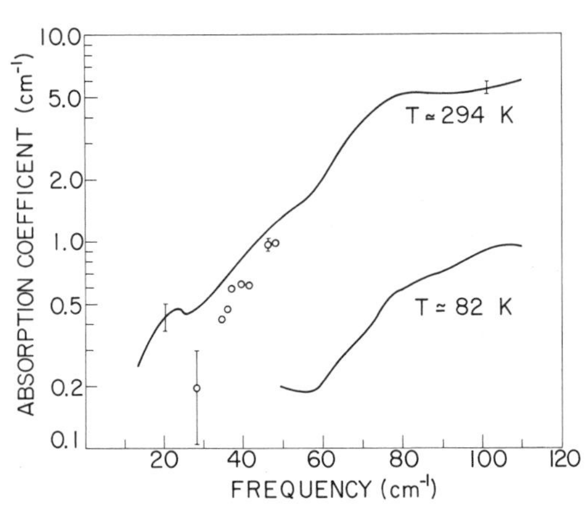

Fig. 2.3. Far infrared absorption coefficient of high-resistivity GaAs at 294 K and 82 K (after *Stolen* [2.54]). The data points indicated by the circles were obtained for GaAs at room temperature by *Rosenbluh* [2.87] with a cw FIR difference-frequency source

absorption spectra obtained by *Christensen* et al. [2.56] for the best available material from several laboratories grown and compensated by different techniques. All samples, with the exception of those from EMC, yield the same spectra with absorption coefficients ranging from 0.005 cm^{-1} at 9.2 μm to 0.010 cm^{-1} at 10.69 μm. Similarity in both the amplitude and structure of these spectra suggests that the absorption is not extrinsic but arises from an intrinsic mechanism. In order to identify the loss mechanism, *Christensen* et al. [2.56] also studied the temperature dependence of the absorption in the low-loss samples of Fig. 2.5. Data obtained at 9.28 μm are shown in Fig. 2.6. These results indicate that for temperature below ~250 °C the absorption is due to multiphonon processes. At higher temperatures the free carrier contribution to absorption becomes increasingly important.

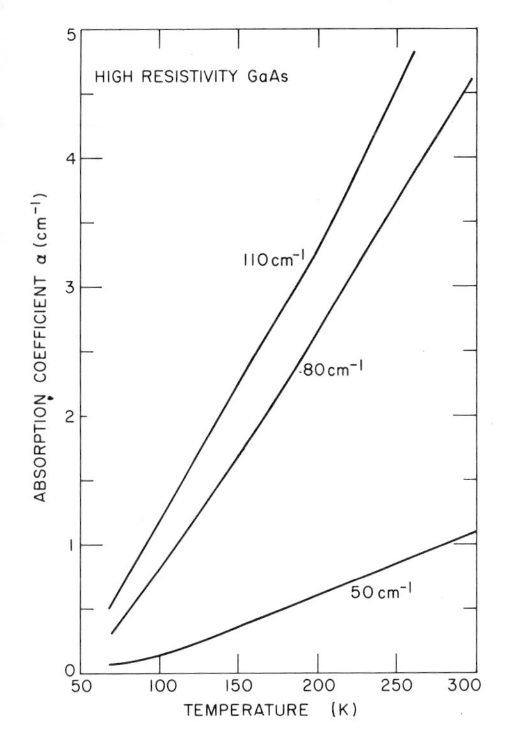

Fig. 2.4. Far-infrared absorption coefficient of high-resistivity GaAs as a function of temperature at 110 cm^{-1} (91 μm), 80 cm^{-1} (125 μm), and 50 cm^{-1} (200 μm) (after *Stolen* [2.55])

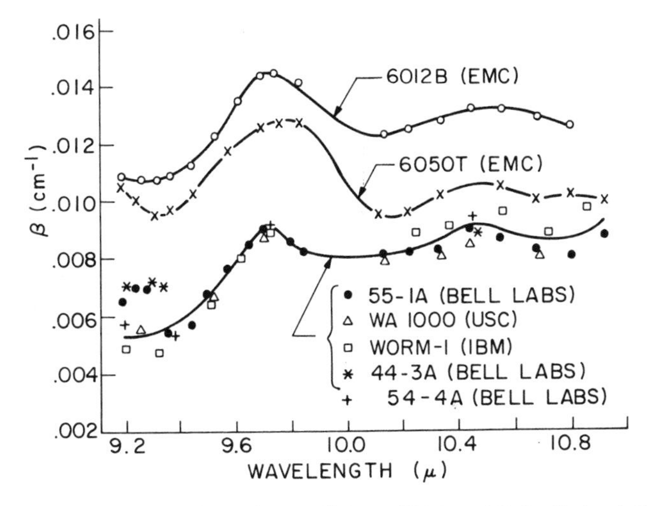

Fig. 2.5. Infrared absorption coefficient of high-resistivity GaAs at $T \simeq 300$ K in the 9 to 11 μm wavelength region (after *Christensen* et al. [2.56])

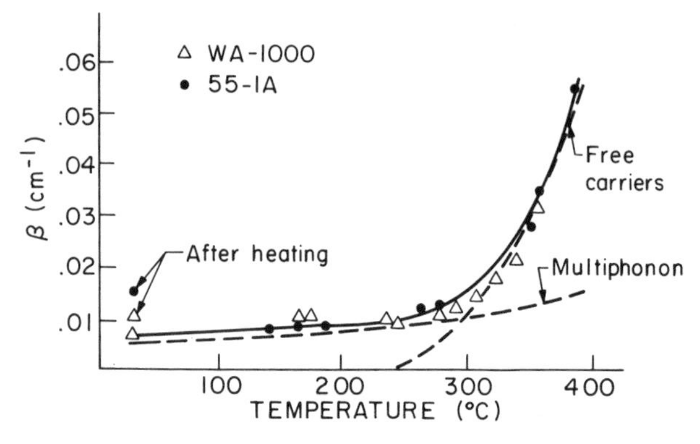

Fig. 2.6. Temperature dependence of the absorption coefficient at 9.28 μm for Bridgman-grown sample #WA-1000 (USC) and float-zone sample #55-1A (Bell Labs.) of high-resistivity GaAs (see Fig. 2.5) (after *Christensen* et al. [2.56])

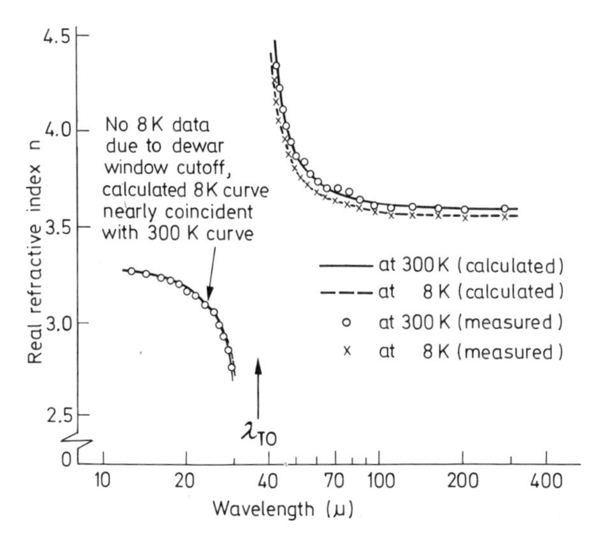

Fig. 2.7. Infrared refractive index of high-resistivity GaAs at 300 K and 8 K. λ_{TO} denotes the wavelength of the zone-center transverse optical phonon (after *Johnson* et al. [2.53])

The FIR dielectric properties of GaAs have been studied by several investigators. Figure 2.7 shows the dispersion of the refractive index measured at 300 K and 8 K by *Johnson* et al. [2.53]. The observed dispersion is fitted to a single classical oscillator model which yields

$$n(\omega)=\left[\varepsilon_\infty+\frac{(\varepsilon_0-\varepsilon_\infty)}{\left(1-\frac{\omega}{\omega_{TO}}\right)^2}\right]^{1/2} \tag{2.63}$$

where ε_∞ is the high-frequency dielectric constant, ε_0 is the low-frequency dielectric constant, and ω_{TO} is the frequency of the TO phonon. The calculated curves in Fig. 2.7 correspond to $\varepsilon_\infty=10.9\pm0.4$ at 300 K and 8 K, $\varepsilon_0=12.8\pm0.5$ at 300 K, 12.6 ± 0.5 at 8 K, $\omega_{TO}=269\ \mathrm{cm}^{-1}$ at 300 K and $273\ \mathrm{cm}^{-1}$ at 8 K. These values are consistent with those deduced at 300 K by

Iwasa et al. [2.57]: $\varepsilon_\infty = 11.10$, $\varepsilon_0 = 13.05$, and $\omega_{TO} = 268.2\ \mathrm{cm}^{-1}$. Refractive index measurements of *Stolen* [2.54] in the frequency range $10\ \mathrm{cm}^{-1}$ to $50\ \mathrm{cm}^{-1}$ are in better agreement with those of *Iwasa* et al. [2.57].

The electro-optic coefficient of GaAs at 10.6 μm has been measured by *Yariv* et al. [2.58] and *Kaminow* [2.59]. Using modulation frequencies well below the acoustic resonances of the sample, they obtained "unclamped" or "free" electro-optic coefficient $r_{14} = 1.6 \times 10^{-12}$ m/V. Using modulation frequencies well above the acoustic resonances of the sample, *Turner* [2.60] measured the "clamped" electro-optic coefficient $r_{14} = (1.5 \pm 0.15) \times 10^{-12}$ m/V at a wavelength of 3.39 μm. Similar values have been obtained by *Mooradian* and *McWhorter* [2.61], $r_{14} = (1.5 \pm 0.3) \times 10^{-12}$ m/V, and by *Johnston* and *Kaminow* [2.62], $r_{14} = (1.5 \pm 0.1) \times 10^{-12}$ m/V from an analysis of Raman scattering efficiencies of LO and TO phonons at a wavelength of 1.06 μm.

As pointed out by *Boyd* et al. [2.41], the nonlinear coefficient to be considered in the generation of FIR by difference-frequency mixing of two near-IR lasers is the one deduced from the electro-optic effect. The reasoning for this goes as follows: In both the electro-optic effect and FIR generation, three frequencies are involved. In either case, two of these frequencies lie above and the third frequency lies below the Restrahlen band. Therefore, the same nonlinearity is involved in both cases. This electro-optic nonlinear coefficient is related to the electro-optic coefficient as

$$d^{eo} = \tfrac{1}{4} n^4 r \tag{2.64}$$

where n is the refractive index at the higher frequencies. Using $n = 3.27$ at 10.6 μm and $r_{14} = (1.5 \pm 0.1) \times 10^{-12}$ m/V, one gets $d_{14}^{eo} = (42.9 \pm 2.9) \times 10^{-12}$ m/V for 10.6 μm.

It is interesting to compare d_{14}^{eo} with the optical nonlinear coefficient, d_{14}^{o} obtained from measurements of second-harmonic generation or sum-frequency generation where all the frequencies involved lie above the Restrahlen band. Using a CO_2 laser, d_{14}^{o} was first measured by *Patel* [2.63] in 1966 as $(387 \pm 126) \times 10^{-12}$ m/V and later by *Wynne* and *Bloembergen* [2.64] in 1969 as $(188.5 \pm 19) \times 10^{-12}$ m/V, and by *McFee* et al. [2.65] in 1970 as $(134 \pm 42) \times 10^{-12}$ m/V. In an attempt to determine a better set of absolute values for the nonlinear optical coefficients of a number of materials, *Levine* and *Bethea* [2.66] have selected the over-all best value of $(90.1 \pm 5.4) \times 10^{-12}$ m/V for d_{14}^{o} of GaAs.

CdTe

Like GaAs, CdTe is another important infrared electro-optic modulator. The FIR absorption coefficient and refractive index of CdTe have been measured by *Johnson* et al. [2.53] at 300 K and 8 K, and by *Danielewicz* and *Coleman* [2.67] at 300 K. Figure 2.8 shows the results of absorption measurements which presumably give an upper limit on the absorption coefficient since it

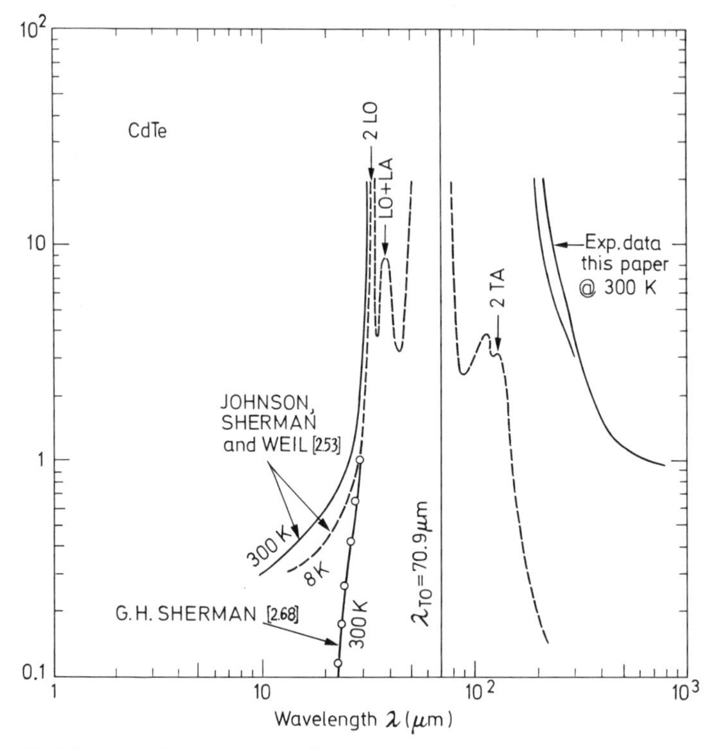

Fig. 2.8. Infrared absorption coefficient of high-resistivity CdTe (after *Danielewicz* and *Coleman* [2.67])

may vary with the quality of the crystal. Included in Fig. 2.8 are some recent data of *Sherman* [2.68] obtained from good quality crystals which reportedly had an absorption coefficient of less than 0.001 cm^{-1} in contrast to that of 0.3 cm^{-1} for the crystal used by *Johnson* et al. [2.53]. A comparison of Figs. 2.3 and 2.8 shows that the FIR absorption loss in CdTe, say, at wavelengths longer than 100 μm, is appreciably higher than that in GaAs. A detailed analysis of the FIR absorption on the long wavelength of the TO phonon similar to that carried out by *Stolen* [2.54] for GaAs, in terms of multiphonon processes, has not been made in CdTe. Nevertheless, the fact that the TO phonon frequency of 141 cm^{-1} (70.9 μm) in CdTe is smaller than that of 268.2 cm^{-1} (37.2 μm) in GaAs accounts, at least, qualitatively for the higher FIR absorption of CdTe.

The refractive index data of CdTe have been fitted to the single oscillator model of (2.63). In this manner *Johnson* et al. [2.53] deduced the following values: $\varepsilon_\infty = 6.7 \pm 0.3$ at 300 K and 8 K, $\varepsilon_0 = 9.4 \pm 0.4$ at 300 K and 9.0 ± 0.4 at 8 K, and $\omega_{TO} = 140.1$ cm^{-1} (71.4 μm) at 300 K and 144.9 cm^{-1} (69.0 μm) at 8 K. The more recent measurements of *Danielewicz* and *Coleman* [2.67] at 300 K yield $\varepsilon_0 = 7.194$, $\varepsilon_\infty = 10.294$, and $\omega_{TO} = 141.0$ cm^{-1} (70.9 μm). Figure 2.9

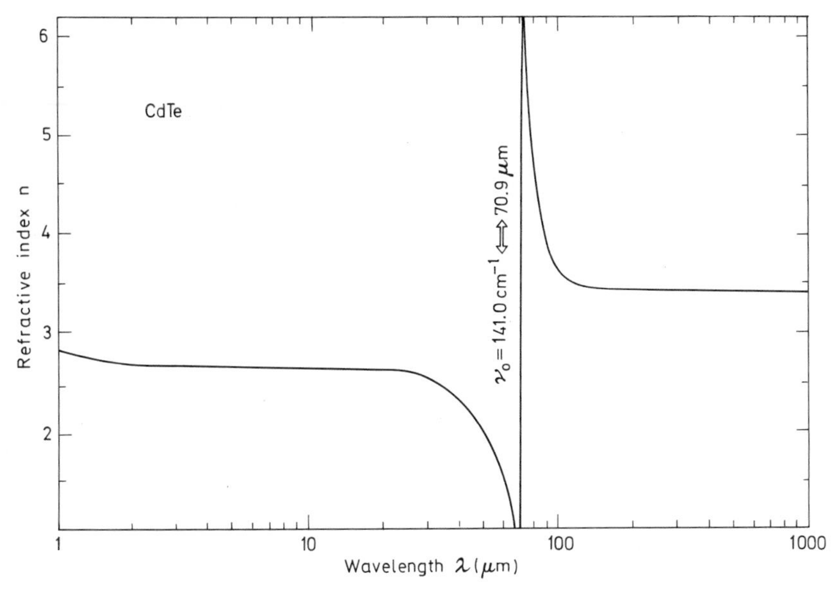

Fig. 2.9. Infrared refractive index of high-resistivity CdTe at room temperature. ν_0 denotes the frequency of the zone-center transverse optical phonon. (after *Danielewicz* and *Coleman* [2.67])

shows the calculated refractive index in the 1–1000 μm range for the latter set of parameters.

In 1969 *Kiefer* and *Yariv* [2.58a] measured the electro-optic properties of high-resistivity CdTe at 10.6 μm. They obtained a value of $(12 \pm 1) \times 10^{-11}$ m/V for the unclamped electro-optic characteristic $n^3 r_{14}$. Using $n = 2.6$, they deduced $r_{14} = (6.8 \pm 0.6) \times 10^{-12}$ m/V. However, if we use the presently accepted value of 2.69 for n at 10.6 μm, we get $r_{14} = (6.16 \pm 0.5) \times 10^{-12}$ m/V. Substituting this value of r_{14} in (2.64), we obtain $d_{14}^{eo} = (80.7 \pm 6.7) \times 10^{-12}$ m/V. In a more recent paper, *Boyd* et al. [2.41] quote a value of 5.6×10^{-12} m/V for r_{14} and a corresponding value of 73×10^{-12} m/V for d_{14}^{eo} which is to be compared with a value of 168×10^{-12} m/V for d_{14}^{o}. The ratio of d_{14}^{eo} to d_{14}^{o} is, therefore. ~0.43 in CdTe. It is interesting to note that this ratio in GaAs (taking $d_{14}^{eo} = 42.9 \times 10^{-12}$ m/V, and $d_{14}^{o} = 90.1 \times 10$ m/V) turns out to be ~0.48 which is quite comparable to the value of 0.43 in CdTe.

ZnSe

ZnSe is one of the low-loss composites developed for use as window material for high-power CO_2 lasers. *Skolnik* et al. [2.69] have reported room-temperature absorption coefficient as low as $0.003\ \mathrm{cm}^{-1}$ at 10.6 μm. Figure 2.10 shows measurements of the absorption coefficient in chemical-vapor deposited (CVD) ZnSe at temperatures between 300 K and 550 K. Over this temperature range, there is no observable increase in the absorption

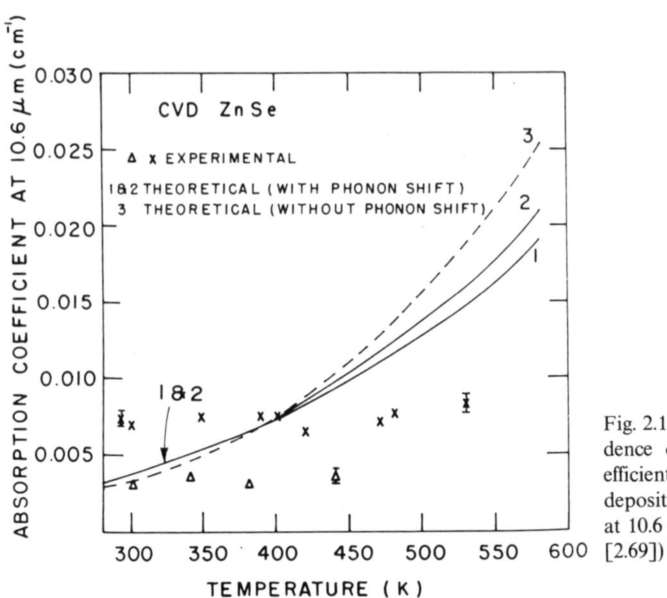

Fig. 2.10. Temperature dependence of the absorption coefficient of chemical vapor deposit (CVD) grown ZnSe at 10.6 μm (after *Skolnik* et al. [2.69])

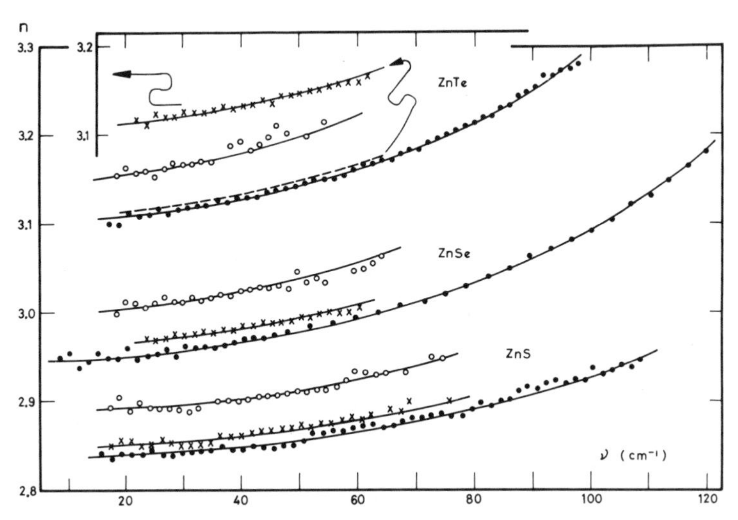

Fig. 2.11. Far-infrared refractive index of ZnS, ZnSe, and ZnTe at 300 K (○), 80 K (×), and 2 K (●) (after *Hattori* et al. [2.71])

coefficient with temperature. On the basis of this observation, *Skolnik* et al. [2.69] have suggested that the surface absorption is the major contribution to the residual absorption at 10.6 μm in ZnSe.

The FIR optical constants of ZnSe were reported by *Manabe* et al. [2.70] in 1967 and by *Hattori* et al. [2.71] in 1973. Figure 2.11 shows the refractive index data of *Hattori* et al. for Bridgman-grown ZnSe at 300 K, 80 K, and

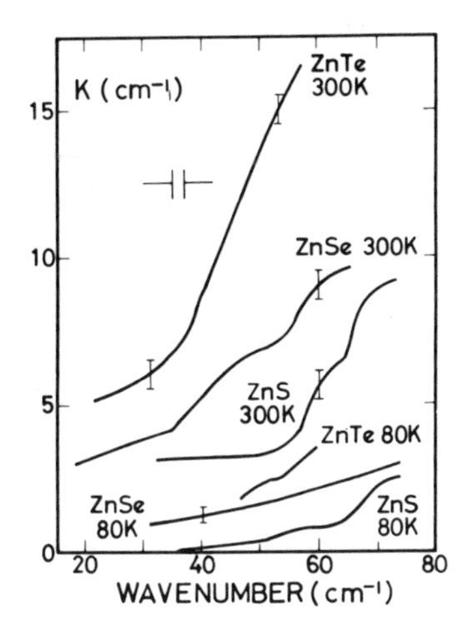

Fig. 2.12. Far-infrared absorption coefficient of ZnTe, ZnSe, and ZnS at 300 K and 80 K (after *Hattori* et al. [2.71])

2 K. By fitting these data to the undamped single oscillator model of (2.63), they have deduced $\varepsilon_\infty = 5.3$ at 300 K, 5.5 at 80 K, 5.6 at 2 K; $\varepsilon_0 = 8.99 \pm 0.05$ at 300 K, 8.76 ± 0.05 at 80 K, and 8.68 ± 0.03 at 2 K, $\omega_{TO} = 203\ \mathrm{cm}^{-1}$ at 300 K, $207\ \mathrm{cm}^{-1}$ at 80 K, and $210\ \mathrm{cm}^{-1}$ at 2 K. The FIR absorption coefficient of ZnSe is shown in Fig. 2.12 for $T = 300$ K and 80 K. A comparison of Figs. 2.3 and 2.12 shows that the FIR absorption in ZnSe is considerably higher than that in GaAs over the same frequency range. Since these FIR absorption measurements were made in the crystals grown by the Bridgman method, it would be interesting to repeat these measurements in CVD ZnSe which exhibits extremely low absorption coefficient at 10.6 μm.

The electro-optic coefficient of Bridgman-grown ZnSe at 10.6 μm was reported by *Kojima* et al. [2.72] in 1969. They found a value of 26×10^{-12} m/V for $n^3 r_{14}$ and a value of 2.3 for n. This yields $r_{14} = 2.1 \times 10^{-12}$ m/V. But if we use a currently accepted value of 2.42 for n, we would get $r_{14} = 1.8 \times 10^{-12}$ m/V and $d_{14}^{eo} = 15.7 \times 10^{-12}$ m/V. This value of electro-optic nonlinear coefficient is about a factor of 5 smaller than that of CdTe. Therefore, the latter would be preferred over ZnSe for FIR generation by difference-frequency mixing of CO_2 lasers.

$ZnGeP_2$ and Other Ternary Compounds

The class of nonlinear ternary compound semiconductors of the type II-IV-V_2 and I-III-VI_2 are tetrahedrally bonded analogues of the III-V and II-VI binary systems. These ternary compounds, e.g., $ZnGeP_2$, $CdGeP_2$, $CdGeAs_2$, $AgGaS_2$, $AgGaSe_2$, etc., usually crystallize in the chalcopyrite structure which has the point group symmetry $\bar{4}2$ m. Both the linear and nonlinear properties of a large number of ternary compounds in the infrared region have been studied by *Boyd* and coworkers [2.73–2.76] and similar measurements in $CdGeAs_2$ have also been reported by *Byer* et al. [2.77]. The FIR

Table 2.1. Electro-optic and nonlinear optical coefficients

Material	Point group symmetry	n ($\lambda = 10.6\ \mu m$)	ε_∞ (high freq.)	ε_0 (low freq.)	r_{14} (10^{-12} m/V)	d^{eo}_{14} (10^{-12} m/V)	d^{o}_{14} (10^{-12} m/V)
GaAs	$\bar{4}3$ m	3.27	11.10 [a]	13.05 [a]	−1.5 [b]	+42.9	90.1 ± 5 [c]
CdTe	$\bar{4}3$ m	2.69	7.194 ± 0.08 [d]	10.294 ± 0.01 [d]	−5.6 [e]	+73.3	168 [e]
ZnSe	$\bar{4}3$ m	2.42	5.90 [f]	8.99 ± 0.05 [g]	2.2 [h]	15.5	19.1 ± 7.2 [i, j]
ZnTe	$\bar{4}3$ m	2.69	7.28 [f]	9.92 ± 0.05 [g]	1.4 [b]	18.3	22.5 ± 8.2 [i, j]
InAs	$\bar{4}3$ m	3.49	11.8 ± 0.1 [k]	14.55 ± 0.3 [k]			200 ± 20 [j, l]
$ZnGeP_2$	$\bar{4}2$ m	3.09 [m]	10.07 [n]	11.48 [n]	1.58 [o]	38.8 [o]	74.8 ± 11.2 [l, p]
$CdGeP_2$	$\bar{4}2$ m	3.15 [m]	9.85 [n]	12.02 [n]			109.0 ± 16.4 [l, p]
$CdGeAs_2$	$\bar{4}2$ m	3.55 [m]					236.1 ± 35.4 [l, p]
$AgGaS_2$	$\bar{4}2$ m	2.32 [q]	5.70 [r]	8.37 [r]			12.1 ± 1.8 [p, q]
$AgGaSe_2$	$\bar{4}2$ m	2.57 [s]					33.2 ± 3.3 [p, s]
$ZnSiAs_2$	$\bar{4}2$ m	3.19 [m]					73.0 ± 10.8 [l, p]
$AgInSe_2$	$\bar{4}2$ m	2.62 [s]					37.6 ± 3.8 [p, s]

a [2.57]
b [2.59, 60]
c [2.66]
d [2.67]
e As given in Table 1 of [2.41]
f D.T.M. Marple: J. Appl. Phys. **35**, 539 (1964)
g [2.71]
h [2.72]
i [2.63]
j Measured values of d_{14} have been scaled down so as to force the agreement with a value of 215 (esu) = 90.1×10^{-12} m/V for d_{14} of GaAs given in [2.66]
k O.G. Lorimer, W.G. Spitzer: J. Appl. Phys. **36**, 184 (1965)
l [2.64]
m [2.75]
n [2.79]
o See Ref. 15 in [2.46]
p The values measured relative to GaAs have been converted to absolute values by using d^{o}_{14} (GaAs) = 90.0×10^{-12} m/V from [2.66]
q [2.74]
r [2.81]
s [2.76]

optical constants: refractive index n, extinction coefficient k, and the real and imaginary parts of the dielectric constant ε' and ε'' have been obtained for several crystals from an analysis of the polarized reflectivity spectra. Such FIR measurements have been made by *Holah* [2.78] in $ZnSiP_2$, by *Miller* et al. [2.79] in $ZnGeP_2$ and $CdGeP_2$, by *Baars* and *Koschel* [2.80] in $CuGaS_2$, and by *van der Ziel* [2.81] in $AgGaS_2$. With the exception of $ZnGeP_2$ [2.46] detailed FIR transmission measurements in these ternary compounds are not yet available.

Since a number of these nonlinear ternary crystals are transparent in the 10 μm region of the CO_2 laser, they have a potential for FIR generation by difference-frequency mixing of CO_2 laser lines. In fact, *Boyd* et al. [2.46] have already reported the birefringent phase-matched mixing of CO_2 laser lines in $ZnGeP_2$ for FIR generation between 70 cm^{-1} and 110 cm^{-1}. With future advances in the quality and size of crystals, it is expected that these ternary semiconductors will play an important role as nonlinear difference-frequency mixers for FIR generation.

In Table 2.1 we summarize the electro-optic and nonlinear optical coefficients of a number of crystals including ternary chalcopyrites which appear promising or have already been used for FIR generation by difference-frequency mixing of CO_2 laser radiation. Table 2.1 also gives the point group symmetry, the refractive index at 10.6 μm and the high- and low-frequency dielectric constants.

2.4 Noncollinear Mixing Geometries

2.4.1 Simple Noncollinear Geometry

Figure 2.13 shows the typical cut of a crystal for noncollinear phase-matched difference-frequency FIR generation at $\omega_3 = \omega_1 - \omega_2$, where ω_1 and ω_2 are the frequencies of the input beams. It is assumed that $\omega_1 > \omega_2$. The ω_1-beam is incident along the normal to the input face, and the ω_2-beam is incident at an external angle θ_E so that the angle between the two input beams inside the crystal is

$$\theta = \sin^{-1}\left(\frac{1}{n_2}\sin\theta_E\right) \tag{2.65}$$

where n_2 is the refractive index at the frequency ω_2. Under phase-matched conditions, the ω_3-beam generated in the crystal propagates at an alge φ relative to the ω_1-beam as shown. For CO_2 laser mixing in GaAs, $\varphi \simeq 21°$ which is larger than the critical angle for total internal reflection of $\sim 17°$ in GaAs. Therefore, it is necessary to cut the output of the crystal in order to avoid total internal reflection of the ω_3-beam. In particular, the output face is cut at an angle φ. This allows the ω_3-beam to emerge out of the crystal without suffering any refraction.

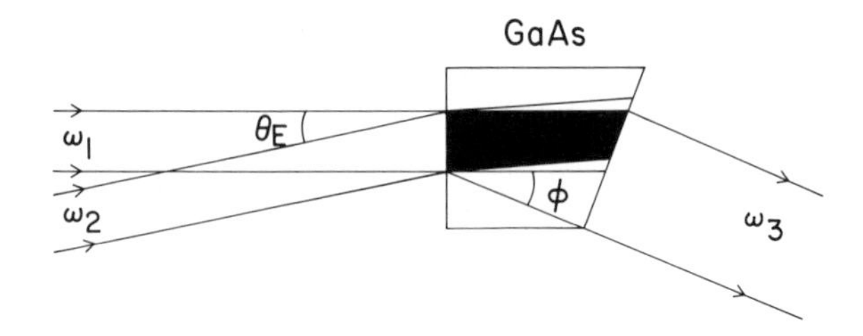

Fig. 2.13. Typical crystal cut for simple noncollinear geometry. ω_1 and ω_2 are the frequencies of the input laser beams and $\omega_3=\omega_1-\omega_2$ is frequency of the far infrared radiation (after *Aggarwal* et al. [2.13])

An expression for the intensity of the FIR difference-frequency radiation generated in the simple noncollinear geometry of Fig. 2.13 can be deduced by solving Maxwell's equations following a procedure similar to that used for collinear mixing as given in the previous section. We [2.14] reduce the problem to two dimensions by assuming that all beams propagate in the x, y plane and are polarized along the z-axis. We further assume that the incident beams at the input frequencies ω_1 and ω_2 are unattenuated plane waves, and that the nonlinear interaction between them occurs in the first quadrant of the x, y plane as shown in Fig. 2.14. We are interested in the component of the frequency $\omega_3=\omega_1-\omega_2$ and express it in the form

$$\boldsymbol{E}_3(\boldsymbol{r}, t)=\mathrm{Re}\{\hat{z}E_3(x, y)\mathrm{e}^{\mathrm{i}(\omega_3 t-\boldsymbol{k}_3\cdot\boldsymbol{r})}\}. \tag{2.66}$$

Neglecting losses and the term $\nabla^2 E_3(x, y)$, $E_3(x, y)$ is obtained from the solution of the following equation (in MKS units):

$$\mathrm{i}(\boldsymbol{k}_3\cdot\nabla)E_3(x, y)=\tfrac{1}{2}\varepsilon_0\mu_0\omega_3^2(2d_{\mathrm{eff}})E_1E_2^*\mathrm{e}^{\mathrm{i}(\boldsymbol{k}_3-\boldsymbol{k}_1+\boldsymbol{k}_2)\cdot\boldsymbol{r}} \tag{2.67}$$

where E_1 and E_2 are the amplitudes of the fields at ω_1 and ω_2, respectively, and d_{eff} is the effective nonlinear coefficient.

Transforming to the coordinates axes x' and y' as shown in Fig. 2.14, (2.67) can be written as

$$k_3\frac{\partial}{\partial x}E_3(x', y')=-\frac{\mathrm{i}}{2}\varepsilon_0\mu_0\omega_3^2(2d_{\mathrm{eff}})E_1E_2^*\mathrm{e}^{\mathrm{i}(\Delta k_{x'}x'+\Delta k_{y'}y')} \tag{2.68}$$

where

$$\Delta k_{x'}=-k_1\cos\varphi+k_2\cos(\theta+\varphi)+k_3\,, \tag{2.69a}$$

$$\Delta k_{y'}=k_1\sin\varphi-k_2\sin(\theta+\varphi)\,. \tag{2.69b}$$

For $\Delta k_{x'}=\Delta k_{y'}=0$, i.e., under phase-matched conditions, one obtains the following solution of (2.68) at the output face of the crystal:

$$E_3(x', y')=-(\mathrm{i}/2k_3)\varepsilon_0\mu_0\omega_3^2(2d_{\mathrm{eff}})E_1E_2^*F(y') \tag{2.70}$$

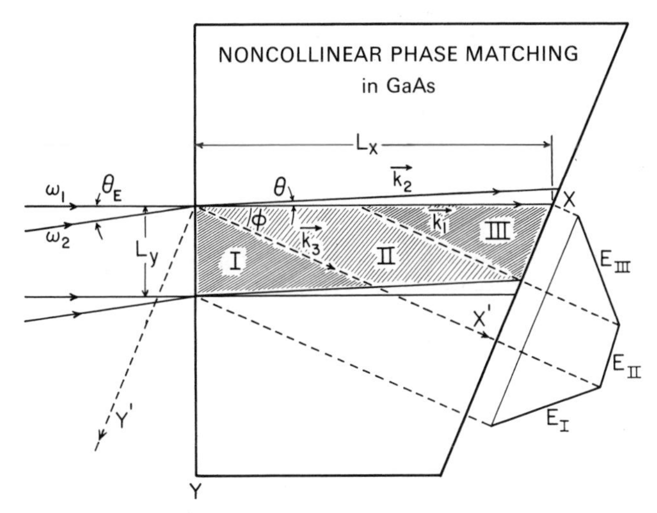

Fig. 2.14. Schematic diagram showing the interaction region for noncollinear mixing and the variation of the electric field amplitude E_3 for the difference-frequency radiation across the output face of the crystal (after *Lax* et al. [2.14])

where the function $F(y')$ is given by the following expressions for the three regions labelled I, II, and III:

$$\text{I)}\quad F(y') = -y'[\cot(\theta+\varphi)+\tan\varphi] + L_y \sin\varphi + L_y \cos\varphi \cot(\theta+\varphi) \tag{2.71}$$

$$\text{II)}\quad F(y') = -y'[\cot(\theta+\varphi)-\cot\varphi] + L_y \sin\varphi + L_y \cos\varphi \cot(\theta+\varphi) \tag{2.72}$$

$$\text{III)}\quad F(y') = y' \cot\varphi + L_x \cos\varphi\,. \tag{2.73}$$

Here L_y is the dimension of the input beams along the y-axis. In deducing (2.71–2.73) it is assumed that $L_y \leqq L_x \sin\varphi \cos\varphi$. The variation of E_3 across the output as predicted by (2.70–2.73) is shown schematically on the right-hand side of Fig. 2.14. We can now solve for the total power at ω_3 in the near-field or far-field approximation. From the conservation of energy these two are equivalent. Mathematically it is simpler to calculate the power in the near-field approximation. In that case

$$P_{\omega_3} = \frac{1}{2}\sqrt{\frac{\varepsilon}{\mu_0}}\, L_z \int E_3(y')E_3^*(y')dy'\,, \tag{2.74}$$

if the summation is carried out for all the filaments parallel to k_3 in the three regions. By using the fact that $\theta \ll \varphi$ and neglecting θ, we obtain the approximate result for a trapezoidal amplitude

$$P_{\omega_3} = \frac{1}{2}\left(\frac{\mu_0}{\varepsilon_0}\right)^{1/2} \frac{(2d_{\text{eff}})^2\omega_3^2}{n_1 n_2 n_3 c^2} \frac{P_{\omega_1} P_{\omega_2}}{A} T_1 T_2 T_3 L_{\text{eff}}^2 \tag{2.75}$$

with

$$L_{\text{eff}} = \left(\frac{L_y}{\sin\varphi}\right)\left(\frac{L_x}{L_y}\sin\varphi - \frac{1+\sin^2\varphi}{3\cos\varphi}\right)^{1/2} \tag{2.76}$$

where A denotes the area of the input beams and T_1, T_2, and T_3 are the single surface transmission coefficients at the frequencies ω_1, ω_2, and ω_3. Equation (2.75) is analogous to (2.28) obtained in the collinear phase-matched configuration except that the crystal length L is replaced by L_{eff}. In other words, L_{eff} represents the equivalent effective interaction length in noncollinear phase-matched mixing. In the limiting case $L_x \gg L_y$, (2.76) yields

$$L_{\text{eff}} = \left(\frac{L_x L_y}{\sin\varphi}\right)^{1/2} \tag{2.77}$$

which shows that for a given value of L_y, L_{eff} increases only as the square-root of the crystal dimension L_x instead of increasing linearly with L_x as would be the case in collinear mixing. Therefore, the simple noncollinear mixing geometry of Fig. 2.13 is not most optimal for obtaining large values of L_{eff}. It has been shown by *Lee* et al. [2.16] that this drawback of the simple noncollinear geometry of Fig. 2.13 can be overcome by using new crystal geometries which make use of multiple total internal reflections. Since the total internal reflections are equivalent to folding a long crystal into a short one, these geometries are called "folded" geometries.

2.4.2 Folded Noncollinear Geometry

The idea of using multiple internal reflections in nonlinear optics was first proposed by *Armstrong* et al. [2.36] in 1962 and later used by *Boyd* and *Patel* [2.11] in 1966 for achieving phase-matching in second-harmonic generation. In noncollinear mixing multiple internal reflections are used to increase L_{eff} since the phase-matching condition is already satisfied by the noncollinear mixing.

The principle of the folded geometry proposed by *Lee* et al. [2.16] is illustrated in Fig. 2.15 which shows two laser beams of frequencies ω_1 and ω_2 with corresponding wave vectors $\boldsymbol{k}_1$ and $\boldsymbol{k}_2$ being incident on a polished crystal surface from inside. The angles θ and φ are chosen to satisfy the conditions for noncollinear phase matching given by (2.59) and (2.60). Under these phase-matched conditions, the wave vector $\boldsymbol{k}_3$ of the FIR beam at the difference-frequency $\omega_3 = \omega_1 - \omega_2$ will be parallel to the crystal surface. If this surface is coated with metal, the reflected laser beams of wave vectors $\boldsymbol{k}_1'$ and $\boldsymbol{k}_2'$ will also generate a FIR beam of frequency ω_3 with wave vector $\boldsymbol{k}_3' = \boldsymbol{k}_3$. The phase of the FIR radiation generated under phase-matched conditions is determined only by the relative phase difference of the two input laser beams, which does not change on reflection from a metal surface since both the laser beams undergo the same phase shift of 180° on reflection. This implies

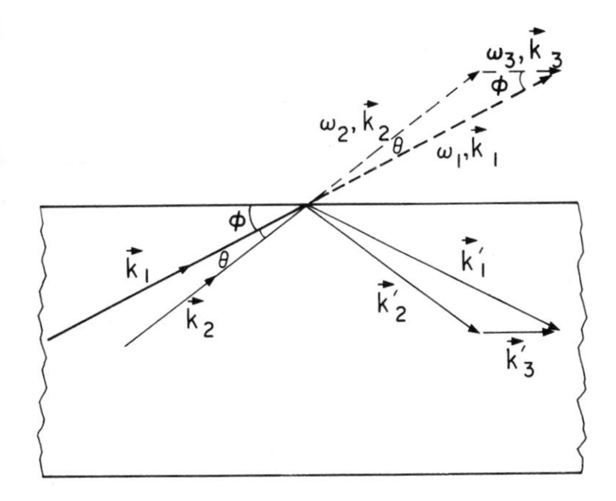

Fig. 2.15. Illustration of the principle of noncollinear folded crystal geometry (after *Lee* et al. [2.16])

that the FIR beam generated by the reflected laser beams propagates in the same direction and in phase with the FIR beam generated by the primary laser beams. In other words, the electric fields of the two FIR beams can be added up in phase. Consequently, the total FIR power output will vary as L^2 where L is the long dimension of the crystal along the reflecting surface.

In case the crystal surface is not coated with metal, the input laser beams will be reflected due to total internal reflection provided the angle φ is larger than the critical angle, which is the case, at least, in GaAs and CdTe. Upon total internal reflection, there occurs a small change in the phase difference of the reflected laser beams relative to that of the input laser beams. Therefore, the FIR beam generated by the reflected laser beams will acquire a small phase shift [2.82].

$$\delta = 2\tan^{-1}\left\{\frac{(n_1^2\cos^2\varphi - 1)^{1/2}}{n_1 \sin\varphi}\right\} - 2\tan^{-1}\left\{\frac{[n_2^2\cos^2(\varphi+\theta) - 1]^{1/2}}{n_2\sin(\varphi+\theta)}\right\} \quad (2.78)$$

relative to the FIR beam generated by the primary laser beams. Since $n_1 \simeq n_2$ and $\theta \ll \varphi$, δ is negligibly small. For multiple reflections, the power loss due to these small shifts can be substantially reduced by a slight deviation from the phase-matching conditions such that the two phase shifts tend to cancel each other. In any case, the resulting loss in power for not coating the crystal surface with metal will be less than a few percent in most cases.

Figure 2.16 shows the simplest case of a folded crystal geometry. Here the ω_2-beam is incident at an angle θ_E to the ω_1-beam $(\omega_1 > \omega_2)$ such that upon refraction into the crystal the angle θ between the two beams satisfies the noncollinear phase-matching condition. These two laser beams propagate zigzag to the right due to reflections between the two plane parallel surfaces. As shown in Fig. 2.15, the generated FIR beam propagates parallel to the long dimension L of the crystal, and that the FIR generated by each input beam

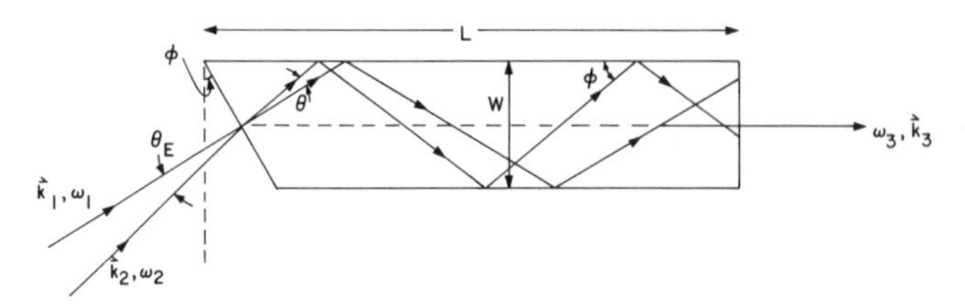

Fig. 2.16. A simple version of the noncollinear folded crystal geometry showing the propagation of the input laser beams of frequencies ω_1 and ω_2 and that of the difference-frequency output beam at ω_3 (after *Lee* et al. [2.16])

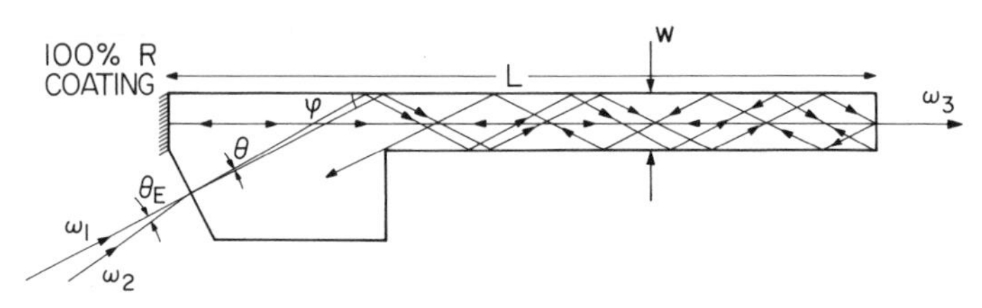

Fig. 2.17. A more sophisticated noncollinear folded geometry with two end faces perpendicular to the propagation direction of the far-infrared difference-frequency beam (after *Lee* et al. [2.16])

segment between reflections can be added up in phase. Thus, under phase-matched conditions, the FIR output P_{ω_3} from this crystal is given by (2.75) with

$$L_{\text{eff}} = gL \tag{2.79}$$

where g is the geometrical factor determined by the overlap of the input laser beams and their reflections. Assuming that the width W of the crystal is chosen to match the dimensions of the input laser beams, simple geometrical considerations show that

$$g = 1 + \tan^2\varphi \quad \text{for} \quad 0 \leqq \varphi \leqq 45^\circ\,, \tag{2.80a}$$

$$g = 2 \qquad \text{for} \quad \varphi \geqq 45^\circ \tag{2.80b}$$

provided that walk-off of the two laser beams as a result of the angle θ between them is neglected. In that case the FIR output from the noncollinear folded geometry will always be larger compared to the collinear case.

Let us now consider the walk-off problem. In the case of the simple (nonfolded) noncollinear geometry shown in Fig. 2.13, the two laser beams propagating at an angle θ between them would walk-off by an amount $l\theta$ from each other in a distance l. However, the situation is quite different in the case of the folded geometry. All those segments of the laser beams which have undergone an odd number of reflections propagate parallel to each other in the same phase. Similarly segments of the laser beams which have undergone an even number of reflections also propagate parallel to each other and in the same phase. While the ω_1-beam walks away from its initial overlapping segment of the ω_2-beam, it walks into another ω_2-segment which has suffered an even number of additional reflections. The maximum fraction f of the beam segments which do not overlap is given by

$$f=(2-g)/g\,. \tag{2.81}$$

Thus, in effect, there is no walk-off for $g=2$.

A somewhat more sophisticated folded geometry is shown in Fig. 2.17. It can provide an even greater enhancement for the FIR generation compared with the simple folded geometry of Fig. 2.16 for which g is determined by φ according to (2.80a, b). But for the folded geometry of Fig. 2.17, it is possible to obtain the maximum value of $g=2$ since the maximum size of beam that can be brought into the crystal is no longer determined by the width W of the crystal. As shown in Fig. 2.17, one end of the crystal is coated with metal and both ends of the crystal are perpendicular to the FIR beam. Thus, the end surfaces of the crystal can form a resonant cavity for the FIR beam. When the length L of the cavity satisfies the resonance condition

$$L_{\text{eff}}=(m+\tfrac{1}{2})\frac{\lambda_3}{2}\,, \qquad m=0, 1, 2, \text{etc.}; \tag{2.82}$$

for the FIR difference-frequency wavelength λ_3, the effective interaction length would become

$$L_{\text{eff}}=\frac{4L}{1-r} \tag{2.83}$$

where

$$r=\frac{n_3-1}{n_3+1} \tag{2.84}$$

is the amplitude reflection coefficient at λ_3. If we consider the case of GaAs, $r\simeq 0.56$ so that $L_{\text{eff}}\simeq 9L$. Therefore, P_{ω_3} could be, in principle, 80 times higher than that obtained with collinear mixing from a crystal of length L. However, the actual gain would not be as great due to the finite absorption of the FIR radiation as well as that of the input laser radiation at ω_1 and ω_2. Secondly, it is not so easy to achieve the resonance condition of (2.82). The simplest method requires a very fine tuning of either ω_1 or ω_2. A more complicated technique would involve placing the nonlinear crystal in an externally tunable cavity.

2.5 Experimental Systems Using CO_2 Lasers

It is well known [2.83] that a CO_2 laser can be tuned to any one of a number of rotational transitions in the *P* and *R* branches of the $\Sigma_u^+(00^\circ1)-\Sigma_g^+(10^\circ0)$ vibrational band centered at 10.4 μm or those of the $\Sigma_u^+(00^\circ1)-\Sigma_g^+(02^\circ0)$ vibrational band centered at 9.4 μm. Figure 2.18 shows a typical spectrum of a grating-tuned CO_2 laser. As can be seen from Fig. 2.18 there are over 100 laser lines which span the spectral region from $\sim 915\,\mathrm{cm}^{-1}$ (10.85 μm) to $1095\,\mathrm{cm}^{-1}$ (9.05 μm). A table of wavelengths and frequencies with absolute

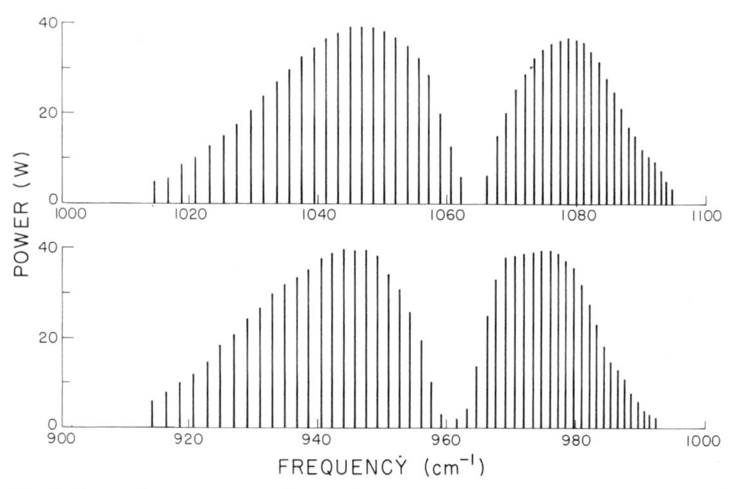

Fig. 2.18. Typical spectrum of a grating-tuned 40 W continuous wave (cw) CO_2 laser (after *Rosenbluh* [2.87])

accuracy of $\sim 10^{-6}$ for 120 transitions was published by *Chang* [2.84] in 1970. Since then *Evenson* et al. [2.85] have reported new frequency measurements and *Baird* et al. [2.86] have made new absolute measurements of wavelength accurate to $\sim 2\times10^{-8}$. The results of these new frequency and wavelength measurements are reproduced in Tables 2.2 and 2.3 for quick reference. Difference-frequency mixing of more than 100 CO_2 laser lines should provide more than 5000 difference frequencies covering the FIR region from $\sim 2\,\mathrm{cm}^{-1}$ (5000 μm = 5 mm) to $180\,\mathrm{cm}^{-1}$ (55 μm) with an average spacing of $\sim 0.04\,\mathrm{cm}^{-1}$. However, there are wide variations in the frequency spacing as well as the relative power between the neighboring FIR lines. This point is well illustrated [2.87] in Fig. 2.19 which shows some of the calculated frequencies and their relative power ($P_{\omega_3} \propto P_{\omega_1}, P_{\omega_2}\omega_3^2$) assuming perfect phase matching and no absorption losses. Only FIR lines with relative power in excess of 0.05 are included in Fig. 2.19.

Table 2.2. Transitions of the $\Sigma_u^+ (00^\circ 1) - \Sigma_g^+ (10^\circ 0)$ band

P branch				*R* branch			
Transition	Wavelength (vac) λ (μm)	Wavenumber (vac) ν (cm^{-1})	Frequency f (MHz)	Transition	Wavelength (vac) λ (μm)	Wavenumber (vac) ν (cm^{-1})	Frequency f (MHz)
P (60)	11.0715398	903.21674	27077756.7	*R* (0)	10.3978980	961.73284	28832025.4
P (58)	11.0434948	905.51046	27146520.8	*R* (2)	10.3813797	963.26310	28877901.4
P (56)	11.0159150	907.77752	27214485.7	*R* (4)	10.3651761	964.76894	28923045.5
P (54)	10.9887923	910.01811	27281656.7	*R* (6)	10.3492850	966.25032	28967456.1
P (52)	10.9621191	912.23238	27348039.0	*R* (8)	10.3337043	967.70719	29011132.0
P (50)	10.9358877	914.42051	27413637.3	*R* (10)	10.3184319	969.13950	29054071.6
P (48)	10.9100910	916.58264	27478456.5	*R* (12)	10.3034659	970.54720	29096273.4
P (46)	10.8847218	918.71893	27542500.8	*R* (14)	10.2888045	971.93022	29137735.1
P (44)	10.8597735	920.82952	27605774.7	*R* (16)	10.2744460	973.28848	29178454.9
P (42)	10.8352393	922.91455	27668282.3	*R* (18)	10.2603891	974.62191	29218429.9
P (40)	10.8111131	924.97413	27730027.1	*R* (20)	10.2466323	975.93040	29257657.6
P (38)	10.7873887	927.00840	27791013.0	*R* (22)	10.2331742	977.21390	29296135.8
P (36)	10.7640602	929.01747	27851243.3	*R* (24)	10.2200138	978.47226	29333860.5
P (34)	10.7411219	931.00144	27910721.3	*R* (26)	10.2071501	979.70539	29370829.0
P (32)	10.7185683	932.96042	27969449.9	*R* (28)	10.1945821	980.91319	29407037.8
P (30)	10.6963942	934.89449	28027431.9	*R* (30)	10.1823089	982.09552	29442483.3
P (28)	10.6745944	936.80374	28084669.8	*R* (32)	10.1703300	983.25226	29477161.5
P (26)	10.6531641	938.68825	28141166.1	*R* (34)	10.1586449	984.38327	29511068.1
P (24)	10.6320984	940.54810	28196922.9	*R* (36)	10.1472529	985.48839	29544199.0
P (22)	10.6113930	942.38334	28251942.1	*R* (38)	10.1361538	986.56750	29576549.9
P (20)	10.5910433	944.19404	28306225.4	*R* (40)	10.1253474	987.62043	29608115.9
P (18)	10.5710453	945.98024	28359774.3	*R* (42)	10.1148336	988.64701	29638891.9
P (16)	10.5513950	947.74198	28412590.1	*R* (44)	10.1046124	989.64707	29668872.9
P (14)	10.5320883	949.47931	28464673.9	*R* (46)	10.0946838	990.62043	29698053.6
P (12)	10.5131217	951.19226	28516026.8	*R* (48)	10.0850481	991.56691	29726428.4
P (10)	10.4944916	952.88085	28566649.3	*R* (50)	10.0757058	992.48631	29753991.1
P (8)	10.4761946	954.54507	28616541.6	*R* (52)	10.0666570	993.37844	29780736.5
P (6)	10.4582276	956.18497	28665704.3	*R* (54)	10.0579026	994.24307	29806657.6
P (4)	10.4405874	957.80052	28714137.3	*R* (56)	10.0494431	995.08002	29831748.6
P (2)	10.4232711	959.39172	28761840.3	*R* (58)	10.0412793	995.88904	29856002.5

Table 2.3. Transition of the $\Sigma_u^+ (00^\circ 1) - \Sigma_g^+ (02^\circ 0)$ band

P branch				*R* branch			
Transition	Wavelength (vac) λ (μm)	Wavenumber (vac) ν (cm^{-1})	Frequency f (MHz)	Transition	Wavelength (vac) λ (μm)	Wavenumber (vac) ν (cm^{-1})	Frequency f (MHz)
P (60)	9.9454584	1005.48407	30143654.3	*R* (0)	9.3940036	1064.50885	31913172.7
P (58)	9.9228349	1007.77652	30212380.1	*R* (2)	9.3805343	1066.03735	31958996.1
P (56)	9.9005305	1010.04689	30280444.2	*R* (4)	9.3673384	1967.53910	32004017.4
P (54)	9.8785443	1012.29490	30347837.8	*R* (6)	9.3544137	1069.01409	32048236.3
P (52)	9.8568756	1014.52026	30414552.5	*R* (8)	9.3417582	1070.46230	32091652.7
P (50)	9.8355233	1016.72272	30480580.4	*R* (10)	9.3293698	1071.88376	32134266.9
P (48)	9.8144866	1018.90199	30545913.5	*R* (12)	9.3172463	1073.27848	32176079.6
P (46)	9.7937644	1021.05785	30610544.4	*R* (14)	9.3053857	1074.64649	32217091.4
P (44)	9.7733556	1023.19003	30674465.7	*R* (16)	9.2937855	1075.98782	32257303.5
P (42)	9.7532590	1025.29831	30737670.4	*R* (18)	9.2824437	1977.30252	32296717.3
P (40)	9.7334735	1027.38247	30800151.8	*R* (20)	9.2713580	1078.59065	32335334.3
P (38)	9.7139977	1029.44228	30861903.5	*R* (22)	9.2605261	1079.85225	32373156.4
P (36)	9.6948305	1031.47755	30922919.3	*R* (24)	9.2499457	1081.08743	32410186.0
P (34)	9.6759704	1033.48808	30983193.3	*R* (26)	9.2396144	1082.29625	32446425.4
P (32)	9.6574160	1035.47367	31042719.8	*R* (28)	9.2295299	1083.47880	32481877.4
P (30)	9.6391660	1037.43415	31101493.5	*R* (30)	9.2196899	1084.63518	32516544.9
P (28)	9.6212189	1039.36935	31159509.3	*R* (32)	9.2100918	1085.76551	32550431.4
P (26)	9.6035731	1041.27911	31216762.5	*R* (34)	9.2007332	1086.86990	32583540.2
P (24)	9.5862271	1043.16327	31273248.4	*R* (36)	9.1916117	1087.94848	32615875.2
P (22)	9.5691792	1045.02171	31328962.8	*R* (38)	9.1827247	1089.00139	32647440.6
P (20)	9.5524279	1046.85427	31383901.8	*R* (40)	9.1740698	1090.02877	32678240.6
P (18)	9.5359715	1048.66085	31438061.6	*R* (42)	9.1656443	1091.03078	32708280.1
P (16)	9.5198083	1050.44132	31491438.8	*R* (44)	9.1574456	1092.00758	32737563.9
P (14)	9.5039365	1052.19558	31544030.2	*R* (46)	9.1494711	1092.95935	32766097.2
P (12)	9.4883544	1053.92354	31595833.0	*R* (48)	9.1417182	1093.88626	32793885.4
P (10)	9.4730601	1055.62509	31646844.4	*R* (50)	9.1341842	1094.78852	32820934.4
P (8)	9.4580519	1057.30018	31697062.3	*R* (52)	9.1268663	1095.66631	32847250.0
P (6)	9.4433278	1058.94873	31746484.4	*R* (54)	9.1197618	1096.51987	32872838.8
P (4)	9.4288860	1060.57068	31795109.2	*R* (56)	9.1128679	1097.34938	32897707.1
P (2)	9.4147245	1062.16597	31842934.9	*R* (58)	9.1061818	1098.15510	32921861.9

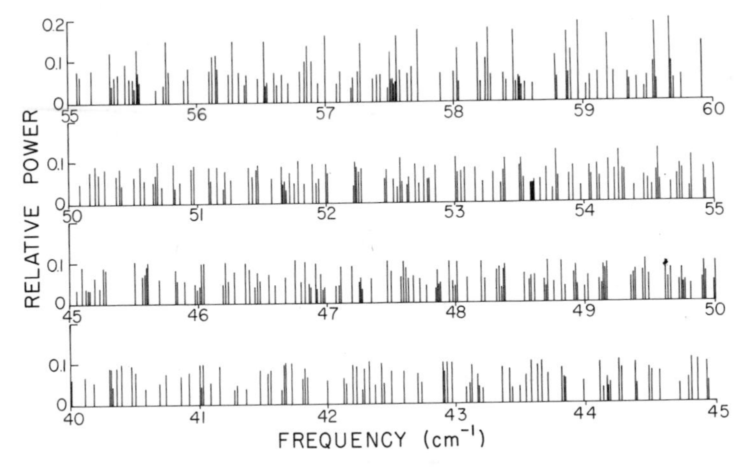

Fig. 2.19a. Typical spectrum from 40–60 cm^{-1} for far-infrared difference frequencies of two CO_2 lasers. Powers were normalized to 1 at the strongest FIR around 129 cm^{-1}, and only lines with relative powers in excess of 0.05 are shown (after *Rosenbluh* [2.87])

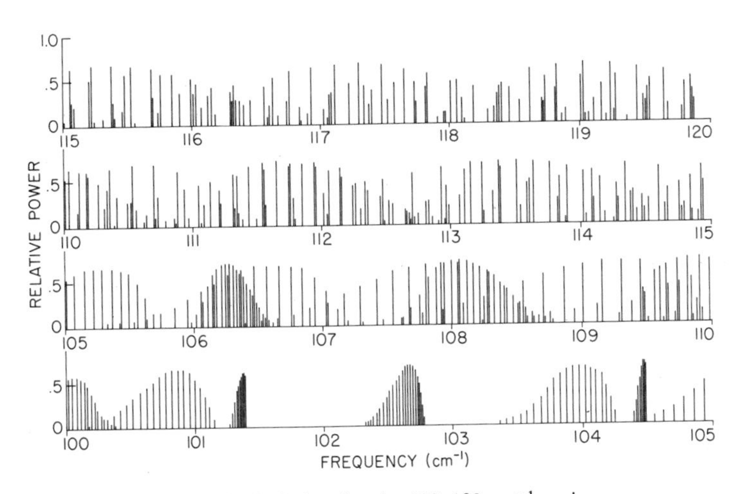

Fig. 2.19b. Same as Fig. 2.19a but for the 100–120 cm^{-1} region

2.5.1 Noncollinear Phase-Matched FIR Generation in GaAs

GaAs is one of the best nonlinear materials for FIR generation by the difference-frequency mixing of CO_2 laser lines for the following reasons:

i) It has a relatively high value for the electro-optic nonlinear coefficient (see Table 2.1).
ii) It is highly transparent in the FIR region beyond 100 μm as well as in the 10 μm region of the CO_2 laser, as can be seen from Figs. 2.26.

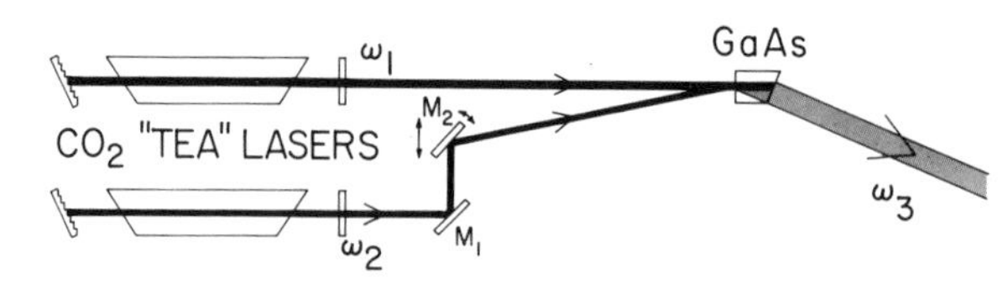

Fig. 2.20. A schematic of experimental arrangement for the generation of pulsed far-infrared radiation by noncollinear phase-matched-mixing of CO_2 laser beams in GaAs (after *Lax* et al. [2.14])

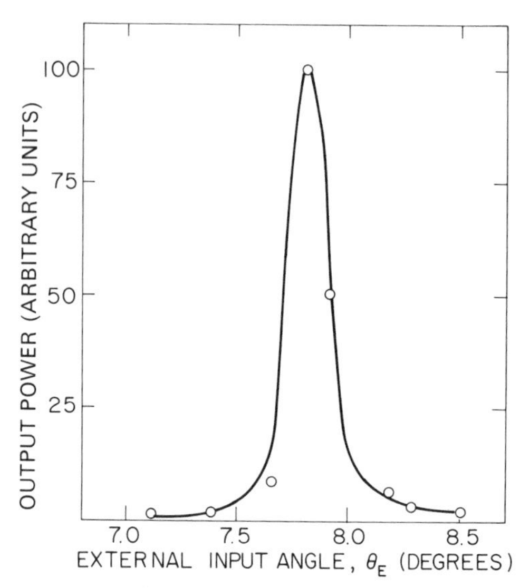

Fig. 2.21. Relative output power of the difference-frequency radiation at 100.8 cm^{-1} vs. external angle θ_E between the 9.57 and 10.59 μm CO_2 laser beams for mixing in GaAs at $T \simeq 20$ K (after *Aggarwal* et al. [2.13])

iii) It possesses anomalous dispersion between 10 μm and FIR. This allows the use of noncollinear phase matching for the difference-frequency mixing, as discussed in Section 2.2.2.

iv) It is commercially available in relatively large sizes. High-quality single crystals with lengths up to 30 cm with a cross section of several cm^2 can be obtained from commercial sources [2.88].

The principle of noncollinear phase matching for the difference-frequency generation in GaAs was verified by *Aggarwal* et al. [2.13] in a crystal of simple noncollinear geometry of Fig. 2.13, using the experimental arrangement similar to that shown in Fig. 2.20. Pulsed laser beams from two transversely excited atmospheric (TEA) type CO_2 lasers were directed toward the input face of the GaAs crystal at an external angle θ_E between them. θ_E can be varied by adjustments of the mirror M_2 in such a way that the two beams maintain an overlap at the input face of the crystal. A plot of the relative FIR output power at the difference-frequency $\omega_3 = 100.8$ cm^{-1} vs. the angle θ_E between the ω_1-beam (1045.0 cm^{-1}) and the ω_2-beam (944.2 cm^{-1}) is shown in Fig. 2.21. It clearly shows that the FIR output is strongly dependent on θ_E. The output is sharply peaked at $\theta_E = 7.82 \pm 0.1°$ which is in good agreement with that calculated from (2.59) and (2.65) using the measured values for the refractive indices of GaAs [2.53, 57].

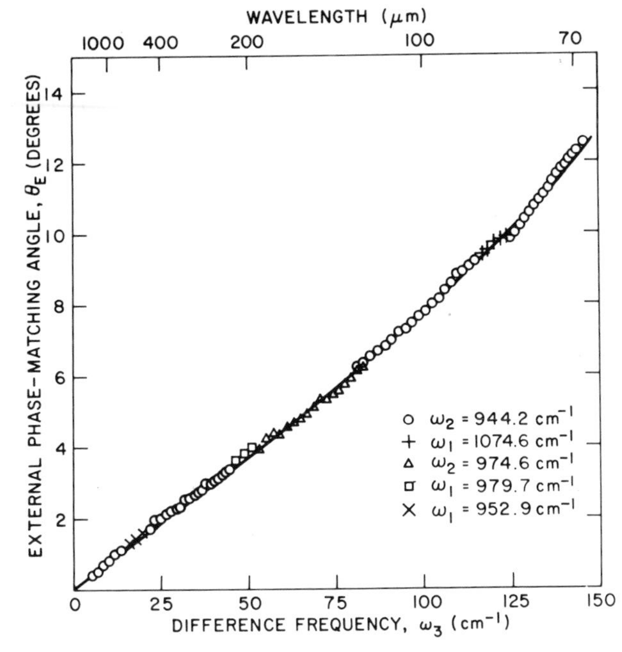

Fig. 2.22. External phase-matching angle θ_E between the two CO_2 laser beams vs. the difference-frequency of the FIR radiation generated in GaAs at $T \simeq 20$ K. The circles, triangles, squares, etc., represent the experimental data, and the solid line represents the calculated curve (after *Lax* et al. [2.14])

Having confirmed the technique of noncollinear phase matching, *Lax* et al. [2.14] tuned the wavelength of the FIR radiation from ~ 70 μm to 2 mm in small steps by selecting the frequencies of the two CO_2 lasers by means of the intracavity diffraction gratings (see Fig. 2.20). The resulting tuning curve is shown in Fig. 2.22 which is a plot of the external phase matching angle θ_E vs. the difference-frequency ω_3. The solid line representing the calculated phase-matching curve is in fact masked by the datum points indicating that the latter are in good agreement with the calculations based on (2.59) and (2.65) over the entire tuning range of Fig. 2.22.

Let us now consider the effect of crystal orientation on the generation of the difference-frequency radiation in GaAs which has the point group symmetry $\bar{4}3$ m for which the nonzero components of the nonlinear coefficient are [2.39] $d_{14}=d_{25}=d_{36}$. For given polarization of the input laser beams, components of the nonlinear polarization can be obtained from (2.5). It is instructive to consider a few cases.

Let us first consider the case where the input laser beams at the frequencies ω_1 and ω_2 are polarized parallel to a [111] crystal axis. That is $\boldsymbol{E}_1 \| \boldsymbol{E}_2 \| [111]$ where $\boldsymbol{E}_1$ and $\boldsymbol{E}_2$ represent the electric vectors of the ω_1 and ω_2 beams, respectively. The components of the nonlinear polarization at the difference-frequency ω_3 are then

$$P_x(\omega_3)=P_y(\omega_3)=P_z(\omega_3)=\tfrac{2}{3}\varepsilon_0(2d_{14})\,E_1E_2^* \tag{2.85}$$

where E_1 and E_2 are the values for the amplitude of the electric field at ω_1 and ω_2, respectively. It is obvious from (2.85) that the resultant nonlinear polarization vector $\boldsymbol{P}_{\mathrm{NL}}(\omega_3)$ will also be parallel to the [111] crystal axis, i.e., $\boldsymbol{P}_{\mathrm{NL}}(\omega_3) \| \boldsymbol{E}_1 \| \boldsymbol{E}_2$. Consequently, the electric vector $\boldsymbol{E}_3$ at ω_3 generated by this nonlinear polarization will also be parallel to $\boldsymbol{E}_1$ or $\boldsymbol{E}_2$. The magnitude of the $\boldsymbol{P}_{\mathrm{NL}}(\omega_3)$ is obtained from (2.85) as

$$|\boldsymbol{P}(\omega_3)| = \sqrt{3}\, P_x(\omega_3) = 2\varepsilon_0 \left(\frac{2}{\sqrt{3}} d_{14}\right) E_1 E_2^* . \tag{2.86}$$

A comparison of (2.86) with (2.17) indicates that the effective nonlinear coefficient

$$d_{\mathrm{eff}} = \frac{2}{\sqrt{3}} d_{14} \quad \text{for} \quad \boldsymbol{E}_1 \| \boldsymbol{E}_2 \| [111] . \tag{2.87}$$

Similarly, it can be shown that

$$d_{\mathrm{eff}} = d_{14} , \qquad \text{for} \quad \boldsymbol{E}_1 \| \boldsymbol{E}_2 \| [110] \tag{2.88}$$

and

$$d_{\mathrm{eff}} = 0 \qquad \text{for} \quad \boldsymbol{E}_1 \| \boldsymbol{E}_2 \| [100] . \tag{2.89}$$

Since the FIR intensity is proportional to $(d_{\mathrm{eff}})^2$, (2.87–89) show that the mixing configuration corresponding to that in (2.87) provides the highest efficiency for FIR generation. It is for this reason that most nonlinear mixing in GaAs and other crystals of the same symmetry class have utilized the configuration $\boldsymbol{E}_1 \| \boldsymbol{E}_2 \| [111]$ with resulting $\boldsymbol{E}_3 \| [111]$ so that all the three beams are polarized along the same [111] direction. In particular, the experimental arrangement shown in Fig. 2.20 corresponds to this case with the [111] axis being perpendicular to the plane of incidence which lies in the plane of the figure.

In order to compare the experimentally observed FIR power output with that expected from theory, (2.75) may be rewritten as

$$P_{\omega_3} = K \left(\frac{P_{\omega_1} P_{\omega_2}}{A}\right) \left(\frac{1}{\lambda_3}\right)^2 L_{\mathrm{eff}}^2 \tag{2.90}$$

where λ_3 is the FIR wavelength and K is a constant given by

$$K = \tfrac{1}{2}\pi^2 \left(\frac{\mu_0}{\varepsilon_0}\right)^{1/2} \left[\frac{32 d_{\mathrm{eff}}}{(n_1+1)(n_2+1)(n_3+1)}\right]^2 . \tag{2.91}$$

For FIR generation with CO_2 laser mixing in GaAs with $\boldsymbol{E}_1 \| \boldsymbol{E}_2 \| [111]$, d_{eff} is given by (2.87). Using $n_1 \simeq n_2 = 3.28$, $n_3 = 3.59$, $d_{14} = 42.9 \times 10^{-12}$ m/V (see Table 2.1) in (2.91) gives $K = 6.61 \times 10^{-19}$ m^2/W.

Using a 1-cm long GaAs crystal of a simple geometry shown in Fig. 2.20, *Aggarwal* et al. [2.89] have observed FIR pulses with peak output power $P_{\omega_3} \simeq 10$ W at $\lambda_3 = 100$ μm with peak input power levels $P_{\omega_1} \simeq P_{\omega_2} \simeq 300$ kW in a beam diameter of about 6 mm. Using $L_x \simeq 1$ cm, $L_y = 0.3$ cm, and $\varphi = 22°$ for the experimental conditions of the above experiment, (2.76) yields $L_{\text{eff}} \simeq 0.75$ cm. Substituting this value of L_{eff} along with the above given values of other parameters, (2.90) predicts $P_{\omega_3} \simeq 12$ W which is in reasonable agreement with the observed value of 10 W. This measurement eliminates the large discrepancy reported previously [2.13, 14] between the observed and calculated values. It has been determined that the discrepancy was largely caused by the erroneous calibration of the FIR detector and the poor specimen of GaAs used in those earlier measurements. It should be pointed out that the experimental results quoted above were obtained with the GaAs crystal cooled to liquid helium temperatures where the FIR absorption at wavelengths longer than 100 μm should be extremely small. The 8 K data of *Johnson* et al. [2.53] shown in Fig. 2.2 only indicate that the absorption coefficient α is considerably smaller than 1 cm^{-1}. The 82 K data of Stolen shown in Fig. 2.3 indicate that the absorption coefficient is approximately 1 cm^{-1} at 100 μm and decreases to approximately 0.2 cm^{-1} at 200 μm. Since the FIR absorption at these long wavelengths arises from multiphonon processes which involve the absorption of a phonon, it is reasonable to expect that the absorption coefficient at liquid helium temperatures is less than 0.1 cm^{-1} for $\lambda_3 \geqq 100$ μm. Therefore, the effect of FIR absorption for $\lambda_3 \geqq 100$ μm is negligible at liquid helium temperatures in crystals with long dimension of the order of 1 cm. But the FIR absorption effects in larger crystals and/or higher temperatures may not be negligible. Under phase-matched generation, the correction factor due to absorption in the collinear mixing geometry is

$$F_{\text{a}} = \mathrm{e}^{-\alpha_3 L} \frac{1 + \mathrm{e}^{-\Delta\alpha} - 2\mathrm{e}^{-\frac{1}{2}\Delta\alpha L}}{(\frac{1}{2}\Delta\alpha L)^2} \tag{2.92}$$

from (2.29) where $\Delta\alpha$ is defined by (2.30). Since α_1 and α_2 are less than 0.01 cm^{-1} in GaAs at room temperature (see Fig. 2.5) it is reasonable to assume that $\alpha_1 = \alpha_2 = 0$ at lower temperatures. Under this assumption, (2.92) becomes

$$F_{\text{a}} = \mathrm{e}^{-\alpha_3 L} \frac{1 + \mathrm{e}^{\alpha_3 L} - 2\mathrm{e}^{-\frac{1}{2}\alpha_3 L}}{(\frac{1}{2}\alpha_3 L)^2}\,. \tag{2.93}$$

For $\alpha_3 L \ll 1$, F_{a} reduces to unity as expected. For $\alpha_3 L \gg 1$, it becomes equal to $(2/\alpha_3 L)^2$. This implies that the effective crystal length is simply $2/\alpha_3$ which is

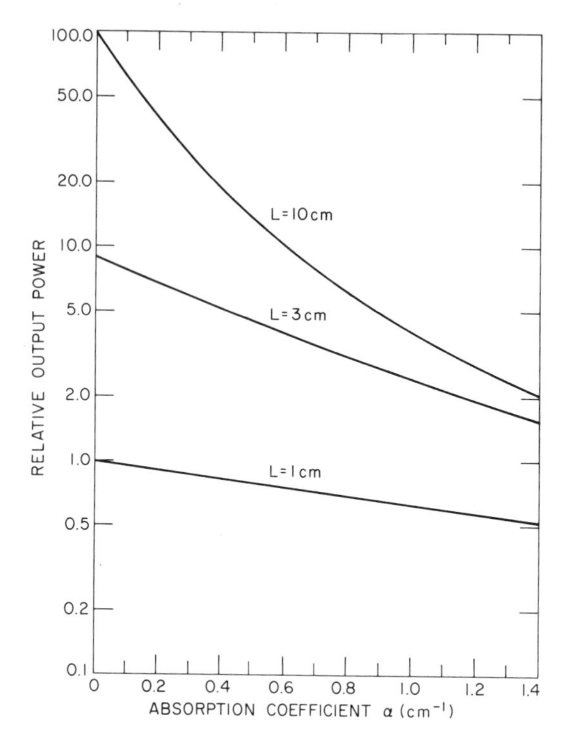

Fig. 2.23. Relative far-infrared difference-frequency output power vs. absorption coefficient for crystal length $L=1$, 3, and 10 cm (after *Rosenbluh* [2.87])

independent of the actual crystal length L as long as $L \gg (1/\alpha_3)$. Using (2.93), *Rosenbluh* [2.87] has calculated the relative FIR output power as a function of the absorption coefficient α_3 for $L=1$, 3, and 10 cm crystals. This is shown in Fig. 2.23. As can be seen, the reduction factor for a 1 cm long crystal is less than a factor of 2 for the highest value of $\alpha_3=1.4$ in Fig. 2.23. In comparison, the FIR power is reduced by a factor of 50 in a 10 cm long crystal for the same value of α_3. In other words, the advantage of using the larger crystal is wiped out to a large extent by absorption effects.

An expression for the correction factor due to absorption in the noncollinear mixing geometry, analogous to that of (2.92) or (2.93) for the collinear case, has not been worked out. Therefore, an approximate correction due to the FIR absorption may be made by substituting for the crystal length L in (2.93) an effective interaction length L_{eff} deduced on the basis of no absorption.

The principle of the folded geometry for FIR difference-frequency generation has been verified by *Aggarwal* et al. [2.89] by mixing in a crystal of GaAs with the simple geometry shown in Fig. 2.16. The dimensions of the GaAs sample were $L \simeq 25$ mm, $W=6$ mm, and $H \simeq 10$ mm with the input face cut at an angle $\varphi \simeq 22°$. In addition to polishing the input and output faces of this sample, the two reflecting faces were carefully polished to be plane parallel to each other. In this experiment the GaAs sample was cooled only to liquid nitrogen temperature. The rest of the experimental setup was similar to that of

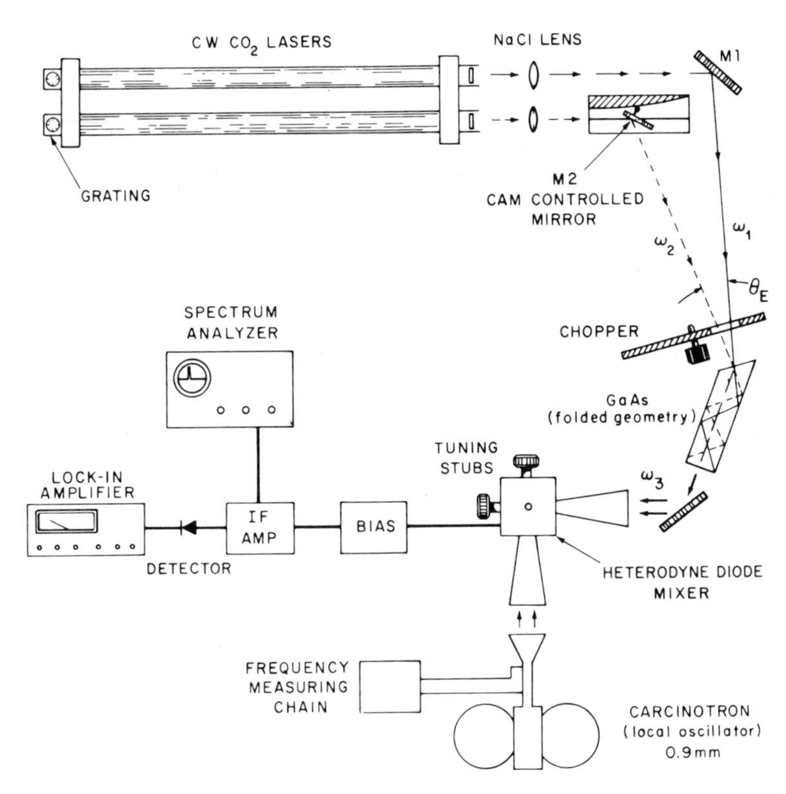

Fig. 2.24. A schematic of the experimental setup for continuous wave (cw) generation of step-tunable far-infrared radiation in the 70 μm to 1 mm wavelength region using noncollinear difference-frequency mixing of two CO_2 lasers in GaAs at $T \simeq 80$ K. Also shown are the components of the heterodyne detection system for measuring the FIR frequency and linewidth (after *Aggarwal* et al. [2.15])

Fig. 2.20. The FIR output around 200 μm was found to be about 20 to 30 times larger than that obtained from the 1 cm long sample of simple noncollinear geometry of Fig. 2.13. Using $\varphi = 22°$ in (2.80a), one gets $g = 1.16$ so that (2.79) yields $L_{\text{eff}} = 2.9$ cm for the sample with the folded noncollinear geometry in comparison to $L_{\text{eff}} = 0.75$ cm for the sample with simple noncollinear geometry. Consequently, the former sample should provide an increase of $(2.9/0.75)^2 = 15$ in the FIR output power over that of the latter. This calculated enhancement factor of 15 is consistent with the observed enhancement factor of 20 to 30 considering that there was appreciable jitter in the FIR output due to the jitter in the temporal overlap of the two CO_2 laser pulses. Using 10 cm long crystals of folded geometry similar to that of Fig. 2.17, we have obtained 0.4 mJ of FIR radiation corresponding to peak powers of approximately 4 kW in 100 ns pulses in the 100 μm region [2.16]. The peak input power was 1.7 MW in the 10.6 μm beam and 3 MW in the 9.6 μm beam with ~ 1 cm^2 cross section.

So far we have discussed the generation of only pulsed FIR radiation by noncollinear mixing of two pulsed TEA CO_2 lasers. Recently *Aggarwal*, *Lax* and coworkers [2.15] have applied this noncollinear phase-matching technique for the continuous wave (cw) generation of tunable FIR radiation in the wavelength region from $\sim 70\,\mu m$ to 1 mm. A schematic of the experimental setup for the cw generation, detection and measurement of the FIR is shown in Fig. 2.24. The CO_2 lasers were fabricated using a somewhat simplified version of *Freed's* design of a sealed-off CO_2 laser [2.90]. Typical single-line TEM_{00} output of these lasers ranged from ~ 20 W to 40 W over most of the tuning range (see Fig. 2.18).

The two laser beams, collimated with long focal length (~ 2 m) NaCl lenses, are directed onto the GaAs crystal. Before entering a 2.5 cm long GaAs crystal of simple folded geometry of Fig. 2.16, the CO_2 laser beams were chopped at a frequency of ~ 100 Hz with a chopper blade providing 20% duty cycle. The low duty cycle was necessary to avoid overheating of the crystal cooled to liquid nitrogen temperature, which acts as a trap for the laser radiation as a consequence of its geometry.

The FIR power was determined using a liquid helium temperature germanium bolometer pumped below the λ-point. The output of the bolometer was observed on an oscilloscope and provided a measurement of the FIR power output. For example, the FIR output at the frequency $\omega_3 = 43.3\ \mathrm{cm}^{-1}$ was determined to be $\sim 0.5 \times 10^{-7}$ W using the CO_2 laser lines at $\omega_1 = 982.10\ \mathrm{cm}^{-1}$, $\omega_2 = 938.70\ \mathrm{cm}^{-1}$ with $P_{\omega_1} \simeq P_{\omega_2} = 25$ W and a beam diameter of about 8 mm. The observed value of 0.5×10^{-7} W is in reasonable agreement with the calculated value of 1.3×10^{-7} W, assuming the absorption losses to be negligible.

By going to heterodyne detection *Aggarwal* et al. [2.15] have measured the frequency and linewidth of the FIR signal. Elements of the heterodyne system developed by *Fetterman* et al. [2.91] are shown in Fig. 2.24. Using the GaAs Schottky diode and a 30 MHz-i.f. strip, the beat signal between the FIR and the second harmonic of the carcinotron was displayed on a spectrum analyzer. The results obtained with the FIR at $22.805\ \mathrm{cm}^{-1}$ corresponding to a frequency of 683.683 GHz, are shown in Fig. 2.25. The sharp peak in Fig. 2.25 is the heterodyne signal, whereas the broad background is due to electrical noise of the detector and the i.f. strip. The observed halfwidth of 100 kHz results not only from the linewidth of the CO_2 laser lines but also from that of the carcinotron. Therefore, this measurement has established an upper limit for the linewidth of the FIR radiation.

2.5.2 Collinear Phase-Matched Generation in GaAs

The coherence length, L_c, for FIR difference-frequency generation with CO_2 laser mixing in GaAs is of the order of the FIR wavelength as can be seen from (2.41) by substituting a value of ~ 0.3 for Δn. Consequently, it is possible to obtain useful FIR power output by collinear CO_2 laser mixing in GaAs

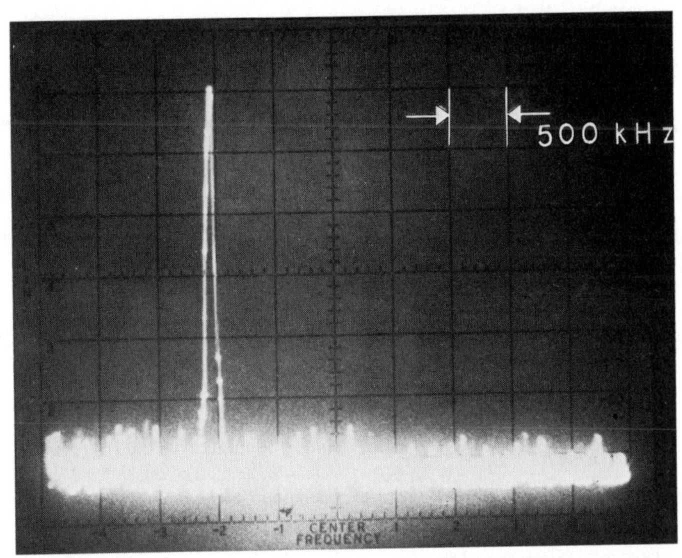

Fig. 2.25. Spectrum analyzer trace of a 30 MHz i.f. strip showing the mixing between a 22.8 cm^{-1} FIR signal and the second harmonic of a carcinotron in a room-temperature GaAs Schottky diode. The heterodyne signal as shown has a linewidth of less than 100 kHz (after *Aggarwal* et al. [2.15])

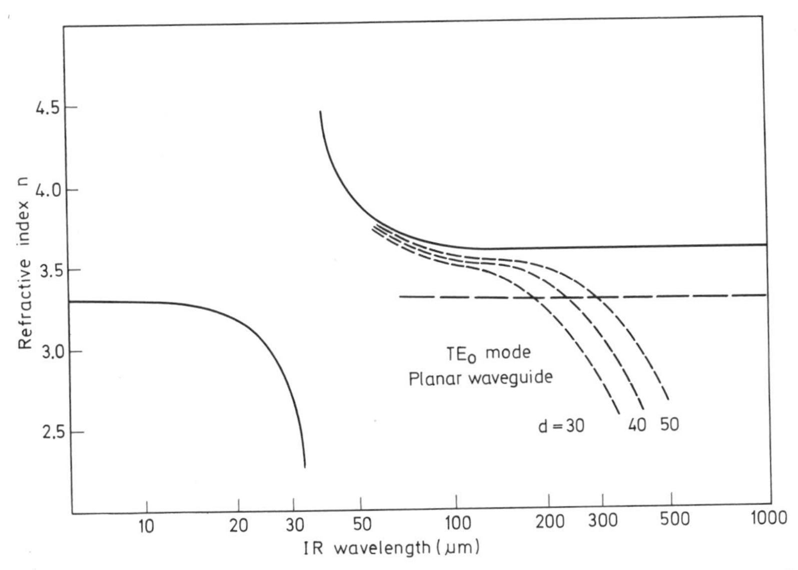

Fig. 2.26. Dispersion of the refractive index of GaAs at room temperature. The solid lines show the refractive index of bulk GaAs. The short-dashed curves show the TE_0 mode waveguide dispersion of planar waveguides for several values of half thickness *d*. The long-dashed curve indicates the required refractive index at the difference-frequency $\omega_3 = \omega_1 - \omega_2$ propagating in a TM_0 mode, with ω_1 in a TM_0 mode, and ω_2 in a TE_0 mode (after *Thompson* and *Coleman* [2.17])

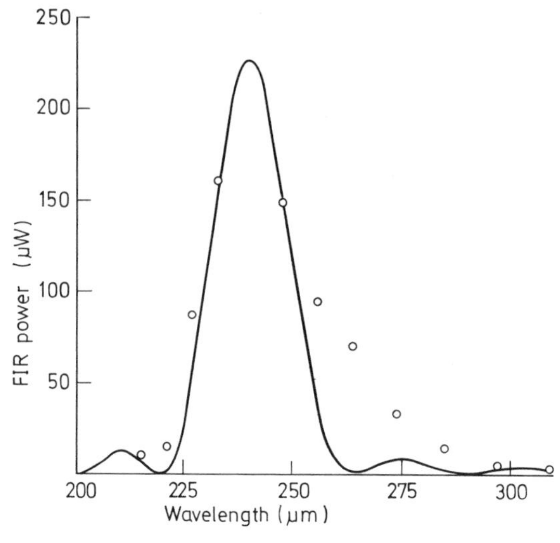

Fig. 2.27. Power output of the infrared difference-frequency radiation vs. wavelength for CO_2 laser mixing in a 85.12 μm thick GaAs waveguide at room temperature. The solid line represents the calculated tuning curve, and the circles denote the data points (after *Thompson* and *Coleman* [2.17])

samples whose thickness is of the order of L_c or less. Following this approach, *Bridges* and *Strand* [2.92] have reported the generation of nonphase-matched FIR radiation in GaAs at 29.7 cm^{-1}. By focussing 1.8 kW Q-switched CO_2 laser beams upon 0.5 mm thick sample of GaAs, they obtained FIR radiation with a peak power of $\sim 4\,\mu W$.

Thompson and *Coleman* [2.17] have reported collinear phase-matched generation of FIR radiation with CO_2 lasers in GaAs. They used dielectric wave-guide phase matching. Figure 2.26 shows dispersion curves for the bulk GaAs and for TE_0 modes of GaAs slab waveguides of thickness $2d$, for several values of d. The long dashed curve in Fig. 2.26 shows the calculated refractive index which is required to phase-match-difference-frequency ω_3 in a TM_0 mode with $\omega_1(>\omega_2)$ in a TM_0 mode and ω_2 in a TE_0 mode.

Figure 2.27 shows the experimental results obtained by *Thompson* and *Coleman* for a 85.12 μm thick waveguide. The solid curve shows the calculated tuning curve while the data points are represented by the circles. By focussing 1–3 kW output of Q-switched CO_2 laser beams, they obtained pulsed FIR signals in the 100–1000 μm range with peak powers in the mW range which represented approximately 20% of the theoretically expected output.

2.5.3 FIR Generation in Other Systems

Aggarwal et al. [2.89] have used CdTe as the nonlinear mixer for the noncollinear phase-matched generation of difference-frequency radiation with CO_2 lasers. A CdTe sample in the form of a 1 cm cube was cut to the shape of the simple noncollinear geometry of Fig. 2.13. However, the angle φ in the case of CdTe was $\sim 32°$ as calculated from (2.62) using the measured values of the refractive indices [2.53, 67]. Figure 2.28 shows the external phase-matching

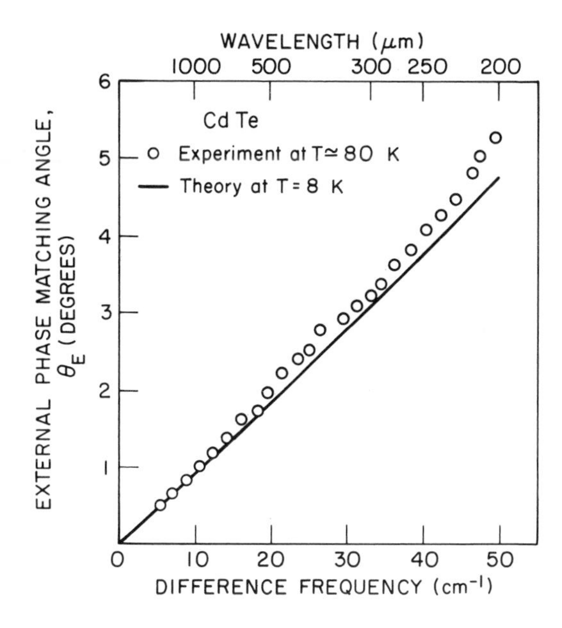

Fig. 2.28. External phase-matching angle θ_E between the two CO_2 laser beams vs. the frequency of the far-infrared radiation generated in CdTe of simple noncollinear geometry of Fig. 2.13 [after *Lee* et al. (unpublished)]

angle vs. the FIR difference-frequency ω_3 obtained from a CdTe crystal at liquid nitrogen temperature. The FIR power output obtained from the CdTe was compared with that from a GaAs crystal of comparable dimensions. For wavelengths longer than $\sim 200\ \mu m$, the FIR output from CdTe was found to be about an order of magnitude larger than that obtained from GaAs. Neglecting absorption losses, the expected ratio of FIR powers is expected to be ~ 7.2 from (2.32) using the values of the electro-optic coefficients given in Table 2.1 along with the values for the refractive indices. The increased absorption in CdTe compared with that in GaAs at wavelengths shorter than $\sim 200\ \mu m$ reduces its advantage over GaAs. In fact, longer crystals of GaAs can more than make up for its lower value of the electro-optic coefficient compared with CdTe.

Boyd et al. [2.46] have employed a ternary birefringent crystal of $ZnGeP_2$ for FIR generation by difference-frequency-mixing of CO_2 laser lines. $ZnGeP_2$ has enough positive birefringence in the 10 μm region of the CO_2 laser to allow collinear phase-matched FIR generation. Figure 2.29 shows the tuning curve obtained for the forward wave mixing in which the input laser beams at ω_1 (as e-wave) and ω_2 (as o-wave) as well as the FIR beam at ω_3 (as o-wave) propagate in the same direction. Peak FIR output in the μW range was obtained in the 70 cm^{-1} to 110 cm^{-1} region by focussing 50 to 400 W beams from *Q*-switched CO_2 lasers on a 0.36 cm thick crystal at room temperature.

Nguyen and *Bridges* [2.93], and *Brown* and *Wolff* [2.94] have shown that it is possible to use the second-order nonlinearity due to the conduction electron spin in InSb for the FIR generation by difference-frequency mixing of CO_2 laser lines or spin-flip Raman laser pumped with a CO_2 laser. This

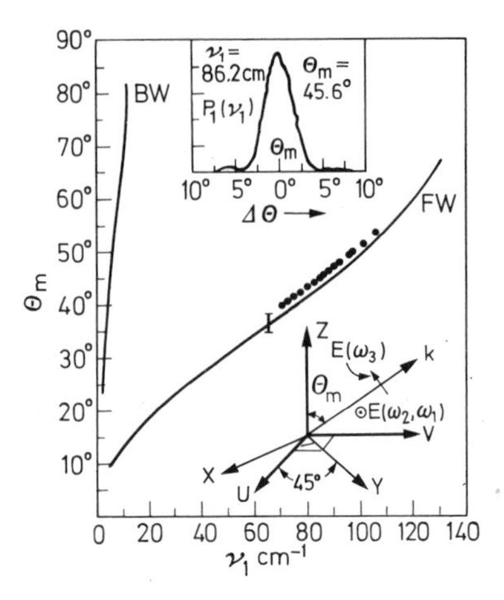

Fig. 2.29. Internal phase-matching angle θ_m vs. far-infrared difference frequency for forward wave (FW) and backward wave (BW) mixing of CO_2 laser lines of frequencies ω_2 and ω_3 ($\omega_3 > \omega_2$) in $ZnGeP_2$ at room temperature. The solid lines represent the calculated results and the dots represent the experimental data. The bottom inset shows geometry for FW mixing. The top inset shows the difference-frequency power vs. internal angle $\Delta\theta$ around θ (after *Boyd* et al. [2.46])

nonlinearity exhibits extremely large resonance enhancement when the difference-frequency ω_3 becomes equal to the spin-flip-frequency ω_s. The FIR radiation in this process may be looked upon as the radiation from the precessing spin system which is driven by an effective magnetic field resulting from the cross product of the vector potentials of the incident laser radiation at ω_1 and ω_2. *Nguyen* and *Bridges* [2.93] mixed 9.6 μm and 10.6 μm CO_2 laser beams in an *n*-type InSb crystal ($n_e \simeq 2.2 \times 10^{15}\,\mathrm{cm}^{-3}$) subjected to an external magnetic field. On tuning the magnetic field to the resonance condition $\omega_3 = \omega_s$, the FIR output was observed to increase by a factor of 88, consistent with the value of 200 predicted from the theory of *Brown* and *Wolff* [2.94]. From the observed resonance enhancement, it was shown that the effective spin-nonlinearity in InSb for a concentration of $2.2 \times 10^{15}\,\mathrm{cm}^{-3}$ is as large as 5.4 times the conventional nonlinear coefficient d_{14}.

Continuously tunable FIR radiation can be generated by difference-frequency mixing of a spin-flip Raman (SFR) laser with a CO_2 laser. If one uses the same CO_2 laser for pumping SFR laser and for mixing, the FIR radiation will be generated at a frequency $\omega_3 = g\beta H$ where g is the spin g-factor, β is the Bohr magneton and H is the strength of the applied magnetic field. Using InSb spin-flip Raman [2.95] laser, one should be able to tune, in principle, over the entire FIR region at the rate of $\sim 2\,\mathrm{cm}^{-1}/\mathrm{kG}$. However, the dependence of the SFR laser output on the magnetic field will determine the achievable FIR frequency range. *Bridges* and *Nguyen* [2.20] have used this technique to obtain tuning in the frequency range from $90\,\mathrm{cm}^{-1}$ to $110\,\mathrm{cm}^{-1}$ with peak power levels in the μW range with a linewidth of $\sim 0.1\,\mathrm{cm}^{-1}$.

Matsumoto and *Yajima* [2.18] were the first to use difference-frequency mixing of dye lasers to generate FIR radiation in the wavelength region above 100 μm. Using the self-beating of a dye laser in ZnTe and ZnSe crystals, they reported the observation of pulsed FIR in the range 300 μm to 2 mm.

More recently *Yang* et al. [2.19] have reported the use of dual-frequency dye laser system shown in Fig. 2.30 for the generation of continuously tunable

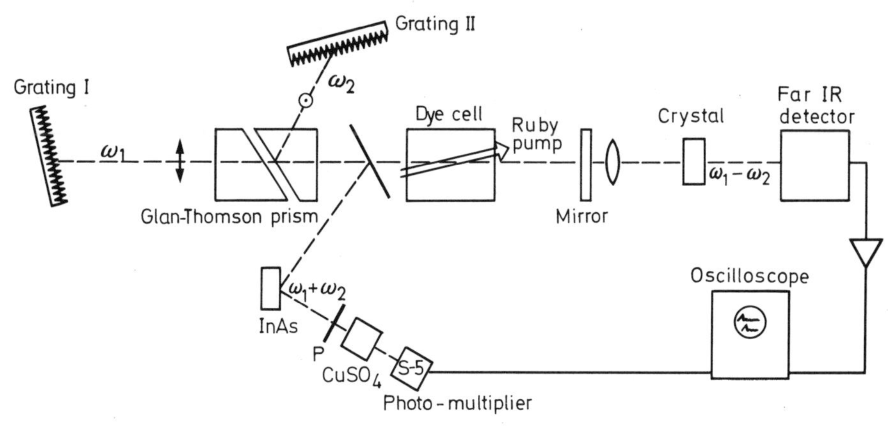

Fig. 2.30. Schematic of the experimental arrangement for the generation and detection of pulsed far-infrared difference-frequency radiation by mixing of radiation from two dye lasers in a non-linear crystal of $LiNbO_3$. The wavelength of the two dye laser beams can be independently varied from 8100 Å to 8400 Å using a single dye cell (after *Yang* et al. [2.19])

FIR radiation in $LiNbO_3$ over the frequency range from 20 cm^{-1} (500 μm) to 190 cm^{-1} (52.5 μm). In the forward collinear phase-matched configuration, this system produced FIR radiation with maximum peak power in the 10 mW range (see Fig. 2.31). In the noncollinear phase-matched mixing configuration, the same system produced a maximum FIR output of 200 mW. The linewidth of the FIR radiation was measured to be ~3 cm^{-1}.

Piestrup et al. [2.21] have reported the generation of continuously tunable FIR radiation by stimulated polariton scattering of 1.06 μm radiation of a Nd:YAG laser from the soft phonon mode of A_1 symmetry in $LiNbO_3$. The experimental arrangement used by *Piestrup* et al. [2.21] is shown in Fig. 2.32. A 5 cm long crystal of $LiNbO_3$ cut parallel to the a axis is placed in an external cavity formed by the mirrors M_1 and M_2 for the Stokes radiation. The laser beam focussed one face of the $LiNbO_3$ crystal and polarized parallel to the c axis is incident at an angle θ to the resonator axis of the cavity. If ω_1, ω_2, and ω_3 represent the frequencies of the incident laser radiation, the Stokes radiation and the FIR radiation, respectively, ω_3 can be tuned by simply adjusting the angle θ at a tuning rate of 80 cm^{-1} per degree in $LiNbO_3$. By focussing 1.7 MW peak power output from the laser into a spot having peak power density of 48 MW/cm^2, *Piestrup* et al. [2.21] obtained FIR output at the kW level over the spectral range from ~15 cm^{-1} to 60 cm^{-1} as shown in Fig. 2.33. As can be seen, the FIR output below 30 cm^{-1} approaches the Manley-Rowe limit indicating quantum conversion efficiency close to 100%. The dropoff in the FIR output at higher frequencies is due to the increasing FIR absorption losses. The linewidth of the FIR radiation generated in this experiment has not been reported. However, it is expected to be appreciably large. A divergence of 1 mrad in the laser beam will contribute ~5 cm^{-1} to the FIR linewidth.

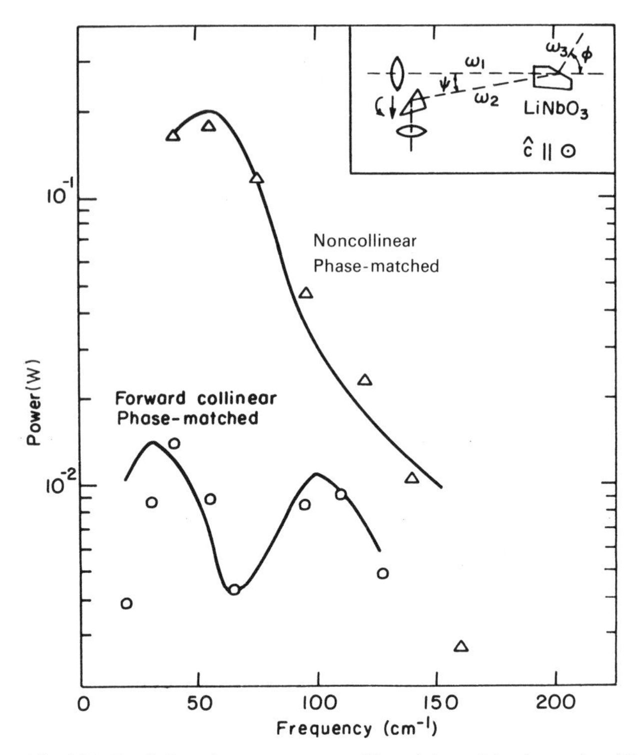

Fig. 2.31. Far-infrared power generated by mixing of dye lasers (see Fig. 2.30) vs. frequency for two phase-matching conditions. The FIR power is normalized by the sum-frequency power. The solid curves are the theoretical calculations based on data of the FIR absorption and the Raman cross section in $LiNbO_3$. The inset shows the experimental geometry for noncollinear phase matching which was achieved by a coupled translation and rotation of the prism (after *Yang* et al. [2.19])

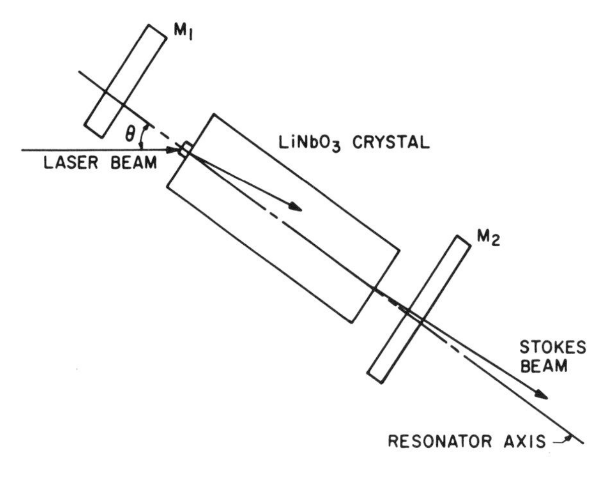

Fig. 2.32. Experimental arrangement for the generation of continuously tunable far-infrared radiation in the 150–700 μm region by stimulated polariton scattering in $LiNbO_3$ of 1.06 μm radiation from a Nd:YAG laser. Mirrors M_1 and M_2 form an interferometer for the Stokes radiation (after *Piestrup* et al. [2.21])

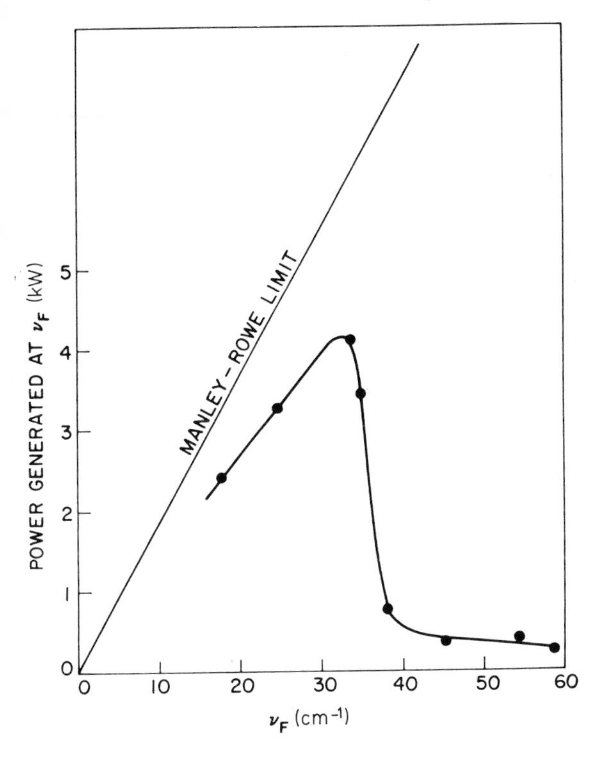

Fig. 2.33. Power generated in the far-infrared radiation as a function of the far-infrared frequency, using the experimental setup of Fig. 2.32. The far-infrared power shown here was actually calculated from the measured Stokes power (after *Piestrup* et al. [2.21])

2.6 Applications

It is not presumptuous to say that tunable high-power monochromatic radiation sources must have numerous applications in spectroscopy and other areas of basic and applied sciences. In particular, we would like to consider applications of the FIR radiation generated by the difference-frequency mixing of CO_2 lasers which provide thousands of discrete frequencies in the spectral region from ~70 μm to several mm. Since the frequency of the FIR radiation is not continuously tunable with the systems developed so far, the CO_2 laser difference-frequency source is not suitable for conventional spectroscopic experiments which require a measurement of the optical response of a medium as a function of frequency or wavelength. However, there is a large class of experiments which do not necessarily demand continuous tunability. All that is required is the ability to select a frequency close to the region of interest. Application of an external tunable perturbation such as magnetic field, electric field, temperature, stress, etc., to the medium can tune the frequency of the optical transition of interest into resonance with the selected FIR frequency. Considering this, the CO_2 laser difference-frequency source is more than adequate for resonance spectroscopy of all kinds, particularly, when high resolution and/or high intensity is a must.

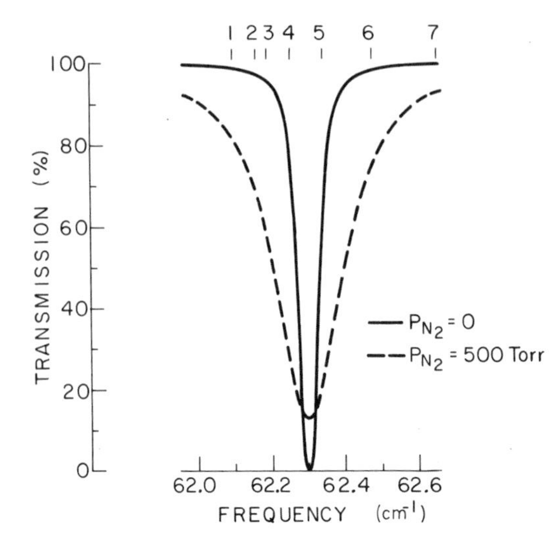

Fig. 2.34. Calculated transmission in the 62 to 63 cm^{-1} region through water vapor for a pressure of 7 Torr, path length of 12.6 cm and temperature of 296 K. The short vertical lines in the top part of the figure indicate the frequencies of FIR lines from a CO_2 laser difference-frequency source which are within the absorption linewidth (after *Mandel* [2.96])

2.6.1 Far-Infrared Molecular and Solid State Studies

Rotational spectra of many molecular gases lie in the FIR region and are therefore well suited for study with a high-resolution source since the natural linewidths, particularly at low pressures, are exceedingly small. *Mandel* [2.96] has shown that it is possible to make useful high-resolution measurements of the parameters for the FIR transitions in water vapor with the quasi-tunable FIR difference-frequency source. As an example, we show in Fig. 2.34 the calculated transmission through water vapor at a pressure of 7 Torr for the transition at 62.3 cm^{-1}. Short vertical lines at the top part of Fig. 2.34 indicate FIR frequencies available from the difference-frequency source within the pressure-broadened linewidth. Using the lines 4 and 5 at the frequencies 62.25008 and 62.33805 cm^{-1}, Mandel measured the transmission T as a function of pressure. Figure 2.35 shows these data in terms of $-(1/lnT)$ vs. $(1/P_{H_2O})^2$ at the above two FIR frequencies. From an analysis of these data, he obtained a value of the half-width parameter for self-broadening. Similar measurements made in the presence of N_2 gas allowed a determination of the foreign-gas broadening parameters.

Rosenbluh [2.87] has applied the quasi-tunable cw FIR source for studying the FIR absorption in high-resistivity GaAs at room temperature. Figure 2.36 shows the transmission through a 0.226 cm thick slab of GaAs measured at a number of FIR frequencies around 28 cm^{-1}. By fitting the data to an absorbing etalon, he was able to determine relatively small absorption coefficient even in a thin sample.

For many years *Button*, *Lax* and coworkers [2.97] have used fixed-frequency submillimeter lasers of HCN, H_2O and its isotopic species for studying cyclotron resonance and Zeeman spectroscopy of shallow impurities in semiconductors. With the larger frequency range and higher powers now

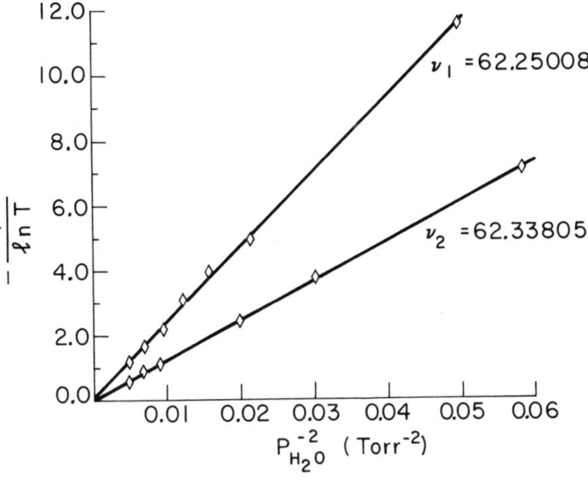

Fig. 2.35. A plot of $-(1/lnT)$ vs. $(1/P_{H_2O})^2$ for water vapor at a temperature of 296 K and path length of 12.6 cm. The transmission T was measured using frequencies $\nu_1 = 62.25008\ cm^{-1}$, and $\nu_2 = 62.33805$. An analysis of this data yields self-broadening half-width parameter $\alpha_0 = 0.413 \pm 0.015\ cm^{-1}/atm$ (after *Mandel* [2.96])

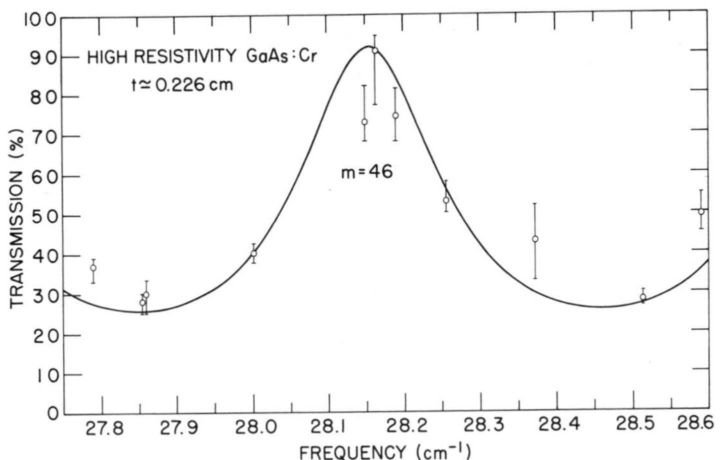

Fig. 2.36. Transmission through a 0.226 cm thick sample of high-resistivity GaAs at room temperature. The error bars indicate observed fluctuations in the transmission values over many trials. The solid curve represents the calculated transmission through an absorbing etalon assuming 46-th order of the Fabry-Perot fringe and absorption coefficient equal to $0.19\ cm^{-1}$ (after *Rosenbluh* [2.87])

provided by the difference frequency quasi-tunable sources and the optically pumped lasers these experiments can be extended in two aspects. Many new materials can be studied and new nonlinear resonant phenomena in the FIR region may be observed. The longer wavelengths, in particular, open up opportunities for studying cyclotron resonance in metals and other materials with higher masses. Spin resonance in magnetic materials can be investigated in the 500 μm to 2 mm range in a convenient fashion with the high magnetic fields available up to 230 kG.

Impurity-level Zeeman studies have been investigated by *Cohn* et al. [2.98] in CdTe to study the polaron effect at high magnetic fields. These experiments can now be extended both in this material and to other materials. The rich variety of lines in the broad spectral range of 70 μm to several millimeters will permit the Zeeman studies of excited states in many semiconductors similar to those examined by *Fetterman* and coworkers in GaAs [2.99].

2.6.2 Plasma Studies with FIR Radiation

A number of applications of submillimeter laser sources for plasma investigations have been considered [2.25]. Among these are cyclotron resonance breakdown and heating of gaseous plasmas, Thomson scattering for measuring ion temperatures of Tokamaks, and their impurity content. Laser excitation of parametric decay instabilities at the plasma frequency is of fundamental interest and is particularly suited to submillimeter studies of relatively lower density plasmas. Cyclotron resonance excitation at harmonic frequencies is also interesting for the study of transport coefficients of Tokamaks.

The theory of laser breakdown and heating at cyclotron resonance has been considered by *Lax* and *Cohn* [2.24] at high magnetic fields using a high-power CH_3F laser as a source. The theory predicts a considerable lowering of the breakdown threshold at lower pressures, well below the power densities available today. In addition they predict efficient heating in the focal volume at resonance. These experiments have recently been successfully demonstrated by *Hacker* and *Temkin* [2.100] in the laboratory. Their results are qualitatively in agreement with the theoretical predictions, but are exhibiting new features which will require more extensive investigations. Analogous experiments of breakdown in extrinsic semiconductors can be performed. At liquid helium temperatures the electrons are "frozen" out of the conduction and valence bands into hydrogenic-like impurity states. These can be readily re-ionized by an avalanche process with a focussed submillimeter laser with and without magnetic field, resonantly at a lower threshold in the former case.

Thomson scattering with high-power submillimeter radiation is perhaps one of the most important applications of these new laser sources to plasma diagnostics. When the wavelength of the laser is longer than the Debye wavelength, λ_D, of the plasma it is scattered by the Debye sphere which surrounds the ions. According to Salpeter's theory the linewidth of the scattered radiation is determined by the Doppler shift of the ions from which the ion temperature can be determined. For Tokamak plasmas the parameters are such that the submillimeter wavelengths are ideal for this type of experiment. Heterodyne techniques will be used to detect the scattered radiation. A GaAs Schottky barrier diode using a stabilized cw submillimeter laser as the local oscillator is expected to provide a signal-to-noise ratio of ~ 100 for the detection of Thomson-scattered radiation. If impurities are present in the plasma additional structure will be present which enables the determination of the effective charge Z_{eff} of the impurities.

The study of the excitation of parametric decay instabilities by submillimeter lasers in a plasma arc has been suggested by *Jassby* [2.101]. Considering a 496 μm CH_3F laser, the threshold intensity for exciting such instabilities in a plasma with a density of 5×10^{15} cm^{-3} would be of the order of $\sim 4 \times 10^4$ W/cm^2, which is readily available at the present time. The continued development of high-power submillimeter sources should also make it possible to study other parametric processes such as stimulated Raman scattering and stimulated Brillouin scattering.

It is well known that Tokamak plasmas emit cyclotron radiation at the fundamental and harmonics of the cyclotron frequency. The same physical mechanism that is responsible for this is also responsible for harmonic absorption of the appropriate frequencies. With tunable and fixed-frequency laser sources two types of experiments are possible. One of these experiments involves absorption measurements for FIR radiation at a number of different frequencies propagating along the horizontal diameter of the Tokamak. Since the major toroidal field is well known in the machine, one can deduce from the absorption of the harmonics and an independent measurement of the density the effective temperature of the plasma as a function of position if it is in local equilibrium, or the deviation of the distribution if the absorption is anomalously high. Such harmonic absorption with a high-power laser can be used to heat the electrons locally at a given position which can then be monitored as a function of time by Thomson scattering of a ruby laser from the same volume. Such scattering would measure the variation of the temperature rise with time after the initial heating and thereby provide a measure of the transverse thermal diffusion. The experiment can explore this transport property as a function of position by varying the magnetic field. The harmonic experiments can in principle also be used to measure the poloidal or transverse field when combined with polarization measurements. Since the resonance measures the total field and the polarization determines its direction, one should be able to deduce the poloidal component. Indeed such an experiment has been attempted by the Princeton group [2.102] with absorption of 33.6 GHz (8.9 mm) radiation at the fundamental at low fields of the order of 14 kG. They then observed the first harmonic emission and its polarization to deduce this important parameter which then provides information about the current distribution in the toroid. However, to extend such measurement to higher-density higher-field Tokamaks like the Alcator at M.I.T. Francis Bitter National Magnet Laboratory, it will be necessary to use submillimeter lasers.

Another important diagnostic technique for plasma studies is the long-wavelength interferometer for measuring the electron density in a machine. Until recently when relatively low-density plasmas were studied these interferometers operating in the millimeter region were very useful in measuring this parameter by observing the change in the refractive index as a function of time along a sector of the plasma. With higher densities it is necessary to go to the submillimeter region and it is desirable to have several convenient frequencies available as the density is varied in these plasma experiments. Furthermore for ease of determining the change in index of refraction and hence density versus time, a two frequency laser system is desirable. Such a system is being developed by *Button* and coworkers [2.103] for measuring electron densities in the 10^{14}–10^{15} cm^{-3} density region of the Alcator by using CO_2 optically pumped lasers at 496 μm and 119 μm.

2.7 Future Outlook

Hopefully there will be a number of advances in the state of the art for FIR generation by difference-frequency mixing of CO_2 lasers. On the one hand, bigger and longer crystals, new crystal geometries, and/or new nonlinear materials will be investigated in an effort to increase the FIR generation efficiency in both the pulsed and cw systems. On the other hand, efforts will be made to achieve continuously tunable FIR radiation with high resolution of 0.001 cm^{-1} or better. This should be possible through the use of interferometric tuning in high-pressure lasers with one or more intracavity Fabry-Perot etalons in conjunction with or without gratings [2.104]. The achievement of high-conversion efficiency, high-resolution, and continuous tunability will undoubtedly widen the range of applications of such FIR sources.

Acknowledgements. The authors wish to express their appreciation to Dr. *N. Lee*, Dr. *H. R. Fetterman* and *M. Rosenbluh* for many stimulating discussions regarding this work. We would like to thank *M. McDowell* for typing the manuscript.

References

2.1 P.A.Franken, A.E.Hill, C.W.Peters, G.Weinreich: Phys. Rev. Lett. **7**, 118 (1961)
2.2 See, for example, N.Bloembergen: *Nonlinear Optics* (W.A.Benjamin, Inc., New York 1965)
2.3 F.Zernike,Jr., P.R.Berman: Phys. Rev. Lett. **15**, 999 (1965)
2.4 J.A.Giordmaine: Phys. Rev. Lett. **8**, 19 (1962)
2.5 P.D.Maker, R.W.Terhune, M.Nissenhoff, C.M.Savage: Phys. Rev. Lett. **8**, 21 (1962)
2.6 C.K.N.Patel: Phys. Rev. Lett. **12**, 588 (1964); Phys. Rev. **136**, A 1187 (1964)
2.7 F.Zernike: Bull. Am. Phys. Soc. **12**, 687 (1967)
2.8 F.Zernike: Phys. Rev. Lett. **22**, 931 (1969)
2.9 R.B.Sanderson: J. Phys. Chem. Solids **26**, 803 (1965)
2.10 V.T.Nguyen, C.K.N.Patel: Phys. Rev. Lett. **22**, 463 (1969)
2.11 G.D.Boyd, C.K.N.Patel: Appl. Phys. Lett. **8**, 313 (1966)
2.12 A.J.Beaulieu: Appl. Phys. Lett. **16**, 504 (1970)
2.13 R.L.Aggarwal, B.Lax, G.Favrot: Appl. Phys. Lett. **22**, 329 (1973)
2.14 B. Lax, R.L.Aggarwal, G.Favrot: Appl. Phys. Lett. **23**, 679 (1973)
2.15 R.L.Aggarwal, B.Lax, H.R.Fetterman, P.E.Tannenwald, B.J.Clifton: J. Appl. Phys. **45**, 3972 (1974)
2.16 N.Lee, R.L.Aggarwal, B. Lax: Opt. Commun. **11**, 339 (1974); to appear in the Summary Intern. Conf. Quant. Electron., Amsterdam (1976)
2.17 D.E.Thompson, P.D.Coleman: IEEE Trans. MTT-**22**, 995 (1974)
2.18 N.Matsumoto, T.Yajima: Japan. J. Appl. Phys. **12**, 90 (1973)

2.19 K.H.Yang, J.R.Morris, P.L.Richards, Y.R.Shen: Appl. Phys. Lett. **23**, 669 (1973)
2.20 T.J.Bridges, V.T.Nguyen: Appl. Phys. Lett. **23**, 329 (1973)
2.21 M.A.Piestrup, R.N.Fleming, R.H.Pantell: Appl. Phys. Lett. **26**, 418 (1975)
2.22 T.Y.Chang, T.J.Bridges: Opt. Commun. **1**, 423 (1970)
2.23 D.L.Jassby, D.R.Cohn, B.Lax, W.Halverson: Princeton Plasma Physics Laboratory Rept. MATT-**1020** (1974)
2.24 B.Lax, D.R.Cohn: Appl. Phys. Lett. **23**, 363 (1973)
2.25 B.Lax, D.R.Cohn: IEEE Trans. MTT-**22**, 1049 (1974)
2.26 M.Garfinkel, W.E.Engeler: Appl. Phys. Lett. **3**, 178 (1963)
2.27 M.Bass, P.A.Franken, J.W.Ward, G.Weinreich: Phys. Rev. Lett. **9**, 446 (1962)
2.28 P.A.Franken, J.F.Ward: Rev. Mod. Phys. **35**, 23 (1962)
2.29 J.A.Giordmaine, R.C.Miller: Phys. Rev. Lett. **14**, 973 (1965)
2.30 R.W.Terhune, P.D.Maker, C.M.Savage: Phys. Rev. Lett. **8**, 404 (1962)
2.31 W.Kaiser, C.G.B.Garrett: Phys. Rev. Lett. **7**, 229 (1961)
2.32 G.Eckhardt, R.W.Hellwarth, F.J.McClung, S.E.Schwarz, D.Weiner, E.J.Woodbury: Phys. Rev. Lett. **9**, 455 (1962)
2.33 R.Y.Chiaco, C.H.Townes, B.P.Stoicheff: Phys. Rev. Lett. **12**, 592 (1962)
2.34 D.I.Mash, V.V.Morozov, V.S.Satrunov, J.L.Fabelinski: JETP Lett. **2**, 22 (1965)
2.35 See, for example, S.Bhagwantam, *Crystal Symmetry and Physical Properties* (Academic Press, New York 1966)
2.36 J.A.Armstrong, N.Bloembergen, J.Ducuing, P.S.Pershan: Phys. Rev. **127**, 1918 (1962)
2.37 D.A.Kleinman: Phys. Rev. **126**, 1977 (1962)
2.38 R.C.Miller: Appl. Phys. Lett. **5**, 17 (1964)
2.39 See, for example, S.Singh: In *Handbook of Lasers*, ed. by R.J.Pressley (Chemical Rubber Co., Cleveland, Ohio 1971) p. 489
2.40 C.G.B.Garrett: J. Quant. Electron. **4**, 70 (1968)
2.41 G.D.Boyd, T.J.Bridges, M.A.Pollack, E.H.Turner: Phys. Rev. Lett **26**, 387 (1971)
2.42 See, for example, A.Yariv: *Introduction to Optical Electronics* (Holt, Rinehart and Winston, Inc., New York 1971) p. 186
2.43 P.P.Bey, J.F.Giuliani, H.Rabin: Phys. Rev. Lett. **19**, 819 (1967)
2.44 J.M.Manley, H.E.Rowe: Proc. IRE **47**, 2115 (1959)
2.45 J.E.Geusic, H.J.Levinstein, S.Singh, R.G.Smith, L.G.Van Vitert: IEEE J. QE-**4**, 352 (1968)
2.46 G.D.Boyd, T.J.Bridges, C.K.N.Patel, E.Buehler: Appl. Phys. Lett. **21**, 553 (1972)
2.47 B.Lax, H.J.Zeiger, R.N.Dexter: Physica **20**, 818 (1954)
2.48 E.D.Palik, G.B.Wright: In *Semiconductors and Semimetals* ed. by R.K.Willardson, A.C. Beer (Academic Press, New York 1967) Vol. 3, p. 421
2.49 C.K.N.Patel, V.T.Nguyen: Appl. Phys. Lett. **15**, 189 (1969)
2.50 N.Bloembergen, A.J.Sievers: Appl. Phys. Lett. **17**, 483 (1970)
2.51 S.Somekh, A.Yariv: Opt. Commun. **6**, 301 (1972)
2.52 Y.Yacoby, R.L.Aggarwal, B.Lax: J. Appl. Phys. **44**, 3180 (1973)
2.53 C.J.Johnson, G.H.Sherman, R.Weil: Appl. Opt. **8**, 1667 (1969)
2.54 R.H.Stolen: Appl. Phys. Lett. **15**, 74 (1969)
2.55 R.H.Stolen: Phys. Rev. B **11**, 767 (1975)
2.56 C.P.Christensen, R.Joiner, S.T.K.Nieh, W.H.Steier: In *3rd Conf. High Power Infrared Laser Window Materials*, Vol. 1, ed. by C.A.Pitha, B.Brendon (Air Force Cambridge Research Laboratories, Bedford, MA 1974) p. 97
2.57 S.Iwasa, I.Balslev, E.Burstein: In *Proc. 7th Intern. Conf. Physics of Semiconductors, Paris*, 1964 (Dunod, 1964) p. 1077
2.58 A.Yariv, C.A.Mead, J.V.Parker: IEEE J. QE-**2**, 243 (1966)
2.58a J.E.Kiefer, A.Yariv: Appl. Phys. Lett. **15**, 26 (1969)
2.59 I.P.Kaminow: IEEE J. QE-**4**, 23 (1968)
2.60 I.P.Kaminow, E.H.Turner: In *Handbook of Lasers*, ed. by R.J.Pressley (Chemical Rubber Co., Cleveland, Ohio 1971) p. 447
2.61 A.Mooradian, A.L.McWhorter: In *Proc. Intern. Conf. Light Scattering Spectra of Solids, New York*, 1968, ed. by G.B.Wright (Springer, New York, Heidelberg, Berlin 1969) p. 297

2.62 W.D.Johnston,Jr., I.P.Kaminow: Phys. Rev. **188**, 1209 (1969)
2.63 C.K.N.Patel: Phys. Rev. Lett. **16**, 613 (1966)
2.64 J.J.Wynne, N. Bloembergen: Phys. Rev. **188**, 1211 (1969)
2.65 J.H.McFee, G.D.Boyd, P.H.Schmidt: Appl. Phys. Lett. **17**, 57 (1970)
2.66 B.F.Levine, C.G.Bethea: Appl. Phys. Lett. **20**, 272 (1972)
2.67 E.J.Danielewicz, P.D.Coleman: Appl. Opt. **13**, 1164 (1974)
2.68 G.H.Sherman: Ph.D. Thesis, University of Illinois, Urbana (1972)
2.69 L.H.Skolnik, H.G.Lipson, B.Bendow, J.T.Schott: Appl. Phys. Lett. **25**, 443 (1974)
2.70 A.Manabe, A.Mitsuishi, H.Yoshinaga: Japan. J. Appl. Phys. **6**, 593 (1967)
2.71 T.Hattori, Y.Homma, A.Mitsuishi, M.Tacke: Opt. Commun. **7**, 229 (1973)
2.72 C.Kojima, T.Shikama, S.Kuninobu, A.Kawabata, T.Tanaka: Japan. J. Appl. Phys. **8**, 1361 (1969)
2.73 G.D.Boyd, E.Buehler, F.G.Storz: Appl. Phys. Lett. **18**, 301 (1971)
2.74 G.D.Boyd, H.Kasper, J.H.McFee: IEEE J. QE-**7**, 563 (1971)
2.75 G.D.Boyd, E.Buehler, F.G.Storz, J.H.Wernick: IEEE J. QE-**8**, 419 (1972)
2.76 G.D.Boyd, H.M.Kasper, J.H.McFee, F.G.Storz: IEEE J. QE-**8**, 900 (1972)
2.77 R.L.Byer, H.Kildal, R.S.Feigelson: Appl. Phys. Lett. **19**, 237 (1971)
2.78 G.D.Holah: J. Phys. C **5**, 1893 (1972)
2.79 A.Miller, G.D.Holah, W.C.Clark: J. Phys. Chem. Solids **35**, 685 (1974)
2.80 J.Baars, W.H.Koschel: Solid State Commun. **11**, 1513 (1972)
2.81 J.P.van der Ziel, A.E.Meixner, H.M.Kasper, J.A.Ditzenberger: Phys. Rev. B **9**, 4286 (1914)
2.82 See, for example, D.Corson, P.Lorrain: *Introduction to Electromagnetic Fields and Waves* (W.H.Freeman and Co., San Francisco 1962) p. 381
2.83 See, for example, C.K.N.Patel: In *Lasers*, Vol. 2, ed. by A.K.Levine (Marcel Dekker, Inc., New York 1968) p. 101
2.84 T.Y.Chang: Opt. Commun. **2**, 77 (1970)
2.85 K.M.Evenson: Conf. Precision Electromagnetic Measurements, Boulder, Colorado (1972)
2.86 K.M.Baird, H.D.Riccius, K.J.Siemsen: Opt. Commun. **6**, 91 (1972)
2.87 M.Rosenbluh: M.S. Thesis, Massachusetts Institute of Technology (1975)
2.88 For example, Crystal Specialties, Inc., Monrovia, Calif. USA
2.89 R.L.Aggarwal, N.Lee, B.Lax: In *Conf. Digest Intern. Conf. Submillimeter Waves and Their Applications Atlanta*, 1974, p. 19
2.90 C.Freed: IEEE J. QE-**4**, 404 (1968)
2.91 H.R.Fetterman, B.J.Clifton, P.E.Tannenwald, C.P.Parker: Appl. Phys. Lett. **24**, 70 (1974)
2.92 T.J.Bridges, A.R.Strand: Appl. Phys. Lett. **20**, 382 (1972)
2.93 V.T.Nguyen, T.J.Bridges: Phys. Rev. Lett. **29**, 359 (1972)
2.94 T.L.Brown, P.A.Wolff: Phys. Rev. Lett. **29**, 362 (1972)
2.95 C.K.N.Patel, E.D.Shaw: Phys. Rev. Lett. **24**, 451 (1970)
2.96 P.D.Mandel: M.S. Thesis, Massachusetts Institute of Technology (1975)
2.97 See, for example, K.J.Button, B.Lax: In *Proc. Symp. Millimeter and Submillimeter Waves, Polytechnic Institute of Brooklyn*, 1970, MRI Symposia Series, Vol. 20
2.98 D.R.Cohn, D.M.Larsen, B.Lax: Phys. Rev. B **6**, 1367 (1972)
2.99 H.R.Fetterman, D.M.Larsen, G.E.Stillman, P.E.Tannenwald, J.Waldman: Phys. Rev. Lett. **26**, 975 (1971)
2.100 M.P.Hacker, R.J.Temkin: (private communication)
2.101 D.L.Jassby: J. Appl. Phys. **44**, 919 (1973)
2.102 R.Cano, I.Fidone, J.C.Hosea: Phys. Fluids **18**, 1183 (1975)
2.103 K.J.Button, D.R.Cohn, Z.B.Drozdowicz, D.L.Jassby, B.Lax, R.J.Temkin, S.Wolfe: Bull. Am. Phys. Soc. **20**, 1228 (1975)
2.104 F.O'Neill, W.T.Whitney: Appl. Phys. Lett. **26**, 454 (1975)

3. Parametric Oscillation and Mixing

R. L. Byer and R. L. Herbst

With 21 Figures

This chapter considers nonlinear interactions primarily in solids and the resultant nonlinear coupling of electromagnetic fields that leads to second-harmonic generation, optical mixing and parametric oscillation. Emphasis is placed on a description of the nonlinear interactions and experiments important for tunable infrared generation.

3.1 Description

When a medium is subjected to an electric field the electrons in the medium are polarized. For weak electric fields the polarization is linearly proportional to the applied field $P=\varepsilon_0\chi^{(1)}E$ where $\chi^{(1)}$ is the linear optical susceptibility and ε_0 is the permittivity of free space with the value 8.85×10^{-12} F/m in mks units. The linear susceptibility is related to the medium's index of refraction n by $\chi^{(1)}=n^2-1$.

A linear polarizability is an approximation to the complete constitutive relation which can be written as an expansion in powers of the applied field

$$\begin{aligned} P_i(\omega_3)=&\varepsilon_0\sum_j\chi_{ij}^{(1)}(\omega_3)E_j(\omega_3)\\ &\varepsilon_0\sum_{jk}\chi_{jk}^{(2)}(-\omega_3,\omega_2,\omega_1);\ \boldsymbol{E}_j(\omega_2)\boldsymbol{E}_k(\omega_1)\\ &+\varepsilon_0\sum_{j,k,l}\tfrac{1}{4}\chi_{ijkl}^{(3)}(\quad)\ldots+\ldots \end{aligned} \tag{3.1}$$

where we have assumed $\omega_3=\omega_2+\omega_1$, and the field components $\boldsymbol{P}(t)$ and $\boldsymbol{E}(t)$ at frequency ω are given by the Fourier relation

$$\boldsymbol{U}(t)=\tfrac{1}{2}[\boldsymbol{U}(\omega)\mathrm{e}^{i(\boldsymbol{k}\cdot\boldsymbol{r}-\omega t)}+cc]\,. \tag{3.2}$$

Like the linear susceptibility the second-order nonlinear susceptibility must display the symmetry properties of the crystal medium. An immediate consequence of this fact is that in centro-symmetric media the second-order nonlinear coefficients must vanish. Thus nonlinear optical effects are restricted to acentric materials. This is the same symmetry requirement for the piezo-electric d tensors and therefore the non zero components of the second-order susceptibility can be found by reference to the listed d tensors. However, the nonlinear coefficient tensors for the crystal point groups are listed in a number of references [3.1–6].

In addition to crystal symmetry restrictions χ_{ijk} satisfies an intrinsic symmetry relation which can be derived for a lossless medium from general energy considerations [3.1, 7]. This relation states that $\chi_{ijk}(-\omega_3, \omega_2, \omega_1)$ is invariant under any permutation of the three pairs of indices $(-\omega_3, i)$; (ω_2, j); (ω_1, k) as was first shown by *Armstrong* et al. [3.8]. Another symmetry relation based on a conjecture by *Kleinman* [3.14], is that in a lossless medium the permutation of the frequencies is irrelevant and therefore χ_{ijk} is symmetric under any permutation of its indices. Kleinman's symmetry conjecture is, however, not generally true.

The driving polarization for optical mixing is usually written in terms of a nonlinear coefficient d defined by

$$\begin{aligned} \boldsymbol{P}_i^{(2)}(\omega_3) = \varepsilon_0 \textstyle\sum_{ijk} d_{ijk}(\omega_3, \omega_2, \omega_1) \boldsymbol{E}_j(\omega) \boldsymbol{E}_k(\omega) \\ \exp i[(\boldsymbol{k}_1 + \boldsymbol{k}_2 - \boldsymbol{k}_3) \cdot \boldsymbol{r}] \,. \end{aligned} \tag{3.3}$$

Comparison of (3.1) and the above definition of d_{ijk} show that

$$\chi_{ijk}(-\omega_3, \omega_2, \omega_1) = d_{ijk}(-\omega_3, \omega_2, \omega_1)$$

for non-degenerate three-frequency processes and that

$$\chi_{ijk}(-\omega_3, \omega_1, \omega_1) = 2 d_{ijk}(-\omega_3, \omega_1, \omega_1)$$

for SHG. The definition of the nonlinear susceptibility has been discussed in detail by *Boyd* and *Kleinman* [3.9] and by *Bechmann* and *Kurtz* [3.4].

We have not yet made an estimate of the magnitude of the nonlinear susceptibility. An important step in estimating the magnitude of $\underset{\approx}{d}$ was taken by *Miller* [3.10] when he proposed that the field could be written in terms of the polarization as

$$E(-\omega_3) = \frac{1}{\varepsilon_0} \textstyle\sum_{jk} 2\Delta_{ijk}(-\omega_3, \omega_1, \omega_2) P_j(\omega_1) P_k(\omega_2) \,. \tag{3.4}$$

Comparing (3.3) and (3.4) shows that the tensors $\underset{\approx}{d}$ and $\underset{\approx}{\Delta}$ are related by

$$d_{ijk} = \varepsilon_0 \textstyle\sum_{lmn} \chi_{il}(\omega_3) \chi_{jm}(\omega_2) \chi_{kn}(\omega_1) \Delta_{lmn}(-\omega_3, \omega_2, \omega_1) \,. \tag{3.5}$$

Miller noted that Δ is remarkably constant for lossless nonlinear materials even though $\underset{\approx}{d}$ varies over four orders of magnitude.

Some insight into the physical significance of Δ can be gained by considering a simple anharmonic oscillator model representation of a crystal similar to the Drude-Lorentz model for valence electrons. The model has been previously discussed by *Lax* et al., [3.11] *Bloembergen* [3.1], *Garrett* and *Robinson* [3.12] and *Kurtz* and *Robinson* [3.13]. For simplicity we neglect the tensor character of the nonlinear effect and consider a scalar model. The anharmonic oscillator satisfies an equation

$$\ddot{x}+\Gamma\dot{x}+\omega_0^2x+\alpha x^2=\frac{e}{m}E(\omega,t) \tag{3.6}$$

where Γ is a damping constant ω_0^2 is the resonant frequency in the harmonic approximation, and α is the anharmonic force constant. Here $E(\omega,t)$ is considered to be the local field in the medium. The linear approximation to the above equation has the well-known solution

$$\chi(\omega)=n^2-1=\omega_p^2/(\omega_0^2-\omega^2-i\Gamma\omega)$$

where $\omega_p^2=Ne^2/m\varepsilon_0$ is the plasma frequency. Substituting the linear solution back into the anharmonic oscillator equation and solving for the nonlinear coefficient d in terms of the linear susceptibilities gives

$$d=\frac{Ne^3\alpha}{\varepsilon_0m^2}\frac{1}{D(\omega_1)D(\omega_2)D(\omega_3)} \tag{3.7}$$

where $D(\omega)$ is the resonant denominator term in the linear susceptibility. Finally, using the relation between Δ and d given by (3.5) we find that

$$\Delta=\left(\frac{m\varepsilon_0\alpha}{N^2e^3}\right). \tag{3.8}$$

On physical grounds we expect that the linear and nonlinear restoring forces are roughly equal when the displacement x is on the order of the inter-nuclear distance a, or when $\omega_0^2\alpha=\alpha a^2$. In addition, if we make the approximation that $Na^3=1$ the expression for Δ simplifies to

$$\Delta=a^2/e\,. \tag{3.9}$$

For $a=2$ Å the value for Miller's delta predicted by our simple model is 0.25 m^2/C. This compares very well with the mean value of 0.45 ± 0.07 m^2/C given by *Bechmann* and *Kurtz* [3.4]. Equation (3.5) shows that the second-order susceptibility to a good approximation is given by

$$\begin{aligned}d&=\varepsilon_0\chi(\omega_3)\chi(\omega_2)\chi(\omega_1)\Delta\\&\approx\varepsilon_0(n^2-1)^3\Delta\approx\varepsilon_0n^6\Delta\,.\end{aligned}$$

The intuitive physical picture inherent in the anharmonic oscillator model gives a remarkably accurate account of the magnitude of a material's non-linear response.

Table 3.1. Nonlinear coefficient, figure of merit, conversion efficiency, burn intensity, and

Material (point group pump wavelength)	$d \times 10^{12}$ (m/V) (References)	n_o n_e	$n_e - n_o$	θ_m	ϱ	$d_{eff} \times 10^{12}$
Te(32) $\lambda_p = 5.3$ µm	$d_{11} = 1089^{(a,d)}$	6.25 4.80	−1.45	14°	0.10	$1065 (d \cos^2 \theta_m)$
$CdGeAs_2$ ($\bar{4}2$ m) $\lambda_p = 5.3$ µm	$d_{36} = 453^{a,g}$	3.51 3.59	+0.086	II 55° I 35°	0.021 0.021	341 $(d \sin\theta)$ 406 $(d \sin 2\theta)$
GaAs ($\bar{4}3$ m)* (10.6 µm)	$d_{36} = 151^a$	3.30	0	—	—	—
GaP ($\bar{4}3$ m) (10.6 µm)	$d_{36} = 58.1^a$	3.00	0	—	—	—
$ZnGeP_2$ ($\bar{4}2$ m) $\lambda_p = 1.83$ µm	$d_{36} = 138^{a,g}$	3.11 3.15	+0.038	II 90° I 62°	0.0 0.01	d_{36} 114 $(d \sin 2\theta)$
GaSe ($\bar{6}$ m2) $\lambda_p = 5.3$ µm	$d_{22} = 88.5^{a,h}$	2.807 2.456	−0.351	12°40′	0.001	86.3 $(d_{22} \cos\theta)$
Tl_3AsSe_3 (3 m) $\lambda_p = 5.3$ µm	$d_+ = 66^{a,e}$	3.34 3.15	−0.182	22°	0.03	d_+
$AgGaSe_3$ ($\bar{4}2$ m) $\lambda_p = 1.83$ µm	$d_{36} = 65^a$	2.62 2.58	−0.32	I 55° I 90°	0.01 0.0	53 $(d \sin\theta)$ d_{36}
CdSe (6 mm) $\lambda_p = 1.83$ µm	$d_{31} = 29.5^a$	2.45 2.47	+0.019	90°	0.0	d_{31}
$AgGaS_2$ ($\bar{4}2$ m) $\lambda_p = 1.06$	$d_{36} = 24.2^{a,f}$	2.42 2.36	−0.054	I 58° I 90°	0.17 0.0	20.5 $(d \sin\theta)$ d_{36}
Ag_3SbS_3 (3 m) $\lambda_p = 5.3$ µm	$d_+ = 20.1^{a,d}$	2.86	−0.19	29°	0.042	d_+
Ag_3AsS_3 (3 m) $\lambda_p = 5.3$ µm	$d_+ = 19$	2.76 2.54	−0.223	22.5°	0.059	d_+
$LiIO_3$(6) $\lambda_p = 0.694$ µm	$d_{31} = 6.8^a$	1.85 1.72	−0.135	23°	0.071	2.75 $(d \sin\theta)$
$LiNbO_3$ (3 m) $\lambda_p = 0.532$	$d_{31} = 5.95^a$	2.24 2.16	−0.081	90°	0.0	d_{31}
ADP ($\bar{4}2$ m) $\lambda_p = 0.266$	$d_{36} = 0.86^{a,d}$	1.53 1.48	−0.0458	90°	0.0	d_{36}
KDP ($\bar{4}2$ m) $\lambda_p = 0.266$	$d_{36} = 0.71^a$	1.51 1.47	−0.0417	90°	0.0	d_{36}
SiO_2 (32) (1.06 µm)	$d_{11} = 0.46^{a,c}$	1.55 1.56	+0.0095	—	—	d_{11}

[a] *M.M.Choy, R.L.Byer*: Phys. Rev. B **14**, 1693 (1976)
[b] *J.Jerphagon, S.K.Kurtz*: Phys. Rev. B **1**, 1739 (1970)
[c] *J.Jerphagnon, S.K.Kurtz*: J. Appl. Phys. **41**, 1667 (1970)
[d] *J.H.McFee, G.D.Boyd, P.H.Schmidt*: Appl. Phys. Lett. **17**, 57 (1970)
[e] *J.D.Feichtner, G.W.Roland*: Appl. Opt. **11**, 993 (1972)

transmission range for nonlinear crystals

$d_{eff}^2/n_o^2 n_3$ $\times 10^{24}$	$l(\varrho)_{eff}$ (cm)	$\Gamma^2 l^2$ (1 W)	$\Gamma^2 l^2$ (1 MW/cm^2)	I_{burn} (MW/cm^2)	Transmission range (μm)
10175	0.011	2.66×10^{-4}	0.50	40–60	4–25
2688	0.34	1.4×10^{-3}	0.12	20–40	2.4–17
3811		1.7×10^{-3}	0.147		
635	$l_{coh} = 107$ μm	—	—	60	0.9–17 1.4–17
125	—	—	—	—	—
625	$l = 1$ cm	2.9×10^{-2}	0.71	> 4	0.7–12
426	0.59	6.1×10^{-3}	0.17		
341	$l = 1$ cm	4.8×10^{-4}	0.045	30	0.65–18
131	0.184	4.2×10^{-5}	7.4×10^{-3}	32	1.2–18
161	0.71	2.14×10^{-3}	0.08	> 10	0.73–17
242	$l = 1$ cm	7.68×10^{-3}	0.271		
58	$l = 2$ cm	3.1×10^{-3}	0.22	60	0.75–25
30.4	0.14	8.3×10^{-4}	0.047	12–25	0.60–13
42.3	$l = 1$ cm	9.2×10^{-3}	0.18		
21.3	0.11	3.5×10^{-6}	1.2×10^{-3}	14–50	0.60–14
22	0.60	1.46×10^{-6}	1.0×10^{-3}	12–40	0.60–13
1.54	0.008	4.6×10^{-6}	4.5×10^{-3}	125	0.31–5.5
3.51	$l = 5$ cm	1.9×10^{-2}	1.15	80–300	0.35–4.5
0.23	$l = 5$ cm	6.6×10^{-3}	0.297	> 500	0.20–1.1
0.11	$l = 5$ cm	4.6×10^{-3}	0.21	> 500	0.22–11
0.015	$l_{coh} = 14$ μ	—	—	> 1000	0.18–35

[f] *G.D.Boyd, H.Kaspar, J.H.McFee*: IEEE J. QE-**7**, 563 (1971)
[g] *D.S.Chemla, R.F.Begley, R.L.Byer*: IEEE J. QE-**10**, 71 (1974)
[h] *G.B.Abdullaev, L.A.Kulevskii, A.M Prokhorov, A.D.Savel'ev, E.Yu Salev, V.V.Smirnov*: JETP Lett. **16**, 90 (1972)
* GaAs is taken as a reference for other infrared nonlinear crystals

Table 3.1 lists the nonlinear coefficient, index of refraction, figure of merit, parametric gain coefficient, transparency range and burn intensity for important nonlinear crystals.

3.2 Nonlinear Interactions

3.2.1 Second-Harmonic Generation

We are now in a position to evaluate the nonlinear interaction in a crystal and to calculate the conversion efficiency and its dependence on phase matching and focusing. Conceptually, it is easier to treat the special case of second-harmonic generation (SHG) and then to extend the principal results to three-frequency interactions.

The starting point for the analysis is Maxwell's equations from which the traveling wave equation is derived in the usual manner.

$$\nabla^2 \boldsymbol{E} - \mu_0 \sigma \dot{\boldsymbol{E}} - \mu\varepsilon \ddot{\boldsymbol{E}} = \mu_0 \ddot{\boldsymbol{P}}^{(2)} . \tag{3.10}$$

Making the usual slowly varying amplitude approximations that $\omega^2 \boldsymbol{P} \gg \omega \dot{\boldsymbol{P}} \gg \ddot{\boldsymbol{P}}$, $\boldsymbol{k}\partial E/\partial z \gg \partial^2 E/\partial z^2$, and $\omega E \gg \dot{E}$, neglecting loss, assuming a steady state solution, and using the definition given (3.3) for the driving polarization, the above equation reduces to a pair of coupled nonlinear equations for the fundamental and second-harmonic waves

$$\frac{d\boldsymbol{E}(\omega)}{dz} = \mathrm{i}\kappa \boldsymbol{E}(2\omega)\boldsymbol{E}^*(\omega)\exp(\mathrm{i}\Delta \boldsymbol{k} z) , \tag{3.11a}$$

$$\frac{d\boldsymbol{E}(2\omega)}{dz} = \mathrm{i}\kappa \boldsymbol{E}(\omega)\boldsymbol{E}(\omega)\exp(-\mathrm{i}\Delta \boldsymbol{k} z) , \tag{3.11b}$$

where $\kappa = \omega d_{\mathrm{eff}}/nc$ and $\Delta \boldsymbol{k} = \boldsymbol{k}(2\omega) - 2\boldsymbol{k}(\omega)$ is the wave vector mismatch. At phase matching $\Delta \boldsymbol{k} = 0$ and $n(\omega) = n(2\omega)$ since $\mathrm{k} = 2\pi n/\lambda$ where n is the index of refraction.

The above coupled equations for second-harmonic generation have been solved exactly [3.8]. It is useful to discuss the solution for second-harmonic generation since the results can be extended to three-frequency interactions. We proceed by considering the low conversion efficiency case for a nonzero Δk, and then discuss the high conversion efficiency case.

In the low conversion limit, the fundamental wave is constant with distance. Therefore, setting $dE(\omega)/dz = 0$ and integrating (3.11b) gives

$$\boldsymbol{E}(2\omega)|_{z=l} = \kappa \boldsymbol{E}^2(\omega)^l \frac{\sin(\Delta \boldsymbol{k} l/2)}{(\Delta \boldsymbol{k} l/2)} .$$

The conversion efficiency is

$$\frac{I(2\omega)}{I(\omega)} = \Gamma^2 l^2 \operatorname{sinc}^2\left(\frac{\Delta kl}{2}\right) \tag{3.12}$$

where we denote $(\sin x)/x$ by $\operatorname{sinc}(x)$, $I = (nc\varepsilon_0/2)|E|^2$ is the intensity; and

$$\begin{aligned}\Gamma^2 l^2 &= \kappa\kappa|E(\omega)|^2 l^2 \\ &= \frac{2\omega^2 |d_{\text{eff}}|^2 l^2 I(\omega)}{n^3 c^3 \varepsilon_0}.\end{aligned} \tag{3.13}$$

Phase matching enters into the nonlinear conversion process through the phase synchronism factor $\operatorname{sinc}^2(\Delta kl/2)$ which is unity at $\Delta kl = 0$. Also, the second-harmonic conversion efficiency is proportional to $|d_{\text{eff}}|^2$ and l^2, as expected, and varies as the fundamental intensity. The above result holds in the plane wave focusing limit where $I = P/A$ and the area $A = \pi w_0^2/2$ with w_0 being the gaussian beam electric field radius.

For high conversion efficiencies, (3.11a) and (3.11b) can be solved by invoking energy conservation such that $E^2(\omega) + E^2(2\omega) = E_{\text{inc}}^2$ where E_{inc} is the incident electric field at $z = 0$. The solution for perfect phase matching is the well-known result

$$\frac{I(2\omega)}{I(\omega)} = \tanh^2(\kappa E_{\text{inc}} z)$$

which for small $\kappa E_{\text{inc}} z$ reduces to the low conversion efficiency result. In theory, second-harmonic generation should approach 100% as a $\tanh^2$ function. In practice, conversion efficiencies of 40–50% are reached under typical experimental conditions.

Phase Matching and Effective Nonlinear Coefficient

We first give our attention to the interaction length which is governed by the phase synchronism factor sinc $(\Delta kl/2)$. When $\Delta kl = \pi$ the interaction is said to have occurred over one coherence length. Using the definition of $\Delta k = k_{2\omega} - 2k_\omega$ for SHG and solving for the coherence length we find

$$l_{\text{coh}} = \frac{\lambda_\omega}{4(n_{2\omega} - n_\omega)}. \tag{3.14}$$

For typical non-birefringent crystals (GaAs for example), l_{coh} varies between 10 μm and 100 μm. Since the SHG efficiency varies as the interaction length squared, nonphase-matched $\Delta kl \neq 0$ interactions are between 10^{-6} and 10^{-4} times less efficient than phase-matched interactions over a 1 cm length crystal. Thus achieving phase matching is of utmost importance for efficient SHG.

Since most materials are positively dispersive over the frequency range of interest, typically $n_{2\omega} > n_{\omega}$ unless special steps are taken. The special step most often taken is to use a birefringent crystal to offset the dispersion [3.14–17].

As a specific example, consider phase matching in the negative ($n_e < n_o$) uniaxial crystal KDP. For a beam propagating at an angle θ_m to the crystal optic axis, the extraordinary index is given by

$$\left[\frac{1}{n_e(\theta_m)}\right]^2 = \frac{\cos^2\theta_m}{n_0^2} + \frac{\sin^2\theta_m}{n_e^2}$$

where n_o and n_e are the ordinary and extraordinary indices of refraction at the wavelength of interest. Two types of phase matching are possible for negative uniaxial KDP:

$$\text{Type I} \quad n_e^{2\omega}(\theta_m) = n_0^{\omega}$$
$$\text{Type II} \quad n_e^{2\omega}(\theta_m) = \tfrac{1}{2}[n_e^{\omega}(\theta_m) + n_o^{\omega}] .$$

For positive uniaxial crystals ($n_e > n_o$) the extraordinary and ordinary indices are reversed in the Type I and Type II phase-matching expressions. Type I phase matching uses the full birefringence to offset dispersion and Type II phase matching averages the birefringence to use effectively half of it in offsetting the dispersion.

A more important factor in choosing the type of phase matching to be used is the nonlinear coefficient tensor. It must be evaluated to see that the effective nonlinear coefficient remains nonzero. Thus for KDP where the $\bar{4}2$ m nonlinear tensor has components d_{xyz}, d_{yxz}, and d_{zxy}, Type I phase matching requires that the second-harmonic wave be extraordinary or polarized along the crystallographic z axis. To generate this polarization, the fundamental waves must be polarized in the E_x and E_y direction and thus be ordinary waves. In addition, the product E_xE_y should be maximized. Therefore, the propagation direction in the crystal should be in the (110) plane at θ_m to the crystal optic or z axis. For this case, the effective nonlinear coefficient becomes $d_{\text{eff}} = d \sin\theta_m$ and it is this coefficient that is used in the SHG conversion efficiency expression. The effective nonlinear coefficients for other crystal point groups have been calculated and listed [3.9]. In addition, the phase-matching angles for a number of nonlinear crystals and pump lasers have been listed by *Kurtz* [3.3].

A second example of some importance is $LiNbO_3$ which belongs to crystal class $3m$. For Type I phase matching the only type of phase matching with adequate birefringence in $LiNbO_3$,

$$d_{\text{eff}} = d_{15} \sin\theta - d_{22} \cos\theta \sin 3\varphi .$$

Now we must in addition to maximizing d_{22} through the correct choice of $\varphi(\varphi = 30°$ or propagation in the $y-z$ plane in trigonal $LiNbO_3$) choose θ correctly so that d_{15} and d_{22} add and not subtract for $\theta \neq 90°$. Recent measurements [3.18] have shown that $|d_{15}| = -|d_{22}|$ so that θ must lie in the negative $y-z$ quadrants to maximize d_{eff}.

Two further aspects of phase matching should be considered. They are double refraction and acceptance angle in the nonlinear interaction.

As the extraordinary wave propagates in a crystal its power flow direction differs by the double refraction angle ϱ from its phase velocity direction. The effect referred to as Poynting vector walk-off leads to a walk-off of the beam energy at an angle ϱ. The double refraction angle is given by

$$\varrho \approx \tan\varrho = \frac{n_{\omega}^{\text{o}}}{2}\left[\frac{1}{(n_{2\omega}^{\text{e}})^2} - \frac{1}{(n_{2\omega}^{\text{o}})^2}\right]\sin 2\theta . \tag{3.15}$$

For typical birefringent crystals at $\theta \sim 45°$, $\varrho \sim 1$–$2°$. Thus for focused laser beams in the crystal the power flow leads to a separation of the extraordinary and ordinary waves after a distance

$$l_{\text{a}} = \frac{\sqrt{\pi}w}{\varrho} \tag{3.16}$$

called the aperture distance [3.19]. Here w is the focused laser beam electric field radius. For example, for KDP $\varrho \sim 1°$ and at $w = 50\ \mu\text{m}$ we find $l_{\text{a}} = 0.5$ cm. This may be less than the full crystal length and thus lead to a substantial reduction in SHG efficiency. Note that at $\theta = 90°$, $\varrho = 0$ and there is no Poynting vector walk-off and thus an infinite aperture length. Therefore, 90° phase matching is desirable when possible.

The acceptance angle of the nonlinear crystal can be calculated by expanding the argument of the phase synchronism factor sinc $(\Delta kl/2)$ in a Taylor series in θ about the phase-matching angle θ_{m}. Carrying out the calculation we find for Type I phase matching

$$\Delta k = \frac{2\pi}{\lambda_{\omega}}(n_{\omega}^{\text{o}})^3\left[\frac{1}{(n_{2\omega}^{\text{e}})^2} - \frac{1}{(n_{2\omega}^{\text{o}})^2}\right]\sin 2\theta\delta\theta .$$

For Type II phase matching we find a similar result but reduced by a factor of two. For crystals with small birefringence this expression simplifies to

$$\Delta k = \frac{2\pi}{\lambda_{\omega}}(n_{2\omega}^{\text{o}} - n_{2\omega}^{\text{e}})\sin 2\theta_{\text{m}}\delta\theta .$$

Finally, to find the acceptance angle we note that the sinc x function reaches its first zero for $x = \pi$. Thus $\Delta kl/2 = \pi$ approximately determines the half width. Using this criteria the half angular acceptance width inside the nonlinear crystal is

$$\delta\theta = \lambda_{\omega}/l(n_{2\omega}^{\text{o}} - n_{2\omega}^{\text{e}})\sin 2\theta_{\text{m}} . \tag{3.17}$$

For a 1 cm crystal of KDP at $\lambda_\omega = 1\ \mu m$, $\Delta n = 0.04$, $\sin 2\theta_m \approx 1$ we find $\delta\theta = 2.5$ mr. The external acceptance angle is $\delta\theta_{ext} = n\delta\theta = 3.75$ mr.

For 90° phasematching $\sin 2\theta$ approaches zero so that the expansion must be carried to the next order. For 90° phase matching we find

$$\delta\theta = \sqrt{\frac{\lambda_\omega}{2l(n_{2\omega}^{o} - n_{2\omega}^{e})}}. \tag{3.18}$$

As an example, for a 1 cm $LiNbO_3$ crystal with $\Delta n = 0.08$ and $\lambda_\omega = 1\ \mu m$, the acceptance angle is $\delta\theta = 25$ mr or an order of magnitude greater than the acceptance angle for critically phase-matching KDP. Thus noncritical phase matching offers the additional advantage of increased acceptance angle compared to critical phase matching.

Focusing

Our analysis to this point has assumed interacting plane waves of infinite extent. In fact, the beams are usually focused into the nonlinear crystal to maximize the conversion efficiency. In addition, the waves are usually generated by laser sources and thus have a gaussian amplitude profile with electric field beam radius w. Historically, the problem of focusing was first solved for SHG [3.19–23] and later generalized to the three-frequency case by *Boyd* and *Kleinman* [3.9] In addition, focusing for parametric oscillators has been discussed in detail [3.24, 25]. Here we summarize the important aspects of the general treatment of focusing by *Boyd* and *Kleinman*,

To include the effects of focusing we introduce the *Boyd* and *Kleinman* focusing factor $h(B, \xi)$. The second-harmonic generation efficiency can be written in terms of $h(B, \xi)$ as

$$\frac{P_{2\omega}}{P_\omega} = \left(\frac{2\omega^2 d_{eff}^2}{\pi n_\omega^2 n_{2\omega} \varepsilon_0 c^3}\right) P_\omega l k_\omega h(B, \xi) \tag{3.19}$$

where B is the double refraction parameter

$$B = \tfrac{1}{2}\varrho (l k_\omega)^{\frac{1}{2}}, \tag{3.20}$$

ξ is the focusing parameter

$$\xi = l/b. \tag{3.21}$$

and b is the confocal distance defined as

$$b = w_0^2 k. \tag{3.22}$$

For second-harmonic generation (3.19) shows that the total second-harmonic efficiency is reduced by $h(B, \xi)$. In parametric processes we are usually interested in the parametric gain coupled into the fundamental mode. For this case $h(B, \xi)$ is reduced further to $\bar{h}(B, \xi)$. Fortunately, $\bar{h}(B, \xi)$ can usually be approximated by one of its limiting forms. In the near field approximation with negligible double refraction

$$\bar{h}(B, \xi) \to \xi \quad (\xi < 0.4, \xi < 1/6B^2). \tag{3.23}$$

Substituting this limit into (3.19) reduces it to the conversion efficiency valid for the near field given by

$$\frac{P_{2\omega}}{P_\omega} = \frac{2\kappa^2 l^2}{nc\varepsilon_0} \cdot \frac{P_\omega}{\pi w_1^2} .$$

In the limit of negligible double refraction but for confocal focusing where $\xi = 1$ the efficiency reduction factor becomes

$$\bar{h}(0, 1) \approx 1$$

which corresponds to the previous near field result first derived by *Boyd* and *Ashkin* [3.26]. In fact, the maximum value of $\bar{h}(B, \xi) = \bar{h}_{mm}(0, 2.84) = 1.068$ occurs at $\xi = 2.84$ instead of $\xi = 1$. However, practical considerations usually limit $\xi < 1$ so that the confocal approximation is an excellent one. This is the limit usually encountered in practice. Figure 3.1 shows $h(B, \xi)$ plotted vs. ξ for various values of the double refraction parameter.

In the tight focusing limit and neglecting double refraction the efficiency reduction factor becomes

$$\bar{h}(0, \xi) \to 1.187\pi^2/\xi \quad (80 < \xi < \pi^2/16B)$$

which shows that $\bar{h}(0, \xi)$ is proportional to b/l in this limit so that the gain reduces to b^2.

When double refraction is important the efficiency reduction factor can be represented by the empirical approximation

$$\bar{h}_{mm}(B) \approx \bar{h}_{mm}(0)/[1 + (4B^2/\pi)\bar{h}_{mm}(0)] \tag{3.24}$$

to better than 10% accuracy over the entire range of B. Here $\bar{h}_{mm}(0) = 1.068$ is the maximum value of $\bar{h}(B, \xi)$ at optimum focusing where $\xi = 2.84$. The condition

$$(4B^2/\pi)\bar{h}_{mm}(0) \approx 1$$

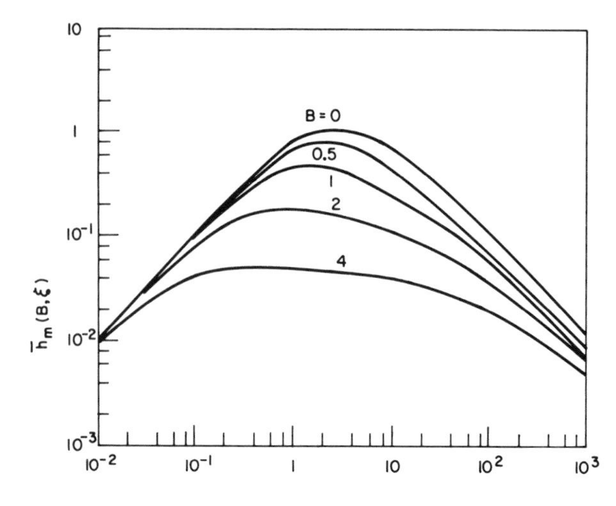

Fig. 3.1. Boyd and Kleinman focusing factor vs. $\xi = l/b$ at various values of the double refraction parameter B (after *Boyd* and *Kleinman* [3.9])

designates the point at which the conversion efficiency begins to limit due to double refraction effects. We use this condition to define an effective useful crystal length

$$l_{\mathrm{eff}}(\varrho) = \frac{\lambda_0}{2n_0\varrho^2 \bar{h}_{mm}(0)} \approx \frac{\lambda_0}{2n_0\varrho^2}. \tag{3.25}$$

For a crystal longer than this length the spot size must be increased to keep $l_a \geqq l$ so that the SHG efficiency per input power remains constant.

In terms of l_{eff} the efficiency reduction factor becomes

$$\bar{h}_{mm}(B) \approx \bar{h}_{mm}(0)/(1 + l/l_{\mathrm{eff}}).$$

In the limit of strong double refraction such that $l_{\mathrm{eff}} \ll l$ or $(4B^2/\pi)\bar{h}_{mm}(0) \gg 1$ the efficiency reduction factor is

$$\bar{h}_m(B, \xi) \rightarrow \frac{\pi}{4B^2} \quad (B^2/4 > \xi > 2/B^2). \tag{3.26}$$

Referring back to the SHG expression given by (3.19) we can see that the strong double refraction limit has the effect of replacing the crystal length l by l_{eff}.

In addition, the large double refraction limit maintains $\bar{h}_m(B, \xi)$ at a nearly constant value over a wide focusing range. Thus to minimize crystal damage problems ξ can be chosen such that $\xi \gtrsim 2/B^2$. This condition corresponds to a focal spot size w_1

$$w_1 \lesssim \frac{1}{2\sqrt{2}} \varrho l \tag{3.27}$$

and an effective area

$$\pi w_1^2/2 \lesssim \frac{\pi}{16} \varrho^2 l^2 \,. \tag{3.28}$$

For larger spot sizes the efficiency is reduced and for smaller spot sizes the efficiency is practically constant but the intensity increases.

The efficiency reduction factor and the effective crystal length are seriously reduced for critical phase matching. We can write a crystal figure of merit for SHG in the limits of noncritical ($\theta_m = 90°$) and critical ($\theta_m \neq 90°$) phase matching as

$$M_{\varrho=0} = \frac{d^2}{n_o^2 n_3} \tag{3.29}$$

and

$$M_{\varrho \neq 0} = \frac{d^2}{n_o^2 n_3} \cdot \frac{l_{\text{eff}}}{l} \,. \tag{3.30}$$

For large double refraction the conversion efficiency is reduced by $\pi/4B^2 \approx l/l_{\text{eff}}$. For $LiNbO_3$ at room temperature with phase matching achieved by angle tuning to $\theta = 43°$ for a 2.12 µm fundamental wavelength, $\varrho = 0.037$ radians and $B = 4.7\, l^{\frac{1}{2}}$. The SHG conversion efficiency is reduced by 28 times for a 1 cm crystal length and 140 times for a 5 cm crystal length compared to the 90° phase-matched case. For this double refraction angle the maximum effective crystal length is only $l_{\text{eff}} \approx 0.36$ mm. For a 1 cm crystal the focusing parameter ξ can vary between $5.5 > \xi > 0.09$ without affecting the conversion efficiency. This corresponds to a confocal parameter variation between 11 cm and 0.2 cm. For experimental ease and reduction of the intensity at the crystal surface the larger value of b would be utilized. For a 1 mm crystal length the gain remains the same but the confocal parameter varies between 1.1 mm $< b <$ 1.8 mm and thus the intensity is considerably increased. The pumping intensity decreases with crystal length as l^2 in agreement with (3.28).

3.2.2 Three-Frequency Interactions

Three-frequency interactions include sum generation and difference frequency generation or mixing in which two waves are incident on the nonlinear crystal and interact to generate a third wave and parametric generation in which a high power pump frequency interacts in a nonlinear crystal to generate two tunable frequencies. The nonlinear polarization now has three components given by

$$\begin{aligned}P^{(2)}(\omega_1)&=\varepsilon_0 2d\,E(\omega_3)E^*(\omega_2)\\P^{(2)}(\omega_2)&=\varepsilon_0 2d\,E(\omega_3)E^*(\omega_1)\\P^{(2)}(\omega_3)&=\varepsilon_0 2d\,E(\omega_2)\ E(\omega_1)\,.\end{aligned}\tag{3.31}$$

Substituting the polarization and electric field expressions at ω_1, ω_2, and ω_3 into the wave equation given by (3.10) gives three coupled equations

$$\frac{d\boldsymbol{E}_1}{dz}+\alpha_1\boldsymbol{E}_1=\mathrm{i}\kappa_1\boldsymbol{E}_3\boldsymbol{E}_2^*\exp(\mathrm{i}\Delta\boldsymbol{k}z)\,,\tag{3.32a}$$

$$\frac{d\boldsymbol{E}_2}{dz}+\alpha_2\boldsymbol{E}_2=\mathrm{i}\kappa_2\boldsymbol{E}_3\boldsymbol{E}_1^*\exp(\mathrm{i}\Delta\boldsymbol{k}z)\,,\tag{3.32b}$$

$$\frac{d\boldsymbol{E}_3}{dz}+\alpha_3\boldsymbol{E}_3=\mathrm{i}\kappa_3\boldsymbol{E}_1\boldsymbol{E}_2\exp(-\mathrm{i}\Delta\boldsymbol{k}z)\,.\tag{3.32c}$$

where $E(\omega_1)$ is now written as E_1, $\kappa_i=\omega_i d/n_i c$ $\Delta\boldsymbol{k}=\boldsymbol{k}_3-\boldsymbol{k}_1-\boldsymbol{k}_2$ and α_i is the loss.

We next investigate the solution of the above coupled equations for the three-frequency processes of interest. The equations have been solved exactly [3.8]; however, we consider only the simplified case of interacting plane waves and weak interactions such that the pump wave is not depleted.

Sum Generation (Up-Conversion)

For the case of sum generation ω_1 is a weak infrared wave that sums with a strong pump wave at ω_2 to generate a high frequency (visible) wave at ω_3. Negligible pump depletion implies that $dE_2/dz=0$ so that the three coupled equations reduce to a pair of (3.32a) and (3.32b). If we assume no loss, $\Delta k=0$, with input boundary conditions that $E_3(z=0)=0$ and $E_1(z=0)=E_1(0)$, the solution of the coupled equations is

$$\begin{aligned}E_1(z)&=E_1(0)\cos\Gamma z\\E_3(z)&=\sqrt{\frac{\omega_3}{\omega_1}\frac{n_1}{n_3}}\,E_1(0)\sin\Gamma z\end{aligned}$$

where

$$\Gamma=\sqrt{\kappa_1\kappa_3|E_2|^2}\,.$$

Thus sum generation has an oscillatory solution with no net gain and a 100% conversion for $\Gamma z=\pi/2$. In the low conversion limit, $\sin\Gamma z\approx\Gamma z$ so that the up-conversion efficiency becomes

$$\frac{I_3}{I_1(0)}=\left(\frac{\omega_3}{\omega_1}\right)\Gamma^2 l^2\,\mathrm{sinc}^2\left(\frac{\Delta kl}{2}\right) \tag{3.33}$$

where $\Gamma^2 l^2=(2\omega_1\omega_3 d^2 l^2 I_2)/(n_1 n_2 n_3 c^3\varepsilon_0)$ is equivalent to the previously derived SHG conversion efficiency factor given by (3.13). In the low conversion efficiency limit the sum generation efficiency equals the SHG efficency as one expects from physical arguments. However, at high conversion efficiencies the sum generation conversion oscillates while the SHG conversion approaches 100% as $\tanh^2\Gamma z$.

Difference Frequency Generation (Mixing)

For difference frequency generation or mixing the pump is the high frequency field at ω_3. Therefore, lack of pump depletion implies $dE_3/dz=0$ so that the coupled equations reduce to (3.32a) and (3.32b). Assuming a solution of the form $\exp(\Gamma z)$ and boundary conditions that $E_2(z=0)=E_2(0)$ and $E_1(0)=0$ results in the solution

$$E_2^*(z)=E_2^*(0)\cosh\Gamma z$$

$$E_1(z)=\sqrt{\frac{\omega_2}{\omega_1}\frac{n_1}{n_2}}\,E_1(0)\sinh\Gamma z\,.$$

Now we notice that exponential functions have replaced the sin and cos functions which occurred in the sum generation case. This implies that both the input field at ω_1 and the generated difference field at ω_2 grow during the nonlinear interaction at the expense of the pump field. Again in the limit of low conversion efficiency where $\sinh\Gamma l\approx\Gamma l$ the mixing efficiency becomes

$$\frac{I_1}{I_2(0)}=\left(\frac{\omega_1}{\omega_2}\right)\Gamma^2 l^2\,\mathrm{sinc}^2\frac{\Delta kl}{2} \tag{3.34}$$

where $\Gamma^2 l^2=(2\omega_1\omega_2 d^2 l^2 I_3)/(n_1 n_2 n_3 c^3\varepsilon_0)$ is the conversion efficiency factor which is again equivalent to the SHG and sum generation efficiency factors. However, now both waves grow and have net gain. This suggests the possibility of parametric oscillation once the gain exceeds the losses.

Parametric Generation

For parametric generation we assume a strong pump field at the highest frequency ω_3 and look for generation of the field at both ω_2 (signal field) and at ω_1 (idler field). We begin by investigating the coupled equations in the limit of no loss.

Multiplying (3.32a–c) successively by

$$\left(\frac{nc\varepsilon_0}{2}\right)\frac{E_1^*}{\omega_1}, \quad \left(\frac{nc\varepsilon_0}{2}\right)\frac{E_2^*}{\omega_2}, \quad \text{and} \quad \left(\frac{nc\varepsilon_0}{2}\right)\frac{E_3}{\omega_3}$$

we note that the right-hand sides of each equation are equal so that

$$\frac{1}{\omega_1}\frac{dI_1}{dz}=\frac{1}{\omega_2}\frac{dI_2}{dz}=-\frac{1}{\omega_3}\frac{dI_3}{dz} \tag{3.35}$$

which is a form of photon conservation or Manley-Rowe relation. Simply stated, parametric generation splits one pump photon into two photons which satisfy conservation of energy at every point in the nonlinear crystal. (This prompted one chemist to call a parametric generator a "photon cutter".) It should also be noted that the signal and idler fields (ω_2 and ω_1) are symmetric under interchange of indices.

We now proceed to solve (3.32a) and (3.32b) including loss and momentum mismatch. We assume, however, no pump depletion so that $dE_3/dz=0$. Again, we look for an exponentially varying solution. The expected solution can be written in the form

$$E_1^*(z)=(E_{1+}^*\mathrm{e}^{gz}+E_{1-}^*\mathrm{e}^{-gz})\mathrm{e}^{-\alpha z}\mathrm{e}^{-\mathrm{i}\Delta kz/2}$$
$$E_2(z)=(E_{2+}\mathrm{e}^{gz}+E_{2-}\mathrm{e}^{-gz})\mathrm{e}^{-\alpha z}\mathrm{e}^{+\mathrm{i}\Delta kz/2}$$

where we have set $\alpha_1=\alpha_2=\alpha$. Substitution into (3.32a) and (3.32b) results in a characteristic equation which gives

$$g=[\Gamma^2-(\Delta k/2)^2]^{\frac{1}{2}}, \tag{3.36}$$

$$\Gamma^2=\frac{\omega_1\omega_2|d|^2|E_3|^2}{n_1n_2c^2}$$
$$=\frac{2\omega_1\omega_2|d|^2I_3}{n_1n_2n_3\varepsilon_0c^3} \tag{3.37}$$

which is identical to the SHG conversion efficiency factor given by (3.13) when $\omega_1=\omega_2=\omega$ which is the degeneracy condition. Of course, for parametric generation, $\omega_1+\omega_2=\omega_3$ is the only restriction the generated frequencies must obey. For parametric amplification we assume input fields at ω_1 and ω_2

of $E_1^*(z=0)=E_1^*(0)$ and $E_2(z=0)=E_2(0)$. These input fields give the boundary conditions that determine the $E_{1\pm}$ and $E_{2\pm}$ coefficients. Carrying out the algebra results in the solutions for the field amplitude at $z=l$

$$E_1(l)e^{\alpha l}=E_1(0)\,e^{i\Delta kl/2}\left(\cosh gl-\frac{i\Delta k}{2g}\,\mathrm{sing}\,gl\right)$$

$$+i\frac{\kappa_1 E_3}{g}\,E_2^*(0)\,e^{i\Delta kl/2}(\sinh gl) \tag{3.38}$$

and

$$E_2(l)e^{\alpha l}=E_2(0)\,e^{i\Delta kl/2}\left(\cosh gl-\frac{i\Delta k}{2g}\sinh gl\right)$$

$$+i\frac{\kappa_2 E_3}{g}E_1^*(0)\,e^{i\Delta kl/2}(\sinh gl)\,, \tag{3.39}$$

where $\kappa_i=\omega_i d/n_i c$ and $\Gamma^2=\kappa_1\kappa_2|E_3|^2$ as before. These expressions are symmetric under the interchange of subscripts $1\leftrightarrow 2$ as expected.

When a single input field at frequency ω_2 is incident on the parametric generator it encounters a single pass power gain

$$G_2(l)=\frac{|E_2(l)|^2}{|E_2(0)|^2}-1$$

$$=\Gamma^2 l^2\,\frac{\sinh^2 gl}{(gl)^2}\,. \tag{3.40}$$

In this case the gain is independent of the phase of the input field $E_2(0)$ relative to the pump field E_3 since the generated idler field assumes the proper phase to maximize the gain. However, if two fields $E_1(0)$ and $E_2(0)$ are simultaneously incident on the amplifier the gain depends on their phases relative to the pump field.

Superfluorescence parametric emission is an extension of the parametric amplifier to the case of extremely high gain such that the input noise fields are amplified to a output intensity on the order of that of the pump field. If we neglect loss and assume negligible pump depletion, then in the high gain limit (3.40) becomes

$$G_2(l)=[1+(\Delta k/2g)^2]\sinh^2 gl$$

or

$$G_2(l)=\tfrac{1}{4}\exp 2\Gamma l \tag{3.41}$$

when $(\Delta k/2)<g$.

The noise input per mode for a parametric amplifier is equivalent to one photon in either the signal or idler channel or 1/2 photon in each. It is interesting to note that (1/2) $\hbar\omega$ input into a laser amplifier also leads to the

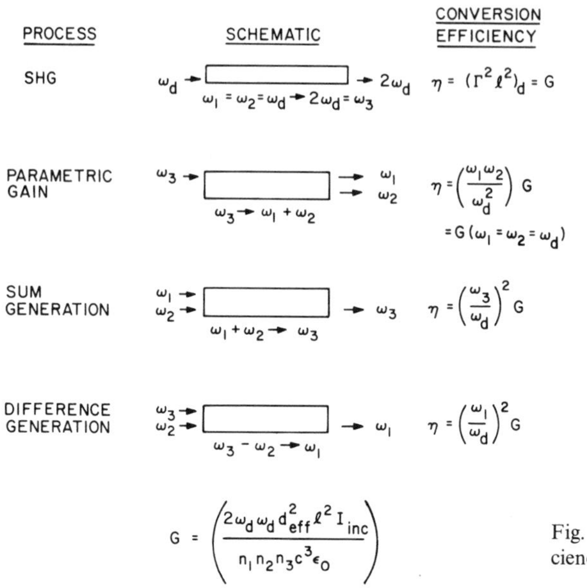

PROCESS	SCHEMATIC	CONVERSION EFFICIENCY
SHG	$\omega_d \rightarrow$ [ ] $\rightarrow 2\omega_d$; $\omega_1=\omega_2=\omega_d \rightarrow 2\omega_d=\omega_3$	$\eta=(\Gamma^2 l^2)_d=G$
PARAMETRIC GAIN	$\omega_3 \rightarrow$ [ ] $\rightarrow \omega_1, \omega_2$; $\omega_3 \rightarrow \omega_1+\omega_2$	$\eta=\left(\frac{\omega_1\omega_2}{\omega_d^2}\right)G$ $=G\ (\omega_1=\omega_2=\omega_d)$
SUM GENERATION	$\omega_1, \omega_2 \rightarrow$ [ ] $\rightarrow \omega_3$; $\omega_1+\omega_2 \rightarrow \omega_3$	$\eta=\left(\frac{\omega_3}{\omega_d}\right)^2 G$
DIFFERENCE GENERATION	$\omega_3, \omega_2 \rightarrow$ [ ] $\rightarrow \omega_1$; $\omega_3-\omega_2 \rightarrow \omega_1$	$\eta=\left(\frac{\omega_1}{\omega_d}\right)^2 G$

$$G=\left(\frac{2\omega_d\omega_d d_{eff}^2 l^2 I_{inc}}{n_1 n_2 n_3 c^3 \epsilon_0}\right)$$

Fig. 3.2. Plane-wave conversion efficiencies for nonlinear processes

correct amplified spontaneous emission output. For efficient superfluorescent operation the input noise field must be amplified by approximately 10^{16} so that $\Gamma l \sim 20$. In practice, very high gains have been achieved [3.27, 28], such that super fluorescence operation is possible [3.29, 30].

In the low gain limit where $\Gamma^2 l^2 < (\Delta k l/2)^2$ the gain of a parametric amplifier is

$$G_2(l)=\Gamma^2 l^2 \operatorname{sinc}^2\{[(\Delta k/2)^2-\Gamma^2]^{\frac{1}{2}} l\}\,. \tag{3.42}$$

Equations (3.40) and (3.42) show that as the gain varies from $\Gamma^2 l^2 > (\Delta k/2)^2$ to $\Gamma^2 l^2 < (\Delta k/2)^2$ the $(\sinh x)/x$ function continuously goes to $(\sin x)/x$. Note that for small gain the argument of the sinc function in (3.42) reduces to the expected $\Delta k l/2$.

The results of this section are summarized in Fig. 3.2 which shows the conversion efficiency for SHG, parametric generation, sum and difference frequency generation. The subscript "d" in the figure refers to the degeneracy case where $2\omega_d=\omega_3$ for SHG and parametric generation.

3.3 Infrared Generation by Mixing

Infrared generation by mixing was first demonstrated by *Zernike* and *Berman* [3.31] in 1965. The development of mixing as a method for obtaining tunable infrared radiation has closely paralleled the progress made in parametric oscillators as a tunable source of infrared radiation. In this section we review infrared generation by mixing. For want of a better alternative, we classify the mixing experiments by the sources used to provide tunable input radiation to the nonlinear crystal.

3.3.1 Dye Laser Sources

Review papers by *Kielich* [3.32] and *Warner* [3.33] have discussed generation of infrared radiation by mixing up to 1971. Since that time, there have been considerable improvements in dye laser sources [3.34] used for the tunable input for frequency mixing.

To be suitable for mixing, the nonlinear crystal must have adequate birefringence for phase matching and be transparent at both the pump wavelength and at the generated infrared wavelength. In general two classes of nonlinear crystals meet the birefringence and transparency requirements; the oxides including $LiNbO_3$ (0.4 – 3.5 μm) and $LiIO_3$ (0.3 – 5.5 μm) and semiconductor crystals including Ag_3AsS_3 (0.7 – 13 μm), $AgGaS_2$ (0.5 – 12 μm) and GaSe (0.6 – 18 μm). The other nonlinear crystals do not have adequate transparency range to include the visible or near infrared dye laser sources, and, except for $LiIO_3$ and $LiNbO_3$, the above materials require near infrared dye lasers to achieve phase matching for mixing.

In 1971 *Dewey* and *Hocker* [3.35] generated 3–4 μm radiation in $LiNbO_3$ by mixing a ruby laser and ruby pumped dye laser output. The output power and linewidth achieved were 6 kW peak power and 3–5 cm^{-1} linewidth. This early result was followed by a number of mixing experiments using the ruby laser and ruby pumped dye laser as a source. Table 3.2 summarizes the experiments, the nonlinear crystals used, generated tuning ranges, and output powers and linewidths. The nonlinear crystals used include $LiNbO_3$ [3.35], $LiIO_3$ [3.36], proustite [3.37], the chalcopyrite $AgGaS_2$ [3.38] and a new material GaSe [3.39, 40]. In addition $AgGaS_2$ [3.41] and $LiIO_3$ [3.42] were pumped by flashlamp pumped dye lasers and $LiIO_3$ [3.43, 44] was pumped by *Q*-switched doubled Nd:YAG laser and dye laser. In a recent experiment, *Pine* [3.45] generated 2.2–4.2 μm in $LiNbO_3$ by cw dye laser source. Of the above experiments, the high repetition rate difference frequency generation in $LiIO_3$ by *Goldberg* [3.44], the mixing in GaSe [3.46] and the very high resolution mixing in $LiNbO_3$ by *Pine* [3.45] represent the state of the art in infrared generation by mixing.

Goldberg [3.44] used a *Q*-switched, internally doubled, Nd:YAG laser source which operated at up to 80 Hz repetition rate and provided an output of 7.5 kW–80 ns pulses at 0.532 μm. The *Q*-switched Nd:YAG laser also provided output of approximately 10 kW peak power at 1.064 μm. The 0.532 μm source pumped a dye laser using rhodamine B in water with 3% ammonyx-LO. The dye laser operated over a tuning range of 0.575–0.640 μm with peak output near 900 W with a 60 ns pulse duration. The $LiIO_3$ crystal used for infrared generation was cut at 23°, antireflection coated, and

Table 3.2 Representative mixing experiments with dye laser sources

Source	Nonlinear crystal	Infrared tuning range (μm)	Output power (peak) and linewidth	Reference
Ruby laser and ruby pumped dye laser (0.84–0.89 μm)	$LiNbO_3$	3–4	6 kW 3–5 cm^{-1}	*C.F.Dewey, Jr, L.O. Hocker*: Appl. Phys. Lett. **18**, 58 (1971)
Ruby laser and ruby pumped dye laser (0.73–0.75 μm)	Ag_3AsS_3 (proustite)	10–13 4.9	0.1 W 6 W	*D.C.Hanna, R.C.Smith, C.R.Stanley*: Opt. Commun. **4**, 300 (1971)
Ruby laser and ruby pumped dye laser (0.8–0.83 μm)	$LiIO_3$	4.1–5.2	100 W	*D.W.Meltzer, L.S. Goldberg*: Opt. Commun. **5**, 209 (1972)
Ruby laser and ruby pumped dye laser (0.79–0.78 μm)	Ag_3AsS_3 (proustite)	3.2–6.5	100 W	*C.D.Decker, F.K. Tittle*: Appl. Phys. Lett. **22**, 41 (1973); Opt. Commun. **8**, 244 (1973)
Ruby laser and ruby pumped dye laser (0.74–0.808 μm)	$AgGaS_2$	4.6–12	0.3 W	*D.C.Hanna, V.V.Rampel, R.C.Smith*: Opt. Commun. **8**, 151 (1973)
Flashlamp pumped dye laser (0.58–0.62 μm)	$AgGaS_2$	8.7–11.6	10^{-4} W	*D.C.Hanna, V.V.Rampel, R.C.Smith*: IEEE J. QE-**10**, 461 (1974)
Q-switched Nd:YAG laser at second-harmonic dye laser (0.635–0.66 μm)	$LiIO_3$	2.8–3.4	80 mW	*H.Tashiro, T.Yajima*: Opt. Commun. **12**, 129 (1974)
Flashlamp pumped dye lasers (0.44–0.62 μm)	$LiIO_3$	1.5–4.8	0.4 W 0.5 cm	*H.Garlach*: Opt. Commun. **12**, 405 (1974)
Q-switched Nd:YAG laser at 1.06 and 0.532 μm (0.575–0.640 μm)	$LiIO_3$	1.25–1.60 3.40–5.65	70 W 0.5 W 0.08 cm^{-1}	*L.S.Goldberg*: Appl. Opt. **14**, 653 (1975)
Ruby laser and ruby pumped dye laser (0.72–0.76 μm)	GaSe	9.5–17	300 W (0.5 cm^{-1})	*G.B.Abdulaev, L.A. Kulevskii, P.V.Nickles, A.M.Drokhorov, A.D. Savel'ev, E. Yu Salaev V.V.Smirnov*: Sov. J. Quant. Elect. (to be published)
Argon ion laser and cw dye laser (0.56–0.62 μm)	$LiNbO_3$	2.2–4.2	1 μW (cw) (5×10^{-4} cm^{-1})	*A.S.Pine*: J. Opt. Soc. Am. **64**, 1683 (1974)

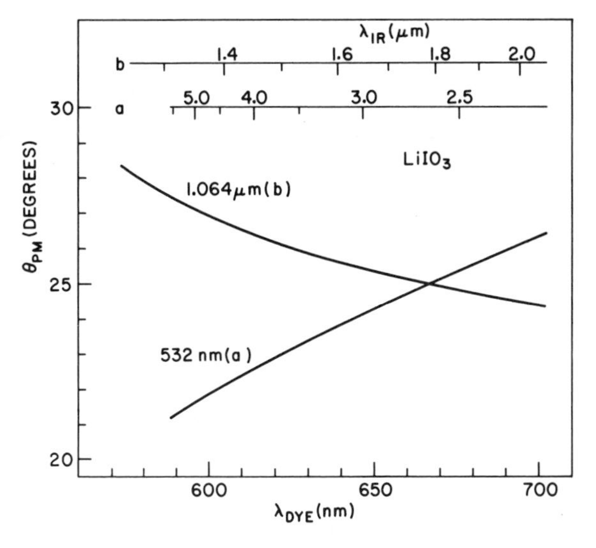

Fig. 3.3. Phase-matching angles for difference frequency mixing of a dye laser against 1.064 μm and 0.532 μm sources in $LiIO_3$ (after *Goldberg* [3.44])

operated within the dye laser cavity. Output in the infrared was obtained over the spectral range 1.25–1.60 μm by mixing the dye laser against 1.064 μm. Output from 3.40–5.65 μm was generated by mixing against the 0.532 μm source. By using an etalon within the Nd:YAG laser cavity and a line-narrowing prism and etalon within the dye laser cavity, output linewidths were kept to less than 0.1 cm^{-1} in the infrared. Peak output powers of 70 W at 1.35 μm and 0.5 W at 5.02 μm were measured in this experiment. Figure 3.3 shows the phase-matching angles for mixing in $LiIO_3$.

The conversion efficiency for infrared mixing in $LiIO_3$ is given by (3.34) with $I_3 = P_3/[\pi(w_2^2 + w_1^2)/2]$ in the near field limit. The effective nonlinear coefficient is $d_{eff} = d_{31} \sin(\theta + \varrho)$ where $d_{31} = 6.8 \times 10^{-12}$ m/V and $\varrho = 0.06$ rad. The large double refraction reduces the conversion efficiency in $LiIO_3$ by requiring off-angle phase matching such that $d_{eff}^2 = 0.18\, d_{31}^2$. More importantly the conversion efficiency is also reduced due to the limited useful interaction length of l_{eff} given by (3.25). For $LiIO_3$, $l_{eff} = \lambda_3/2n_o\varrho^2 = 8 \times 10^{-3}$ cm which is the interaction length for confocal focusing. Goldberg's measured output powers agreed very well with the predicted powers at the input powers and spot sizes used.

$LiIO_3$ is useful for infrared generation by mixing high peak power pulsed laser sources since its large double refraction can be offset by loose focusing at the high peak powers. However, its conversion efficiency is very poor for cw powers on the order of 1 Watt. As shown in Table 3.1, for a 1 Watt incident power, the conversion in $LiIO_3$ is only 4.6×10^{-6} for a 1 cm long crystal. This is two orders of magnitude less conversion efficiency than available from mixing in a 1 cm long 90° phase-matched $LiNbO_3$ crystal.

In an important demonstration of difference frequency mixing *Pine* [3.45] has generated cw output at the 1 μW power level from 2.2–4.2 μm in $LiNbO_3$.

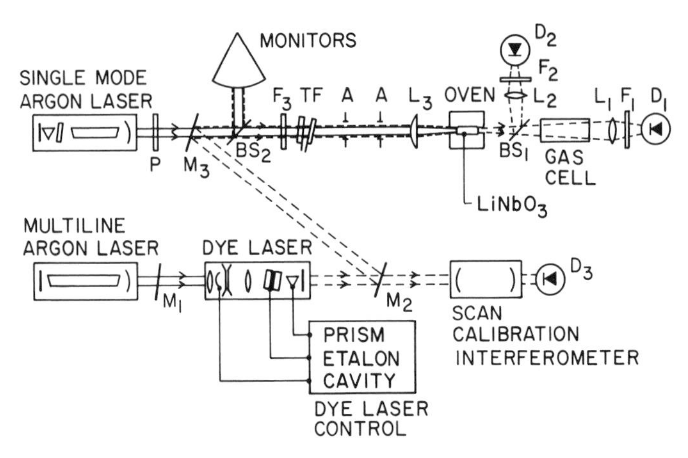

Fig. 3.4. Experimental apparatus used for cw mixing in $LiNbO_3$ (after *Pine* [3.45])

The experiment demonstrated the usefulness of infrared generation by applying the very high resolution source to spectroscopic studies of water vapor, ammonia, methane and nitrous oxide. A resolution of $5\times10^{-4}\,cm^{-1}$ with a continuous scan range of up to $1\,cm^{-1}$ was achieved by mixing a cw dye laser and single mode argon ion laser source in a 90° phase-matched $LiNbO_3$ crystal.

Figure 3.4 illustrates the experimental set up used by Pine in the cw mixing experiment. The rhodamine 6 G dye laser operated in a single axial mode with a linewidth of 15 MHz at 25–50 mW output power. Its output was combined by dielectric mirror with the single mode output of an argon ion laser operating at either 0.5145 or 0.4880 μm with 50–100 mW output power.

For the 5 cm long $LiNbO_3$ crystal used by Pine, the conversion efficiency is found to be 1.9×10^{-2} per watt of input power at optimum focusing. Pine focused somewhat more loosely than the confocal condition using $\xi=0.38$ instead of unity. At 50 mW input power the conversion efficiency reduces to 3.6×10^{-4}. Thus for 50 mW in the dye laser beam the output power is expected to be 2 μW at 3 μm wavelength. Pine measured a cw power of 0.5 μW, without correcting for various surface reflectivity losses, in good agreement with the expected value. This is still 10^4–10^5 times greater than the noise equivalent power of the photovoltaic InSb detector.

Figure 3.5 shows the measured mixing phase-matching curves as a function of crystal temperature for $LiNbO_3$. Figure 3.6 shows the *l*-type doubling in the NO spectrum at a resolution of $0.0005\,cm^{-1}$ or 15 MHz. $LiNbO_3$ is transparent to 4.5 μm in the infrared. Beyond that wavelength other crystals must be used for mixing.

Recently *Abdullaev* et al. [3.39, 40, 46] have investigated GaSe as a nonlinear crystal for infrared use. The nonlinear properties of GaSe are given in Table 3.1. Its important properties include a high nonlinear coefficient, large birefringence and a wide (0.65–18 μm) transparency range. GaSe is therefore

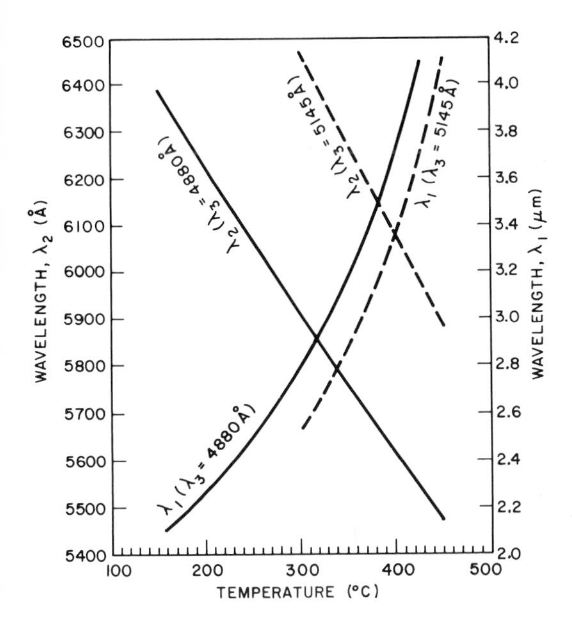

Fig. 3.5. Temperature tuned 90° phase-matching curves for $LiNbO_3$ mixing using 0.488 and 0.514 µm and a rhodamine 6 G dye laser source (after *Pine* [3.45])

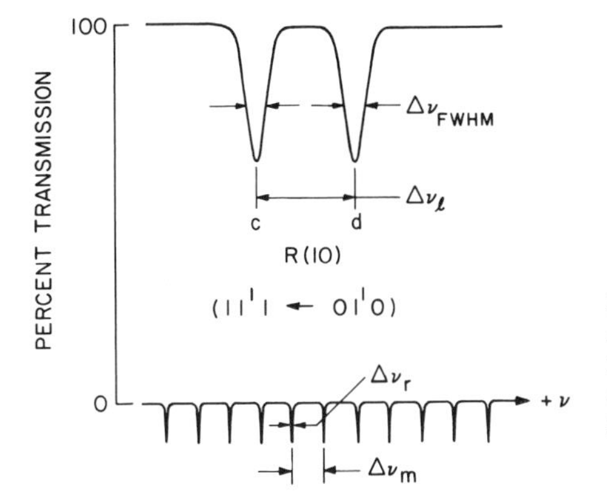

Fig. 3.6 Nitrous oxide spectrum $\nu_1+\nu_3$ combination band for a 30 cm cell length and 2 torr pressure. The frequency markers are spaced $\Delta\nu_m=0.01$ cm^{-1} and the resolution is 0.0005 cm^{-1} or 15 MHz (after *Pine* [3.45])

useful for infrared generation over an extended spectral range by mixing. To demonstrate the potential of GaSe, *Abdullaev* et al. [3.46] have generated infrared in the 9.5–13 µm range by mixing a ruby laser and ruby pumped dye laser. Output powers of 300 W were produced at a 0.5 cm^{-1} linewidth. The tuning range was limited on the short wavelength end by the dye laser range and by the total internal reflection angle of the layered GaSe crystal. Figure 3.7 illustrates the power output and angle tuning vs the infrared output in wave numbers. The large birefringence of GaSe leads to small phase-matching angles and a corresponding small double refraction angle ($\varrho\sim0.001$) and large effective nonlinear coefficient since $d_{\text{eff}}=d_{22}\cos\theta$.

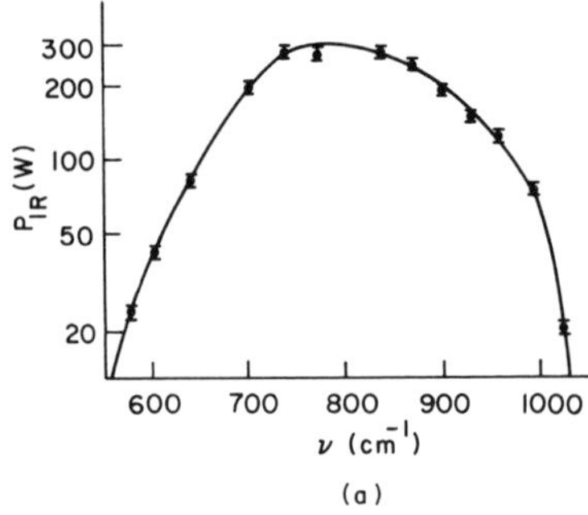

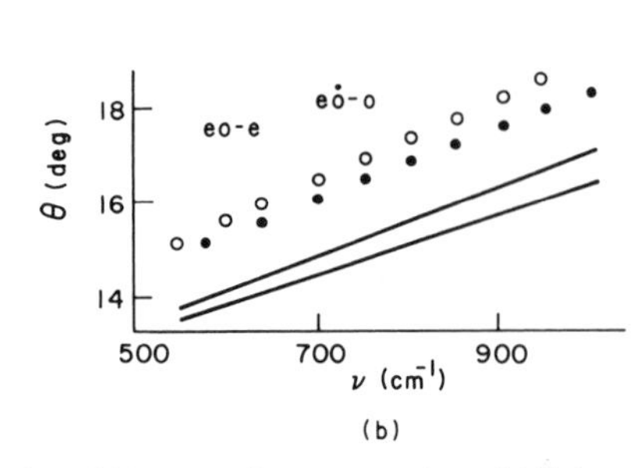

Fig. 3.7. (a) Peak output power (watts) vs generated infrared frequency in wave numbers. (b) Calculated (line) and observed (crosses) phase-matching angles for Type I and Type II phase matching in GaSe. The offset is due to the lack of precise birefringence data at the ruby pump wavelength region (*Kulevskii*, private communication)

The transparency requirement placed on a nonlinear material for infrared generation by mixing using dye laser sources limits the number of useful materials to only a handful. However, recent progress in the growth and perfection of the wide bandgap ternary semiconductor crystals belonging to the chalcopyrite group has opened the possibility of generating coherent radiation over an extended infrared tuning range by mixing. Of the chalcopyrite crystals only four are useful for nonlinear optics: $CdGeAs_2$ (2.4–17 μm), $ZnGeP_2$ (0.7–12 μm), $AgGaSe_2$ (0.73–17 μm), and $AgGaS_2$ (0.6–13 μm). Of these, only $AgGaS_2$ allows phase-matched mixing of near infrared dye laser sources to generate output from 4.6–12 μm [3.38]. Thus, though transparency range severely limits the choice of nonlinear crystals, the wide phase-matching ranges of $LiNbO_3$, $LiIO_3$, Ag_3AsS_3, $AgGaS_2$, and GaSe allow complete coverage of the infrared spectrum from less than 1 μm to beyond 17 μm by mixing of dye laser sources.

3.3.2 Parametric Oscillator Sources

Crystal choices for infrared generation by mixing can be considerably broadened by using longer wavelength tunable pump sources. Since dye lasers are limited to operating wavelengths less than 1 μm, parametric oscillator sources have provided the tunable input radiation.

The choice of crystals suitable for mixing now includes Ag_3AsS_3 (proustite), CdSe, Tl_3AsSe_3, GaSe and the chalcopyrites $AgGaS_2$, $AgGaSe_2$, $ZnGeP_2$, and $CdGeAs_2$. Mixing experiments have been carried out in Ag_3AsS_3, CdSe, and $AgGaSe_2$ using parametric oscillator sources. However, the phase-matching properties of the other crystals are well known and phase-matching curves are available in the review article by *Byer* [3.25].

Mixing was first demonstrated in CdSe by *Herbst* and *Byer* [3.47]. That experiment demonstrated that CdSe phase matched for mixing and was of high

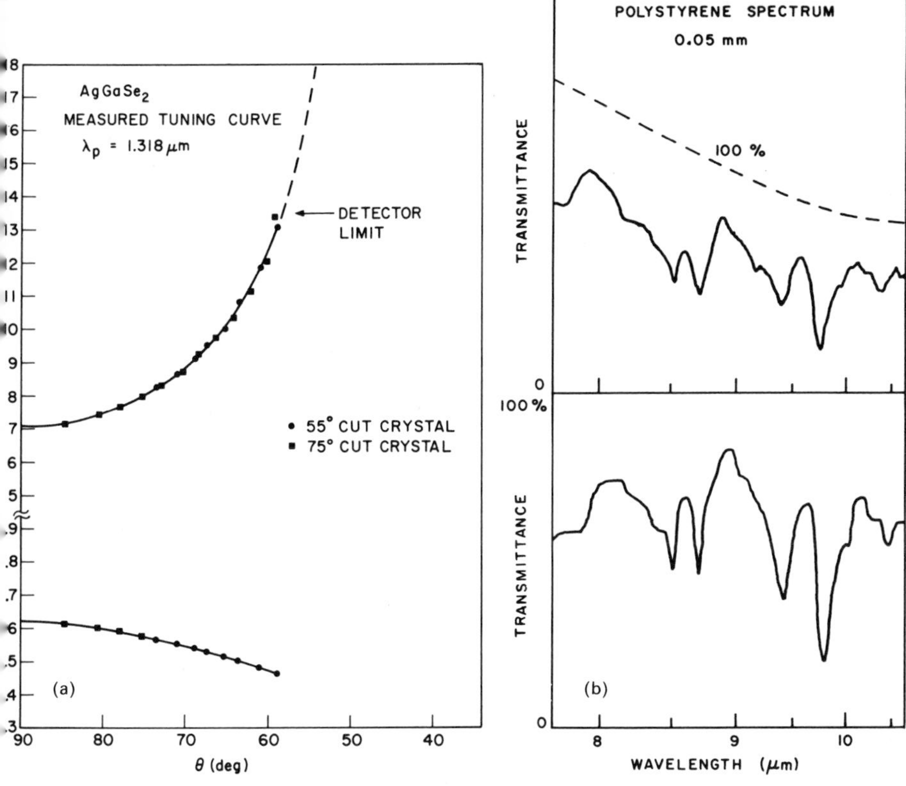

Fig. 3.8. (a) The 7–15 μm angle phase-matched mixing output in $AgGaSe_2$ using a $LiNbO_3$ parametric oscillator source. (b) The spectrum of polystyrene scanned with the mixer source (above) and a Perkin Elmer spectrophotometer (below) (after *Byer* et al. [3.50])

optical quality. By mixing a cw CO_2 laser a Q-switched Nd:YAG laser source operating at 1.833 μm, *Herbst* and *Byer* were able to demonstrate a 35% mixing efficiency in generating 2.2 μm. In a later experiment, the 1.833 μm wavelength was mixed against a tunable $LiNbO_3$ parametric oscillator source near 2.2 μm to generate output from 10.4–13 μm in CdSe [3.28] In the same article, *Byer* [3.28] predicted that CdSe phase matches for far infrared generation between 10 and 130 cm^{-1}.

In a similar experiment, *Bhar* et al. [3.48] used a proustite parametric oscillator source to generate output between 8 and 12 μm by mixing in proustite. The experiment was then extended by *Hanna* et al. [3.49] by mixing in CdSe to tune over the very wide 9.5–24 μm range. In that experiment the output powers ranged from 10 W at 10 μm to 100 W at 22 μm.

A representative mixing experiment using a parametric oscillator source was performed by *Byer* et al. [3.50] in $AgGaSe_2$. In the experiment a $LiNbO_3$ temperature-tuned parametric oscillator was mixed against a *Q*-switched

Table 3.3. Representative mixing experiments with parametric oscillator sources

Source	Nonlinear crystal	Infrared tuning range (μm)	Output power	Reference
$LiNbO_3$ temperature tuned parametric oscillator (2.08–2.22 μm)	CdSe	10.4–13		*R. L. Byer*: Parametric oscillators" in *Laser Spectroscopy*, ed. by R. *G. Brewer*, *A. Mooradian* (Plenum Press, N. Y. 1973)
Proustite parametric oscillator (1.87–2.47 μm)	Ag_3AsS_3 (proustite)	8–12	200 μW	*G. C. Bhar, D. C. Hanna, B. Luther-Davies, R. C. Smith*: Opt. Commun. **6**, 323 (1972)
$LiNbO_3$ temperature tuned parametric oscillator (1.5–1.7 μm)	$AgGaSe_2$	7–13	—	*R. L. Byer, M. M. Choy, R. L. Herbst, D. S. Chemla, R. S. Feigelson*: Appl. Phys. Lett. **24**, 65 (1974)
Proustite parametric oscillator (1.87–2.47 μm)	CdSe	9.5–24	10 W – 0.1 W	*D. C. Hanna, B. Luther-Davies, R. G. Smith*, R. *Wyatt*: Appl. Phys. Lett. **25**, 142 (1974)

Nd:YAG laser operating at 1.318 μm. Output wavelengths between 7 and 13 μm were generated for input wavelengths between 1.45 and 1.62 μm from the $LiNbO_3$ oscillator source. Figure 3.8a shows the phase-matching angle and input and generated infrared wavelengths for mixing in $AgGaSe_2$. Figure 3.8b shows a polystyrene spectrum taken by continuously tuning the $LiNbO_3$ parametric oscillator tunable source and the generated 7.5–10.5 μm wavelength range. $AgGaSe_2$ allows complete coverage of the infrared to 18 μm for pump wavelengths longer than 1.5 μm. Thus *Byer* [3.28] has proposed that $AgGaSe_2$ be used in conjunction with a 1.06 μm Nd:YAG pumped $LiNbO_3$ parametric oscillator to generate 3 μm to 18 μm by mixing the signal and idler waves of the $LiNbO_3$ oscillator source.

Table 3.3 summarizes the mixing experiments performed to date using parametric oscillator sources. The potential for high conversion efficiency, wide tuning range and an all solid state system makes this approach to infrared generation attractive.

3.3.3 Infrared Laser Sources

Infrared generation by mixing infrared laser sources leads to output frequencies in the far-infrared spectral range. Since the generation of far infrared by mixing is the subject of Chapters 2 and 5 only a summary of experiments reported since 1970 is given in this section. For references to earlier work see *Morris* and *Shen* [3.51] and *Bridges* and *Strnad* [3.52].

In 1971, following earlier theoretical work [3.51] *Yang* et al. [3.53] reported mixing picosecond Nd:Glass laser pulses in $LiNbO_3$ to generate picosecond broadband output in the 2–16 cm^{-1} far-infrared range. The experiment was followed two years later by a ruby laser, ruby pumped dye laser mixing experiment in $LiNbO_3$ to generate 20–190 cm^{-1} output at 0.1 W peak power [3.54].

At longer infrared wavelengths the coherence length increases, [cf. (3.14)], so that phase matching is somewhat less important. *Bridges* and *Strnad* [3.52] used GaAs to demonstrate last coherence length submillimeter wave generation using a pulsed CO_2 laser source.

In an important demonstration of the properties of the chalcopyrite nonlinear crystals, *Boyd* et al. [3.55] demonstrated phase-matched far-infrared generation in $ZnGeP_2$. Using two-step tunable CO_2 lasers, *Boyd* et al. were able to generate 70–110 cm^{-1} output at powers of 1.7 μW. Only recently has another chalcopyrite, $CdGeAs_2$, been used in a mixing experiment. *Kildal* and *Mikkelsen* [3.56] used CO and CO_2 laser sources to generate 11.4–16.8 μm by mixing in $CdGeAs_2$. Output powers of 4 μW were obtained.

The demonstration of tunable stimulated Raman spin-flip radiation in InSb by *Patel* and *Shaw* [3.57] led to the possibility of far-infrared generation in the spin-flip laser itself [3.58–60] or in a following mixing crystal.

Using a second InSb crystal after the CO_2 pumped InSb Raman spin-flip laser source, *Bridges* and *Van Tran Nguyen* [3.61] generated 93–110 cm^{-1} tunable far-infrared output at a 2 μW peak power level. In this mixing experiment the nonlinearity was provided not by the bound lattice electrons, but by the free electron spins. The mixing process was phase matched by adjusting the free electron concentration in the InSb mixing crystal.

In a similar experiment, *Brignall* et al. [3.62] generated 85–105 cm^{-1} radiation by mixing the output of a TEA CO_2 laser and InSb Raman spin-flip laser in InSb. Output powers of up to 10 μW were obtained.

In an approach designed to take advantage of the longer coherence lengths for far-infrared generation, *Lax* et al. [3.63] generated output step tunable from 70 μm to 2 mm in GaAs pumped by a CO_2 laser source. Phase matching was achieved by propagating the input CO_2 TEA laser beam noncollinearly in the GaAs sample. The far-infrared output was generated at a 21° angle to the input beams and interacted over an effective length of 0.75 cm. In a related experiment *Aggarwal* et al. [3.64] mixed cw CO_2 lasers in GaAs using a "folded" geometry to achieve phase matching. Output wavelengths between 70 μm and 1 mm were obtained at linewidths of less than 100 KHz. The folded geometry utilized total internal reflections in the GaAs sample to maintain phase velocity synchronism. An alternate approach to phase matching employs a stack of plates with every other plate inverted to achieve a 180° phase reversal in the sign of the nonlinearity. This method, which has been theoretically analyzed [3.65–68] has been used by *Thompson* [3.69] to frequency double a CO_2 laser in GaAs.

Table 3.4. Representative mixing experiments using infrared sources

Source	Nonlinear crystal	Infrared tuning range	Output power	Reference
Nd: glass laser (mode-locked)	$LiNbO_3$	2–16 cm^{-1}	200 W	*K.H. Yang, P.L. Richards. Y.R. Shen*: Appl. Phys. Lett. **19**, 320 (1971)
CO_2 laser (Q-switched)	GaAs	30 cm^{-1}	3 μW	*T.J. Bridges, A.R. Strnad*: Appl. phys. Lett. **20**, 382 (1972)
CO_2 laser	$ZnGeP_2$	70–110 cm^{-1}	1.7 μW	*G.D. Boyd, T.J. Bridges, C.K.N. Patel, E. Buehler*: Appl. Phys. Lett. **21**, 553 (1972)
CO_2 laser InSb Raman spin- -flip laser	InSb	93–100 cm^{-1}	2 μW	*T.J. Bridges, V.T. Nguyen*: Appl. Phys. Lett. **23**, 107 (1973)
Ruby laser ruby pumped dye lasers	$LiNbO_3$	20–190 cm^{-1}	0.1 W	*K.H. Yang, J.R. Morris, P.L. Richards, Y.R. Shen*: Appl. Phys. Lett. **23**, 669 (1973)
CO_2 laser	GaAs	70 μm–2 mm	20 mW	*B. Lax, R.L. Aggarwal, G. Favrot*: Appl. Phys. Lett. **23**, 679 (1973)
CO_2 laser InSb Raman spin-flip laser	InSb	85–105 cm^{-1}	10 μW	*N. Brignall, R.A. Wood, C.R. Pidgeon, B.S. Wherrett*: Opt. Commun. **12**, 17 (1964)
CO_2 and CO lasers	$CdGeAs_2$	11.4–16.8 μm	4 μW	*H. Kildal, J.C. Mikkelsen*: Opt. Commun. **10**, 306 (1974)

Table 3.4 summarizes mixing experiments using infrared laser sources. Due to restrahlen absorption bands, the frequency region between 25 μm and 70 μm has not been adequately covered by crystal mixing. An experimental approach based on Raman mixing, which does not have this transparency limitation, is considered in the next section. However, within limitations, mixing experiments in nonlinear crystals are straightforward and do provide useful output powers over a very wide spectral range.

3.4 Infrared Generation by Raman Mixing

Stimulated Raman scattering has been considered a potential source of new coherent frequencies since its accidental discovery by *Woodbury* and *Ng* [3.70] and interpretation by *Eckhart* et al. [3.71] in 1962. Since then stimulated Raman scattering has been adequately reviewed in the literature [3.72–76]. With the development of high-power tunable lasers Raman scattering is again being considered as a means of extending the tuning range of available tunable sources.

There are three approaches to generating radiation by the Raman process. They are stimulated Raman scattering, four-wave Raman mixing, which is a parametric process, and stimulated Raman scattering followed by a mixing interaction. In this section only stimulated Raman scattering and the coherent mixing process are discussed since the four-wave parametric mixing process first used to generate tunable infrared radiation by *Sorokin* et al. [3.77] is discussed in Chapter 6.

3.4.1 Stimulated Raman Scattering

Stimulated Raman scattering is most easily described as an interaction third order in the fields. Thus the polarization can be written in the form

$$P_i(\omega_p k_p) = \varepsilon_0 \chi^{(3)}_{ijk\,l}(-\omega_p, \omega_1, \omega_2, \omega_3)$$
$$E_j(\omega_1, k_1) E_k(\omega_2, k_2) E_l(\omega_3 k_3)$$

where in general $\chi^{(3)}$ is a fourth-rank tensor which may have complex components and must obey the symmetry properties of the medium. Unlike $\chi^{(2)}_{ijk}(-\omega_3, \omega_2, \omega_1)$, which is identically zero in centro-symmetric media, $\chi^{(3)}_{ijkl}(-\omega_p, \omega_1, \omega_2, \omega_3)$ is nonzero so that Raman scattering and four-wave mixing processes occur in gases and liquids as well as solids.

To find the Raman gain coefficient we assume that near resonance the Raman susceptibility has a Lorentzian line shape $g_L(\omega_R, \omega_p - \omega_s) = 1/\pi[\omega_R - (\omega_p - \omega_s) + (\Delta\omega_R/2)]$ where $\int_{-\infty}^{\infty} g_L(\omega_R, \omega_p - \omega_s) d(\omega_p - \omega_s) = 1$. On resonance the Raman susceptibility $\chi_R = \chi'_R + i\chi''_R$ is imaginary and less than zero. Here $\Delta\omega$ is the full width at half maximum of the Raman mode and ω_R, ω_p, and ω_s are the Raman, pump and Stokes frequencies. The polarization

$$P(\omega_S, k_S) = i\varepsilon_0 \chi''_R(-\omega_S, \omega_p, -\omega_p, \omega_S)|E_p|^2 E_s\,.$$

This driving polarization substituted into the wave equation shows that the generated Stokes field must satisfy

$$\frac{\partial E_s}{\partial z} + \alpha_s E_s = \frac{\omega_s \chi''_R}{2n_s c}(\omega_R = \omega_p - \omega_s)|E_p|^2 E_s \tag{3.43}$$

where $\Delta k = k_R - (k_p - k_s) = 0$ for stimulated Raman scattering since the polarization in the Raman medium is generated at the proper phase.

The Raman Stokes field grows exponentially with a gain coefficient

$$\Gamma = \frac{\omega_s \chi''_R |E_p|^2}{2n_s c}\,. \tag{3.44}$$

The intensity growth rate is $2\Gamma = g_s$ or

$$g_s = \frac{\omega_s \chi_R'' |E_p|^2}{2n_s c} = \frac{2\omega_s \chi_R'' I_p}{n_s n_p c^2 \varepsilon_0}. \tag{3.45}$$

The gain coefficient for the Stokes wave can also be written in terms of the differential Raman cross section $(d\sigma/d\Omega)$ using the relation [3.78]

$$\chi_R'' = \frac{(2\pi)^3 n_p c^4 \varepsilon_0 N (d\sigma/d\Omega)}{\pi n_s \omega_p \omega_s^3 \hbar \Delta\omega_R}. \tag{3.46}$$

Thus

$$\begin{aligned} g_s &= \frac{16\pi^2 c^2 N (d\sigma/d\Omega) I_p}{n_s^2 \omega_p \omega_s^2 \hbar \Delta\omega_R} \\ &= \frac{4\lambda_s^2 N (d\sigma/d\Omega) I_p}{n_s^2 \hbar \omega_p \Delta\omega_R} \end{aligned} \tag{3.47}$$

which is the form of the Raman gain most often seen in the literature. Equation (3.45) shows that the Raman gain decreases as ω_s. In addition, the pump intensity decreases as ω_p at constant peak power due to the frequency dependence of the beam area. Thus in practice stimulated Raman scattering becomes progressively more difficult as frequencies move from the visible to the infrared range.

Raman cross sections have been accurately measured by spontaneous Raman scattering [3.79]. On a scale where the N_2 cross section at 0.4880 µm of $d\sigma/d\Omega = 3.31 \pm 1.1 \times 10^{-31}\,\mathrm{cm}^2$ is unity, $H_2[Q(1)] = 1.6$ at $4155\,\mathrm{cm}^{-1}$, $CH_4(\nu_1) = 6.0$ at $2914\,\mathrm{cm}^{-1}$, and $C_6H_6(\nu_3) = 1.6$ at $992\,\mathrm{cm}^{-1}$ shift. The measured Raman gain factors g_s/I_p in units of cm/GW are for liquid N_2 at $2326.5\,\mathrm{cm}^{-1}$ shift 16 ± 5, for benzene 2.8, and for H_2 gas 1.5. The gain factors calculated using (3.47) are subject to the uncertainty in the linewidth $\Delta\omega_R$. In addition, if the laser linewidth is greater than $\Delta\omega_R$, which may be the case for scattering in gases, then the gain is reduced by the ratio of the laser to Raman linewidth.

The threshold for Raman oscillation within a resonator with mirror reflectivities R_1 and R_2 and loss α is $1 = R_1 R_2 e^{2g_s l - 2\alpha l}$. For pulsed operation the Raman gain must be "on" long enough for the Stokes field to build up to powers that are on the order of the pump power. Finally, superfluorescent Raman amplification occurs when the net gain over the cell length exceeds approximately e^{30}. Details of Raman oscillation, conversion efficiency and focusing are discussed in [3.74, 76, 80]. However, as an example, H_2 gas at 20 atm pressure pumped at 1.06 µm has a gain of $g_s/I_p = 4.0 \times 10^{-3}$ cm/MW. Thus to achieve a reasonable gain, pump intensities of 100 MW/cm^2 over interaction lengths of 100 cm are required. Since gas breakdown limits the

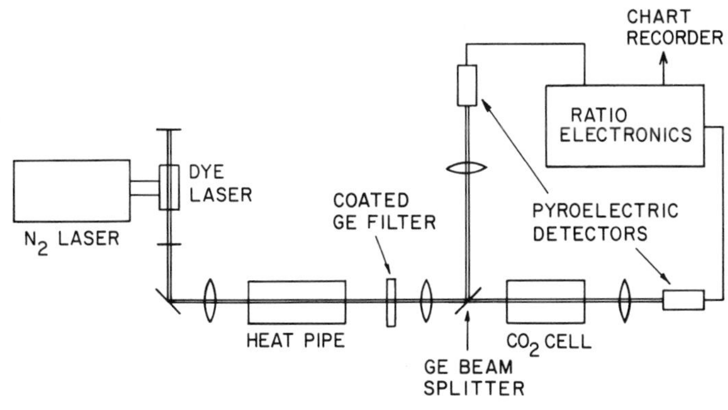

Fig. 3.9. Arrangement for the measurement of the CO_2 absorption spectrum by stimulated Raman down-conversion in potassium (after *Cotter* et al. [3.87])

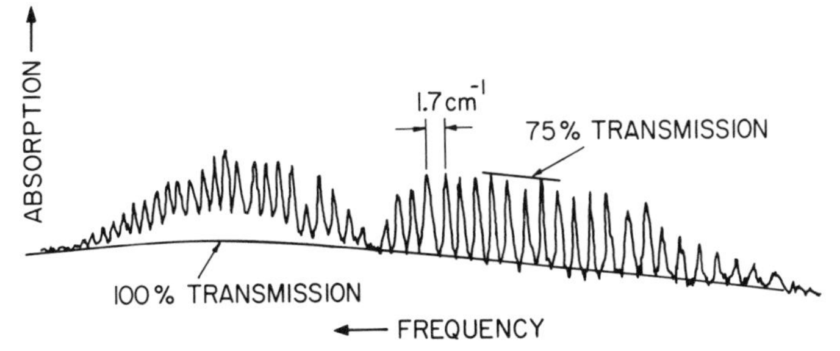

3.10. Absorption spectrum of CO_2 gas (50 Torr 10 cm cell) around 2.7 μm (after *Cotter* et al. [3.87])

intensity to less than 1 GW/cm^2, cell lengths must be long enough to achieve the desired net exponential gain. For confocal focusing in a 1 meter cell, the required power at 1.06 μm to generate 100 MW/cm^2 is 250 kW. Thus stimulated Raman scattering in molecular gases is limited to relatively high peak power laser sources.

In 1972 *Schmidt* and *Appt* [3.81] reported generating tunable Raman anti-Stokes emission in H_2 pumped by a ruby laser pumped dye laser. In later work [3.82] they reported generating tunable infrared radiation using the same apparatus. In an extension of the experiment by *Schmidt* and *Appt*, *Frey* and *Pradere* [3.83] generated tunable infrared radiation over the range from 14000 to 2300 cm^{-1} by pumping H_2 and CH_4 with a ruby pumped dye laser-amplifier system. Output powers of 20–60 MW between 14000 and 9700 cm^{-1}, 5–20 MW between 9700 and 5500 cm^{-1} and 0.5–2 MW between 5500 and 2300 cm^{-1} were achieved at linewidths of 0.07 cm^{-1}.

To obtain the relatively high conversion efficiencies to the first Stokes of 10–20%, Frey and Pradere used a second Raman amplifier cell after the first cell. In this way the kW output powers from the first Raman cell were

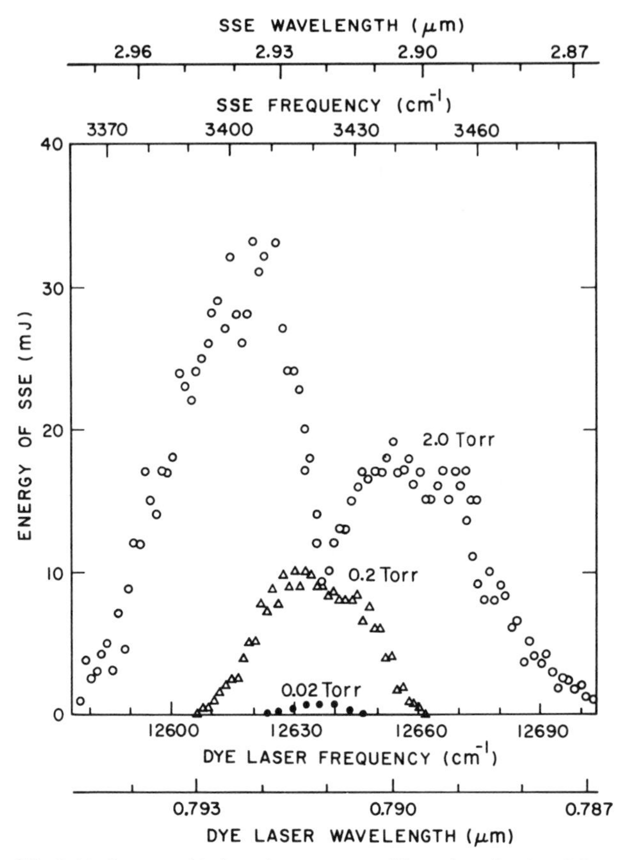

Fig. 3.11. Generated infrared energy near 2.9 μm by stimulated Raman emission in barium vapor (after *Carlsten* and *Dunn* [3.89])

amplified to the MW peak power levels. For wavelengths between 2.6 and 4.3 μm the low Raman gain was overcome by using a 4 meter cell length with three fluorine lenses to refocus the pump and Stokes beams. The high conversion efficiency and operation of the Raman cell without an external cavity illustrate the potential for stimulated Raman scattering as a near-infrared tunable source.

To overcome the low Raman gain in molecular gases, Raman scattering near electronic resonances in atomic systems may be used. The advantage of higher gains in electronic Raman scattering is achieved at the expense of decreased tuning range due to local resonances. However, the generally larger electronic Raman frequency shifts are more suitable for generating infrared with dye laser pump sources.

Early work in stimulated electronic Raman scattering [3.84–86] was concerned with the process itself. Only recently has stimulated electronic Raman scattering been studied as a source of tunable radiation [3.80, 87–89].

Using a nitrogen pumped dye laser *Cotter* et al. [3.87] were able to generate tunable Stokes emission near 2.7 μm in potassium vapor with a tuning range of near 1000 cm^{-1}. Output powers of up to 1 kW were obtained for 20–30 kW input power near 0.404 μm at 0.2 cm^{-1} linewidth. The Raman shift in potassium between the 5 *S* and 4 *S* levels is 21026.8 cm^{-1}. Figure 3.9 shows the apparatus for Raman scattering in potassium and Fig. 3.10 illustrates a CO_2 spectrum at 2.7 μm taken with the generated radiation. The potassium was maintained in a heat pipe at 10–20 Torr vapor pressure with an interaction length of 30 cm.

In a similar experiment *Carlsten* and *Dunn* [3.89] generated infrared radiation near 2.9 μm by Raman scattering in barium vapor. They used a ruby pumped dye laser/amplifier system [3.89] to generate over 30 mJ dye laser of output at up to 40% conversion efficiency. Figure 3.11 shows the generated output energy vs wavelength near the 2.9 μm range. The conversion efficiency saturated at a 40% photon efficiency at 10 mJ input dye laser energy and at 0.1 Torr of barium vapor pressure. This experiment very clearly demonstrates both the advantages and limitations of stimulated electronic Raman scattering as a source of tunable infrared radiation.

3.4.2 Coherent Raman Mixing

An alternate method to generating tunable radiation by Raman process is to coherently scatter an input beam from a stimulated Raman cell. In this process, which is responsible for anti-Stokes generation for example [3.76] a pump beam generates stimulated Raman scattering in the Raman medium. The input beam at frequency ω_i scatters from the generated Raman polarization to generate output frequencies at $\omega_i \pm \omega_R$. Thus the Raman medium acts like a mixer with a local oscillator frequency ω_R.

Scattering from coherently driven Raman oscillation was first discussed by *Shen* and *Bloembergen* [3.72] and by *Giordmaine* and *Kaiser* [3.90] who demonstrated the process in calcite. Earlier *Garmire* et al. [3.91] correctly interpreted higher order Stokes and anti-Stokes generation by the mixing mechanism. Later, *Duardo* et al. [3.92] carried out experiments in H_2 and CH_4 and described the generation of new frequencies in terms of mixing from coherently driven Raman oscillation.

Harris and *Byer* [3.93] first proposed using mixing from coherently driven Raman oscillation as a source of tunable radiation. They pointed to the advantage of a fixed Raman threshold for the pump while still obtaining tunable output with good conversion efficiency due to the mixing process. Later *Fleming* [3.94] showed that under proper excitation conditions phase matching is not required. Similar results were arrived at by *Venkin* et al. [3.95] who demonstrated frequency conversion by mixing.

To evaluate the conversion efficiency for coherent Raman mixing we assume four fields are present: the pump, the generated Stokes, the tunable input field at ω_i and the generated output field at ω_o. Substituting the fields and polarizations into the wave equation yields four simultaneous equations for the fields given by

$$\frac{\partial \boldsymbol{E}_{\mathrm{p}}}{\partial z} = - \frac{\omega_{\mathrm{p}}}{2cn_{\mathrm{p}}} \chi_{\mathrm{R}}''(|\boldsymbol{E}_{\mathrm{s}}|^2 \boldsymbol{E}_{\mathrm{p}} + \boldsymbol{E}_{\mathrm{i}} \boldsymbol{E}_{\mathrm{o}}^* \boldsymbol{E}_{\mathrm{s}} \mathrm{e}^{\mathrm{i}\Delta kz}), \tag{3.48a}$$

$$\frac{\partial \boldsymbol{E}_{\mathrm{s}}}{\partial z} = + \frac{\omega_{\mathrm{s}}}{2cn_{\mathrm{s}}} \chi_{\mathrm{R}}''(|\boldsymbol{E}_{\mathrm{p}}|^2 \boldsymbol{E}_{\mathrm{s}} + \boldsymbol{E}_{\mathrm{i}}^* \boldsymbol{E}_{\mathrm{o}} \boldsymbol{E}_{\mathrm{p}} \mathrm{e}^{-\mathrm{i}\Delta kz}), \tag{3.48b}$$

$$\frac{\partial \boldsymbol{E}_{\mathrm{i}}}{\partial z} = - \frac{\omega_{\mathrm{i}}}{2cn_{\mathrm{i}}} \chi_{\mathrm{R}}''(|\boldsymbol{E}_{\mathrm{o}}|^2 \boldsymbol{E}_{\mathrm{i}} + \boldsymbol{E}_{\mathrm{p}} \boldsymbol{E}_{\mathrm{s}}^* \boldsymbol{E}_{\mathrm{o}} \mathrm{e}^{-\mathrm{i}\Delta kz}), \tag{3.48c}$$

$$\frac{\partial \boldsymbol{E}_{\mathrm{o}}}{\partial z} = - \frac{\omega_{\mathrm{o}}}{2cn_{\mathrm{o}}} \chi_{\mathrm{R}}''(|\boldsymbol{E}_{\mathrm{i}}|^2 \boldsymbol{E}_{\mathrm{o}} + \boldsymbol{E}_{\mathrm{p}}^* \boldsymbol{E}_{\mathrm{s}} \boldsymbol{E}_{\mathrm{i}} \mathrm{e}^{\mathrm{i}\Delta kz}), \tag{3.48d}$$

where $\Delta k = -(k_{\mathrm{p}} - k_{\mathrm{s}}) + (k_{\mathrm{i}} - k_{\mathrm{o}})$.

In general the solution of these coupled equations requires numerical techniques. However, assuming minimal depletion of the pump and input waves the problem reduces to the solution of the two coupled equations, (3.48b) and (3.48d). If we assume solutions of the form

$$\boldsymbol{E}_{\mathrm{s}} = E_{\mathrm{s}} \mathrm{e}^{\left(\Gamma - \frac{\mathrm{i}\Delta k}{2}\right) z} \quad \text{and} \quad \boldsymbol{E}_{\mathrm{o}} = E_{\mathrm{o}} \mathrm{e}^{\left(\Gamma + \frac{\mathrm{i}\Delta k}{2}\right) z}$$

we then find

$$\Gamma_{\pm} = \frac{\Gamma_{\mathrm{p}} + \Gamma_{\mathrm{i}}}{2} \pm \tfrac{1}{2} \left[(\Gamma_{\mathrm{p}} - \Gamma_{\mathrm{i}}) - 4 \left(\frac{\Delta k}{2} \right)^2 + 4 \left(\frac{\mathrm{i}\Delta k}{2} \right) (\Gamma_{\mathrm{p}} - \Gamma_{\mathrm{i}}) \right]^{\frac{1}{2}} \tag{3.49}$$

where

$$\Gamma_{\mathrm{p}} = \frac{\omega_{\mathrm{s}}}{2n_{\mathrm{s}} c} \chi_{\mathrm{R}}'' |E_{\mathrm{p}}|^2, \tag{3.50a}$$

$$\Gamma_{\mathrm{i}} = \frac{\omega_{\mathrm{o}}}{2n_{\mathrm{o}} c} \chi_{\mathrm{R}}'' |E_{\mathrm{i}}|^2 \tag{3.50b}$$

are the field gain coefficients.

At exact phase matching the gain coefficient becomes $\Gamma = \Gamma_{\mathrm{p}} + \Gamma_{\mathrm{i}}$. Then, from (3.48), we find

$$\frac{E_{\mathrm{o}}}{E_{\mathrm{i}}} = \frac{\omega_{\mathrm{o}}}{\omega_{\mathrm{s}}} \frac{n_{\mathrm{s}}}{n_{\mathrm{o}}} \frac{E_{\mathrm{s}}}{E_{\mathrm{p}}} \tag{3.51}$$

which results in a conversion efficiency for the coherent Raman mixing process given by

$$\frac{I_o}{I_i} = \left(\frac{\omega_o}{\omega_s}\right)^2 \frac{n_s n_p}{n_o n_i} \frac{I_s}{I_p} \tag{3.52}$$

a result first derived by *Giordmaine* and *Kaiser* [3.90]. Thus the coherent Raman mixing efficiency equals the conversion efficiency from the pump to the Stokes field times a frequency factor $(\omega_o/\omega_s)^2$.This is the same conversion efficiency for mixing in nonlinear crystals if we identify I_s/I_p with the small signal gain or conversion efficiency for mixing.

In stimulated Raman scattering processes the photon conversion efficiency to the Stokes field is a maximum of 40%. Thus under optimum conditions the coherent Raman mixing process should allow generation of tunable down-shifted output with a corresponding 40% photon efficiency.

The advantage of coherent Raman mixing is twofold. First a strong fixed frequency pump beam can be used to efficiently generate the stimulated Raman scattering and, secondly the mixing process efficiency is independent of the intensity of the tunable input beam to a first approximation.

The coherent mixing process does require phase matching to achieve optimum conversion efficiency. However, if the gain constant is large compared to the phase mismatch factor Δk, then the conversion efficiency is reduced to approximately

$$\frac{I_o}{I_i} = \left(\frac{\omega_o}{\omega_s}\right)^2 \frac{n_s n_p}{n_o n_i} \frac{I_s}{I_p} \left(\frac{1}{1 + \frac{\Delta k^2}{\Gamma_p^2}}\right) \tag{3.53}$$

where $\Gamma_p \gg \Gamma_i$ is also assumed. Thus in H_2 gas pumped at 1.06 μm at 20 atm pressure, $\Gamma_p/I_p \sim 4 \times 10^{-3}$ cm/MW. At 200 MW/cm^2 input intensity $\Gamma_p \sim 0.8$ cm^{-1}. If ω_i varies from 4155–7000 cm^{-1} (the tuning range of a 1.06 μm pumped $LiNbO_3$ parametric oscillator) and the generated output frequency varies from $0-2845$ cm^{-1}, then $\Delta k = 0.4-0.6$ cm^{-1} for collinear beam propagation. At 50 atm Δk increases to 1.2–1.5 cm^{-1}. Thus for the above input intensity $\Delta k/\Gamma_p \sim 0.5$–0.75 and lack of phase matching does result in some conversion efficiency reduction. Since Δk is relatively constant over a wide range of generated frequencies, deviation from exact phase matching can be kept relatively small without adjusting the cell length or gas pressure. Thus phase matching is not a serious problem in coherent Raman mixing.

Figure 3.12 shows the tuning range available by coherent mixing in H_2 pumped at 1.064 μm with a 1.4 μm→2.40 μm tunable $LiNbO_3$ parametric oscillator as the source at ω_i. Preliminary experiments have demonstrated a Raman conversion efficiency to the first Stokes of 20% energy conversion and 40% photon conversion at two times above threshold. The pumping energy

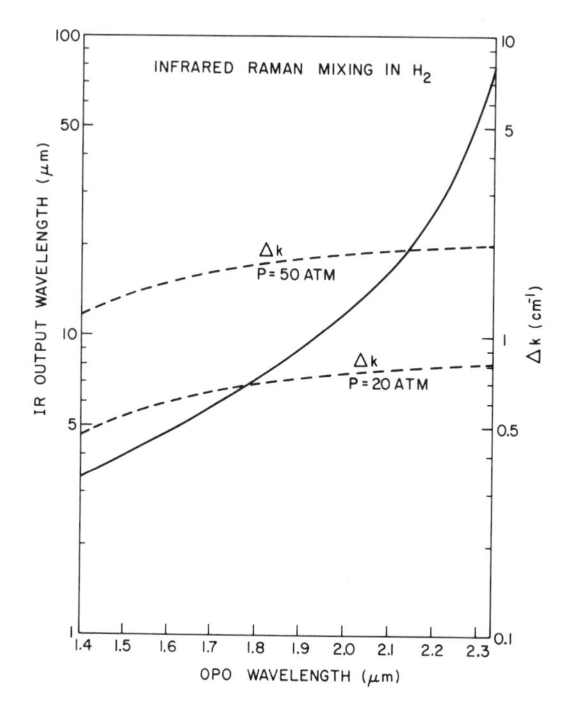

Fig. 3.12. Available tuning range in H_2 by coherent Raman mixing with a 1.06 μm pumped $LiNbO_3$ parametric oscillator source

required at 1.06 μm to achieve the peak conversion efficiency was 30 mJ in an 80 cm long cell. The corresponding input intensity was 200 MW/cm^2 to give a total gain $2\Gamma l$ of 128 [3.94].

Coherent Raman mixing offers the possibility of generating tunable infrared radiation without the limitations inherent in nonlinear crystal mixing of phase matching, low damage threshold and difficulty in obtaining adequate size and quality crystals. Experiments are in progress to demonstrate the full capability of this method.

3.5 Infrared Parametric Oscillators

3.5.1 Introduction

A parametric oscillator is schematically represented by a nonlinear crystal within an optical cavity. The nonlinear crystal when pumped proivdes gain at the two frequencies ω_2 and ω_1 (the signal and idler fields). When the gain exceeds the loss the device reaches threshold and oscillates. At threshold the output power increases dramatically, similar to the behavior of a laser. The generated output is coherent and collinear with the pump laser beam. Once above threshold, the parametric oscillator efficiently converts the pump radiation to continuously tunable signal and idler frequencies.

Wang and *Racette* [3.96] first observed parametric gain in a three-frequency interaction. It remained for *Giordmaine* and *Miller* [3.97] in 1965 to achieve adequate parametric gain in $LiNbO_3$ to overcome the losses and reach threshold for parametric oscillation. Giordmaine and Miller's early success led to considerable activity in the study of parametric oscillators as tunable coherent light sources.

To date parametric oscillators have been tuned across the visible and near infrared in KDP [3.98, 99] and ADP [3.100] when pumped by the second and fourth harmonic of a Nd:YAG laser and in RDA [3.101]. $LiNbO_3$ parametric oscillators have operated over an extended wavelength range from 0.6 μm to 4.4 μm [3.102–109] in the far infrared [3.110]. $LiIO_3$ parametric oscillation has been obtained out to 5.5 μm [3.11–115]. In the infrared range CdSe [3.116–118] and proustite [3.119–121] have operated as parametric oscillators out to 12 μm. In 1969 *Harris* [3.24] reviewed the theory and device aspects of parametric oscillators. More recently *Smith* [3.122] and then *Byer* [3.25] discussed parametric oscillators in their review articles. Table 3.5 lists representative parametric oscillators and their parameters.

There are a number of configurations for an optical parametric oscillator (OPO). The first distinction is between cw and pulsed operation. Due to the much higher gains, we consider only pulsed operation where the pump is typically generated by a Q-switched laser source. Parametric oscillators have also operated internal to the pump laser cavity, but the external configuration is more common. Finally, there are different parametric oscillator cavity configurations. The two most important are the doubly resonant oscillator (DRO), where both the signal and idler waves are resonated by the cavity mirrors, and the singly resonant oscillator (SRO), where only one wave is resonant.

3.5.2 Threshold and Rise Time

In a typical parametric oscillator the pump makes a single pass through the nonlinear crystal. The generated idler and signal waves $E_1(l)$ and $E_2(l)$ grow according to (3.38) and (3.39) during the single pass in the pump wave direction. Following reflection and the backward trip in the cavity, the waves are again traveling with the pump and are amplified. The gain is thus single pass and the corresponding loss is the round trip electric field loss α.

For the small gain case the threshold condition for low loss at both waves, or the doubly resonate oscillator case (DRO) is

$$\Gamma^2 l^2 \operatorname{sinc}^2\left(\frac{\Delta k l}{2}\right) \approx \alpha_1 l \alpha_2 l = a_1 a_2 . \quad (3.54)$$

Table 3.5. Representative parametric oscillators

Pump laser	Nonlinear material	Output power & pulse width	Conversion efficiency	Tuning range	Reference
Nd:Glass and second and third harmonics at 0.532 μm and 0.35 μm	KDP	100 kW (20 ns)	3%	0.957–1.17 μm (0.532 μm pump) 0.48–0.58 μm and 0.96–1.16 μm (0.35 μm pump)	*A.A. Akmanov, A.I. Kov V. A. Kolosov, A. S. Pisk kas, V. V. Fadeav, R. V. Khoklov*: JETP Lett. **3**, (1966) *S.A. Akhmanov, O.N. Ch sev, V. V. Fadeav, R. V. Kh lov, D.H. Klyshko, A.I. Kovrigin, A.S. Piskarska* Presented at Symp. Mo Optics, Brooklyn, 1967
Nd:YAG operating at fourth harmonic harmonic 0.266 μm	ADP	100 kW (2 ns)	25%	0.42–0.73 μm	*J.M. Yarborough, G.A. Massey*: Appl. Phys. Le 438 (1971)
Nd:YAG operating at second harmonic wavelengths 0.472, 0.532, 0.579, 0.635 μm	$LiNbO_3$	0.1–10 kW (200 ns)	45%	0.55–3.65 μm	*R. W. Wallace*: Appl. Phy Lett. **17**, 497 (1970)
Nd:YAG 1.06 μm	$LiNbO_3$	0.1–1 MW (15 ns)	40%	1.4–4.4 μm	*R.L. Herbst, R.N. Flemi R.L. Byer*: Appl. Phys. L **25**, 520 (1974)
Ruby 0.6943 μm	$LiNbO_3$	3 W	10^{-6}	66–200 μm	*B. C. Johnson, H.E. Puth J. Soo Hoo, S.S. Sussman* Appl. Phys. Lett. **18**, 18 (1971)
Nd:Glass operating at second harmonic 0.53 μm	$\alpha-HIO_3$ $LiIO_3$	10 MW (20 ns)	10%	0.68–2.4 μm	*A. I. Izrailenko, A. I. Ko gin, P. V. Nikles*: JETP L **12**, 331 (1970); *A.I. Kovr P. V. Nikles*: JETP Lett. 313 (1971)

If only one wave is resonant, the singly resonant oscillator (SRO) condition, the threshold increases to

$$\Gamma^2 l^2 \operatorname{sinc}^2\left(\frac{\Delta k l}{2}\right) \approx 2\alpha_2 l = 2a_2 \,. \qquad (3.55)$$

where α_2 is the single pass loss at the resonated wave.

A comparison of (3.54) and (3.55) shows that the singly resonant oscillator has a threshold that is $2/a_1$ times that of the doubly resonant oscillator. For a single pass power loss of 1% the SRO threshold is thus 200 times the DRO threshold. This of course assumes that both waves of the DRO are resonant simultaneously. In practice, this requires a single frequency pump source and

3.5. (continued)

p laser	Nonlinear material	Output power & pulse width	Conversion efficiency	Tuning range	Reference
laser 3 μm	$LiIO_3$	2 kW (20 ns)	1 %	0.77–4 μm	*L.S. Goldberg*: Appl. Phys. Lett. **17**, 489 (1970)
laser 3 μm	$LiIO_3$	100 kW (15 ns)	10 %	1.1–1.9 μm	*A.J. Campillo, C.L. Tang*: Appl. Phys. Lett. **19**, 36 (1971); *A.J. Campillo*: IEEE J. QE-**8**, 809 (1972)
y laser operating cond harmonic μm	$LiIO_3$	10 kW (5 ns)	8 %	0.415–2.1 μm	*G. Nath, G. Pauli*: Appl. Phys. Lett. **22**, 75 (1973)
YAG μm	CdSe	1 kW (100 ns)	40 %	2.2–2.3 μm 10.5–9.7 μm	*R.L. Herbst, R.L. Byer*: Appl. Phys. Lett. **21**, 189 (1972)
$_2$:Dy μm	CdSe	5 kW (30 ns)	0.5 %	3.3 μm 7.86 μm	*A.A. Davydov, L.A. Kulevskii, A.M. Prokhorov, A.D. Savel'ev, V.V. Smirnov*: JETP Lett. **15**, 513 (1972)
aser μm	CdSe	800 W (300 ns)	10 %	4.3–4.5 m 8.1–8.3 m	*J.A. Weiss, L.S. Goldberg*: Appl. Phys. Lett. **24**, 389 (1974)
$CaWO_4$ 5 μm	Ag_3AsS_3	100 W (25 ns)	0.1 %	1.22–8.5 μm	*D. C. Hanna, B. Luther Davies, H.N. Rutt, R.C. Smith*: Appl. Phys. Lett. **20**, 34 (1972); *D. C. Hanna, B. Luther Davies, B. C. Smith*: Appl. Phys. Lett. **22**, 440 (1973)

careful cavity control to assure resonance at both parametric waves. Lack of cavity length control or pump frequency control leads to instabilities in the threshold condition resulting in amplitude fluctuations of the oscillator. For these reasons, the lower threshold advantage of the DRO is not utilized except where necessary such as for cw operation.

The possibility of cw operation was first pointed out by *Boyd* and *Ashkin* [3.26] in 1966. It was not until 1968 that *Smith* et al., [3.123] using a 5 mm crystal of $Ba_2NaNb_5O_{15}$ pumped at 0.532 μm achieved cw operation at degeneracy. The threshold for that experiment was 45 mW. Later, *Smith* [3.124] demonstrated a cw pumped oscillator with only 3 mW threshold.

Using an idea proposed by *Harris* [3.24] that the full power of a multiaxial mode laser may be used to pump a parametric oscillator if the

DRO axial mode interval equals that of the pump, *Byer* et al. [3.125] constructed a visible cw DRO using a 1.7 cm $LiNbO_3$ crystal. This oscillator tuned from 0.68 to 0.71 μm and 1.9 and 2.1 μm for a 0.5145 μm argon ion laser pump. It achieved a threshold of less than 500 mW in a long cavity that included an internal lens to achieve proper focusing.

The low gains and resultant difficulty of constructing cw oscillators have held back work in this area. However, *Laurence* and *Tittel* [3.126] and *Weller* et al. [3.127] have extended work on cw argon-ion laser pumped $Ba_2NaNb_5O_{15}$ parametric oscillators. Weller et al. achieved 60 mW of output at a total efficiency of 15% in a tuning range from 0.93 μm to 1.15 μm.

Thus far we have discussed the threshold of a parametric oscillator under steady state conditions. In fact, most optical parametric oscillators are constructed using a Q-switched laser source with pump pulse lengths that typically vary from 10 ns to 1 μs. When the pump is incident on the nonlinear crystal the initial signal and idler fields are amplified. However, the oscillator is not "on" until the fields are amplified from the initial noise level to a magnitude of the order of the pump field.

The number of round trips in the optical cavity necessary to fully amplify the signal and idler waves multiplied by the cavity round trip time leads to the rise time necessary to achieve threshold. The excess gain which is proportional to the incident pump power, the pump pulse length, and the cavity length are important parameters in the threshold of a pulsed parametric oscillator.

The rise time of a DRO for a step input pump pulse has been discussed by *Byer* [3.25] and *Kreuzer* [3.128]. Recently *Pearson* et al. [3.129] extended the analysis to include an arbitrary input pulse shape for both the DRO and SRO cases.

The rise time of the parametric oscillator can be directly determined by integrating the time-dependent coupled equations

$$\frac{1}{c_1}\frac{\partial E_1}{\partial t}+\alpha_1 E_1=\mathrm{i}\kappa_1 E_3 E_2^* \exp(\mathrm{i}\Delta k r)$$

$$\frac{1}{c_1}\frac{\partial E_2}{\partial t}+\alpha_2 E_2=\mathrm{i}\kappa_2 E_3 E_1^* \exp(\mathrm{i}\Delta k r)$$

where $c_1=c/n_1$ and $c_2=c/n_2$ are the phase velocities. Assuming $\Delta k=0$, $\alpha_1=\alpha_2=a/L_c$ where L_c is the effective cavity length. For a step pump source turned on at $t=0$ the calculated rise time is

$$\tau_R=\frac{(L_c/c)\ln(P_1(t)/P_1(0))}{2a(\sqrt{N}-1)} \tag{3.56}$$

where $N=\Gamma^2 l^2/a^2$ is the number of times above threshold. Equation (3.56) shows that it is important to keep the oscillator cavity length as short as possible for pulsed operation in order to minimize the rise time and maintain high conversion efficiency.

Pulsed parametric oscillator operation also leads to the consideration of a minimum threshold energy density rather than power density. That is, the peak gain must remain "on" long enough to allow the oscillator to build up to its threshold value. The minimum threshold energy density has been discussed in detail by *Byer* [3.25].

3.5.3 Conversion Efficiency

The conversion efficiency for a DRO was first treated by *Siegman* [3.130]. He assumed that the spatial variation of the resonated waves was nearly constant and took into account the back-generated wave at the pump frequency. With these assumptions the calculated signal and idler output powers for the DRO are

$$\frac{\omega_3}{\omega_2}\frac{P_2}{P_{30}}=\frac{\omega_3}{\omega_1}\frac{P_1}{P_{30}}=2\frac{1}{N}(\sqrt{N}-1)\,, \tag{3.57}$$

where N is the number of times above threshold defined by $\Gamma^2 l^2 = a_1 a_2 N$.

The conversion efficiency of the DRO is therefore

$$\eta=\frac{P_1+P_2}{P_{30}}=\frac{2}{N}(\sqrt{N}-1)\,, \tag{3.58}$$

which reaches a maximum of 50% at four times above threshold. The generated power is divided between the two waves in the ratio ω_2/ω_1 since equal signal and idler photons are generated.

The transmitted pump power P_3 limits at the threshold power $P_3(\mathrm{TH})$ which forms the basis for the optical limiter first proposed by *Siegman* [3.130]. Finally, the reflected pump power is given by

$$\frac{P_3(R)}{P_3(\mathrm{TH})}=(N^{\frac{1}{2}}-1)^2 \tag{3.59}$$

The reflected pump power is not ruly "reflected" but is generated by sum generation of the back-traveling signal and idler waves. The back-generated pump acts to reduce the signal and idler powers and simultaneously feed power back into the pumping laser which induces pump laser frequency and power instabilities.

Bjorkholm [3.131] analyzed the DRO without the back-generated pump wave and showed that its efficiency increases from 50% to 100% at $N=4$. One method to ensure no backward pump generation is to construct a ring cavity oscillator. This oscillator is referred to as a ring resonator oscillator (RRO) which implies that it is doubly resonant. Recently *Fischer* [3.132] has treated the RRO in detail.

All of the DRO configurations suffer from the disadvantage of requiring simultaneous resonance at the signal and idler waves within a single cavity. Due to crystal dispersion, the axial mode intervals at the signal and idler frequencies are not the same so that the particular signal and idler cavity modes that exactly sum to the pump frequency may occur off gain center. Furthermore, slight shifts in crystal index or cavity length cause the modes that align to shift about rapidly. Typically a few modes within a "cluster" are aligned or nearly aligned and the next "cluster" of modes is spaced a number of axial mode intervals away. *Giordmaine* and *Miller* [3.102, 103] first discussed the cluster effect which has since been observed by *Bjorkholm* [3.133]. The frequency instability due to double resonance is the main reason for sacrificing the low threshold of the DRO for operation as a single resonant oscillator whenever possible.

The SRO was first demonstrated by *Bjorkholm* [3.104] and later analyzed by *Kreuzer* [3.134]. It has since become the most common parametric oscillator configuration for a number of reasons. They include: lack of the cluster effect, use of the full multiaxial mode power of the pump laser, ease of mirror and cavity design, good conversion efficiency and frequency and output power stability. These advantages more than offset the disadvantage of increased threshold relative to the DRO.

Kreuzer showed that the signal power for the SRO is given by the implicit relation

$$\kappa_1 \kappa_2 |\mathscr{E}_{30}|^2 l^2 \operatorname{sinc}^2 \beta l = 2a_s \,, \tag{3.60}$$

or

$$\operatorname{sinc}^2 \beta l = \frac{1}{N} ,$$

where

$$\beta^2 = \alpha^2 + \left(\frac{\Delta k}{2}\right)^2 , \tag{3.61}$$

and

$$\alpha^2 = \kappa_1 \kappa_3 |\mathscr{E}_{so}|^2 .$$

Here N is the number of times above threshold which for $\beta = 0$ reduces to the previously derived SRO threshold condition $\Gamma^2 l^2 = 2a_s$.

At line center the SRO reaches 100% efficiency for $P_{30} = (\pi/2)^2 P_3(\mathrm{TH})$. As the pumping level increases beyond $(\pi/2)^2$ the pump wave begins to grow at the expense of the signal. Figure 3.13 shows the transmitted pump power for the SRO as a function of the number of times above threshold in the low gain plane-wave approximation.

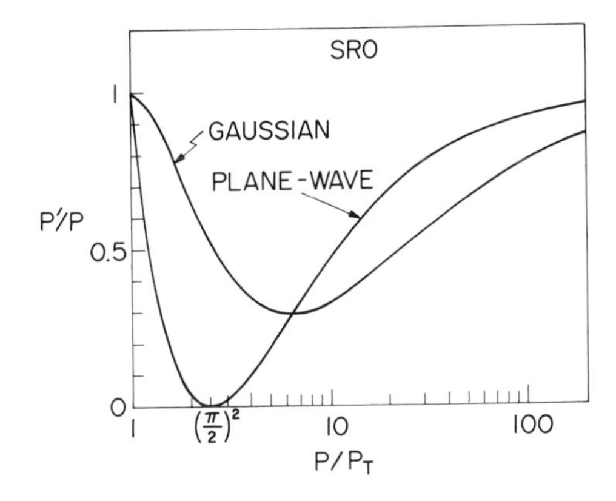

Fig. 3.13. Transmitted pump power P'/P vs number of times above threshold for a SRO for a plane wave and a gaussian intensity wave (after *Bjorkholm* [3.104])

Early comparisons of experimental results to the calculated SRO efficiency and pump depletion did not show close agreement. *Bjorkholm* [3.135] and later *Fischer* [3.132] extended the present plane wave analysis to multimode beams with gaussian intensity profile. Bjorkholm then demonstrated that many previous observations agreed very well with predictions based on the gaussian intensity model.

The DRO conversion efficiency for a gaussian intensity distribution is given by

$$\eta=\frac{P_1+P_2}{P_{30}}=\frac{4}{N}(\sqrt{N}-1-\ln\sqrt{N})\,. \tag{3.62}$$

The conversion efficiency is less than the plane-wave efficiency by $-(4/N)\ln N$. Compared to the plane-wave case, the DRO pumped with a gaussian intensity does not show power limiting and reaches a maximum efficiency of only 41% at $N=12.5$.

The SRO conversion efficiency for a gaussian intensity input has been treated in a similar manner by *Bjorkholm* [3.135]. In this case, however, the solution remains in an integral form since the resonant signal power is not solved for explicitly. Figure 3.13 shows a comparison of the transmitted pump power for the plane wave and gaussian SRO as a function of N. For a gaussian intensity profile beam the maximum efficiency reaches only 71% at 6.5 times above threshold.

The conversion efficiencies calculated for the various OPO configurations apply to operation in the low gain essentially steady state condition. In practice, the steady state condition is achieved only after a finite build-up time. Thus the energy conversion efficiency is less than the power conversion efficiency by approximately the fraction of the pumping time that the oscillator is above threshold. Again, operation of an OPO with a short cavity length to reduce build-up time leads to improvements in the conversion efficiency.

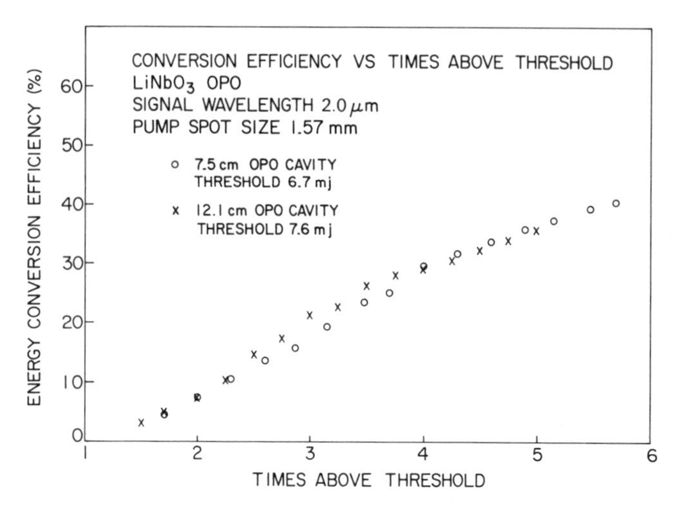

Fig. 3.14. Conversion efficiency vs number of times above threshold for a 1.06 μm pumped $LiNbO_3$ SRO (after *Byer* et al. [3.109])

Up to this point we have considered OPO conversion efficiency with the approximation that the resonant wave is slowly varying within the cavity. The exact solution for a three-wave parametric interaction has been given by *Armstrong* et al. [3.8]. *Bey* and *Tang* [3.136] have extended Armstrong's results to the singly resonant parametric oscillator and to an OPO with intracavity up-converter and doubler.

Experimental measurements of OPO conversion efficiency have been made for both cw and pulsed operation and for DRO, SRO parametric oscillators. *Bjorkholm* [3.104] and *Kreuzer* [3.137] investigated DRO conversion efficiencies. Using a single mode ruby pump Kreuzer obtained 36% conversion efficiency for a DRO and 6% conversion for an SRO. Using a ruby laser pump and a non-collinear SRO cavity arrangement. *Falk* and *Murray* [3.105] obtained 70% power conversion and 50% energy conversion in a $LiNbO_3$ oscillator. The energy conversion was limited by the finite build-up time of the oscillator.

Early conversion efficiency results were not in very good agreement with the conversion efficiency expected from plane-wave theory. As previously discussed, *Bjorkholm* [3.104] investigated the spatial dependence of pump depletion for multimode gaussian amplitude waves and showed that much better agreement was obtained if the nonuniform pumping intensity was taken into account.

Wallace [3.107, 138] described a $LiNbO_3$ SRO pumped by a *Q*-switched Nd:YAG laser. The oscillator operated at typically 50% conversion efficiency with threshold powers near 600 watts. Average output powers up to 70 mW have been obtained.

Using $LiNbO_3$ internal to a Q-switched Nd:YAG laser enabled *Ammann* et al. [3.106] to obtain an 8% conversion efficiency at 2.13 μm with up to 17 mW of average power. Recently these results were improved to over 1.2 W of average power at 2.1 μm and 60 mW at 3.1 μm.

Efficient traveling wave oscillators have been operated in ADP and recently α-HIO_3. *Yarborough* and *Massey* [3.100] reported up to 25% power conversion in an ADP oscillator pumped with the fourth harmonic of Nd:YAG laser. The output pulses were typically 2 ns in duration at 100 kW peak power. *Kovrigin* and *Nikles* [3.139] obtained 57% energy conversion at 1 to 1.10 μm in α-HIO_3 pumped with a doubled Nd: glass laser. The pulse length was 30 ns at $P_p = 20$ mW/cm^2. At higher pump power densities the α-HIO_3 crystal damaged.

Recently *Byer* et al. [3.109] measured a 40% energy conversion efficiency in an angled tuned $LiNbO_3$ SRO pumped by 1.06 μm radiation from a Nd:YAG oscillator/amplifier. The parametric oscillator which uses a special [01.4] growth direction 5 cm long $LiNbO_3$ crystal [3.108, 140] operated at up to 40 mJ input and 16 mJ output energies at 10 pps. Figure 3.14 shows the measured conversion efficiency vs the number of times above threshold. The results agree well with the predictions of a gaussian amplitude SRO conversion efficiency vs N shown in Fig. 3.13, if account is taken of the rise time on peak power reduction.

3.5.4 Tuning and Bandwidth

The spectral properties of a parametric oscillator include tuning range and method of tuning, gain bandwidth and the detailed spectral structure within the gain bandwidth determined by the optical cavity.

Tuning is achieved by varying the crystal birefringence while simultaneously satisfying the frequency and phase-matching conditions

$$\omega_3 = \omega_2 + \omega_1 , \tag{3.63}$$

$$\boldsymbol{k}_3 = \boldsymbol{k}_2 + \boldsymbol{k}_1 + \Delta \mathrm{k} . \tag{3.64}$$

In general, a parametric oscillator is tuned by altering the crystal birefringence. This is usually done by crystal rotation, thus changing the extraordinary index of refraction, or by controlling the crystal temperature and thus the birefringence. As an example, Fig. 3.15 shows the angle tuning curves of a $LiIO_3$ parametric oscillator pumped by 1.08 μm of a Q-switched Nd:$YAlO_3$ laser. The tuning range extends from 1.4 μm to 5 μm [3.115]. Tuning curves for a number of nonlinear crystals have been given in the review article by *Byer* [3.25].

The gain bandwidth of the parametric oscillator is determined by the relation $\Delta kl \approx 2\pi$. If we expand Δk in frequency about the phase-matching frequency we find for the bandwidth that

$$\delta\omega_2 \approx 2\pi / l\beta_{12} \tag{3.65}$$

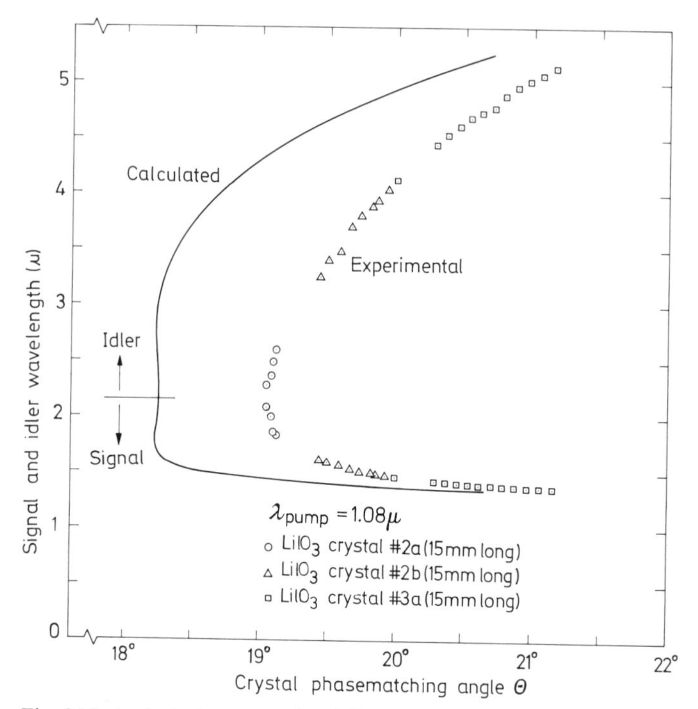

Fig. 3.15. Angle tuning curves for 1.08 μm pumped $LiIO_3$ parametric oscillator (after *Ammann* [3.115])

where

$$\beta_{12}=\left(\frac{\partial k_1}{\partial\omega_1}-\frac{\partial k_2}{\partial\omega_2}\right)$$

and l is the crystal length. For the special case when $\beta_{12}\approx 0$ (near degeneracy) the gain bandwidth is approximated by

$$\delta\omega_2\approx\sqrt{4\pi/l\gamma_{12}} \tag{3.66}$$

where

$$\gamma_{12}=(\partial^2 k_1/\partial\omega_1^2+\partial^2 k_2/\partial\omega_2^2)$$

is the group velocity dispersion. The dispersion constant β_{12} has been calculated for $LiNbO_3$ for various pump wavelengths [3.24]. The calculation of γ_{12} shows that the second-order term is not significant until tuning is within 200 cm^{-1} of degeneracy.

Expanding β_{12} in terms of the crystal index of refraction

$$\beta_{12}=\frac{1}{c}\left(n_2-n_1+\lambda_1\frac{\partial n_1}{\partial\lambda_1}-\lambda_2\frac{\partial n_2}{\partial\lambda_2}\right)\approx\frac{\Delta n}{c}, \tag{3.67}$$

and substituting into the bandwidth expression (3.65) we find

$$\delta\nu(\mathrm{cm}^{-1}) \approx \frac{1}{\Delta n l(\mathrm{cm})}, \tag{3.68}$$

where Δn is the birefringence. Thus crystals with small birefringence have larger bandwidths and tuning rates than crystals with large birefringence. Equation (3.68) predicts a bandwidth of $12\,\mathrm{cm}^{-1}$ for a 1 cm crystal of $LiNbO_3$. The actual bandwidth is $4\,\mathrm{cm}^{-1}$. Therefore the dispersion term can be neglected only to within this accuracy.

Tuning curves for parametric oscillators usually are determined prior to their operation. The straightforward approach is to solve the phase-matching equations (3.63) and (3.64) for signal and idler frequencies at a given pump frequency as a function of the tuning variable. To carry out the calculation the indices of refraction must be given by an analytical expression in the form of a Sellmeier equation. For accurate calculation of tuning curves the indices of refraction must be known over the entire tuning range to an accuracy of 0.1–0.01%.

Tuning curves can be experimentally measured without constructing an oscillator by observing parametric fluorescence [3.141–143]. This method for obtaining tuning and curves was first applied to $LiNbO_3$ [3.142, 144] and then extended to include ADP [3.145], $LiIO_3$ [3.146] and proustite [3.147]. In general the method is more useful in the visible spectral region where photomultiplier detectors or the eye can be used in conjunction with a spectrometer to measure the wavelength.

Unlike other laser sources which show power loss as the bandwidth is narrowed, the OPO remains af full efficiency even for a single mode operation. In this respect it behaves like a homogeneously saturated laser source.

Wallace [3.148] has demonstrated a long-term frequency stability of better than $0.005\,\mathrm{cm}^{-1}$ by use of a temperature-scanned infrared etalon internal to the oscillator cavity. In this experiment the OPO remained at 80% of its wideband power as expected showing a slight decrease in output power due to the etalon insertion loss.

The tilted etalon in only useful for mode control in oscillators with relatively wide beam cross sections. This led *Pinard* and *Young* [3.149] to study an alternate interferometric mode control technique based on an interferometer proposed by *Damaschini* [3.150] which is a dual of the better known *Smith* [3.151] interferometer. Pinard and Young obtained a bandwidth of $0.001\,\mathrm{cm}^{-1}$ in the $2.5\,\mu\mathrm{m}$ spectral region tunable over a $0.015\,\mathrm{cm}^{-1}$ tuning range before a second cavity mode appeared.

For operation with an etalon it is desirable to have an OPO with a narrow gain bandwidth to avoid multiple etalon transmission bands within the gain linewidth. Recently *Byer* et al. [3.109] used a thin tilted etalon, two-element birefringent filter and a grating for primary line narrowing elements within a $LiNbO_3$ parametric oscillator.

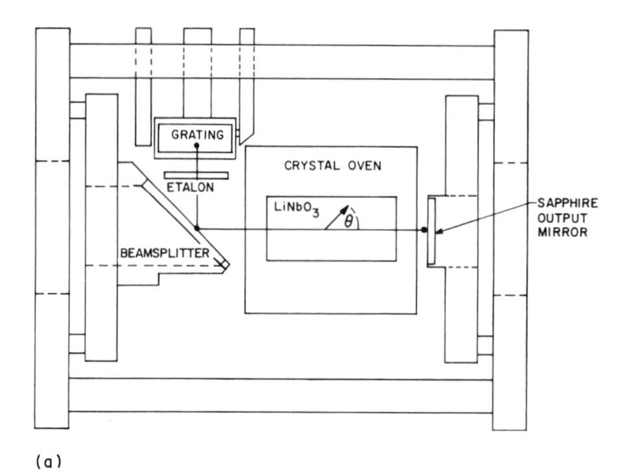

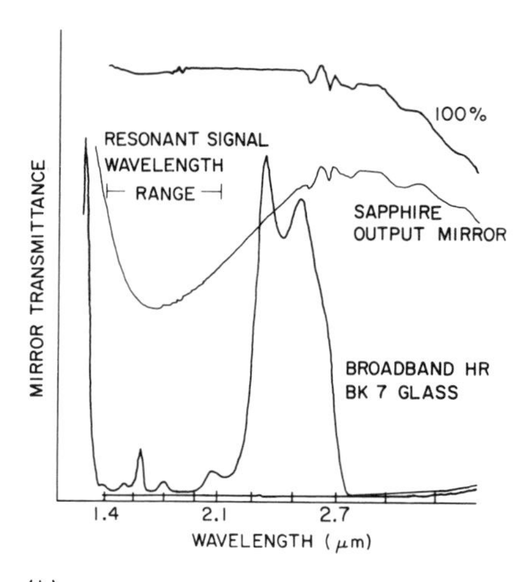

Fig. 3.16. (a) $LiNbO_3$ SRO cavity design for a widely tunable high energy angle tuned parametric oscillator source, (b) Transmittances for the 45° input beamsplitter and the sapphire output coupler over the signal wavelength range of 1.4–2.1 μm (after *Byer* et al., [3.109])

Figure 3.16 shows the $LiNbO_3$ SRO cavity design which utilizes an input dichroic beam splitter to form the cavity. This cavity allows the insertion of line narrowing elements into the resonant signal beam but out of the 1.06 μm pump beam. The beam splitter transmits up to 90% of the incident "p" polarized pump beam and reflects 99% of the "s" polarized signal beam over a 1.4–2.1 μm wavelength range. The output sapphire mirror is between 50% and 70% transmitting over the same signal wavelength range which enables SRO operation over the very wide 1.4–4.2 μm tuning range. When operated with a thin etalon or birefringent filter the grating is replaced by a broad band high reflector mirror which also reflects at the signal wavelength.

Figure 3.17 shows the tuning range vs crystal angle for the 1.06 μm pumped $LiNbO_3$ SRO. Also shown is the gain bandwidth. A single layer SiO_2 anti-reflection coating centered at 1.7 μm is sufficient to reduce to crystal reflection

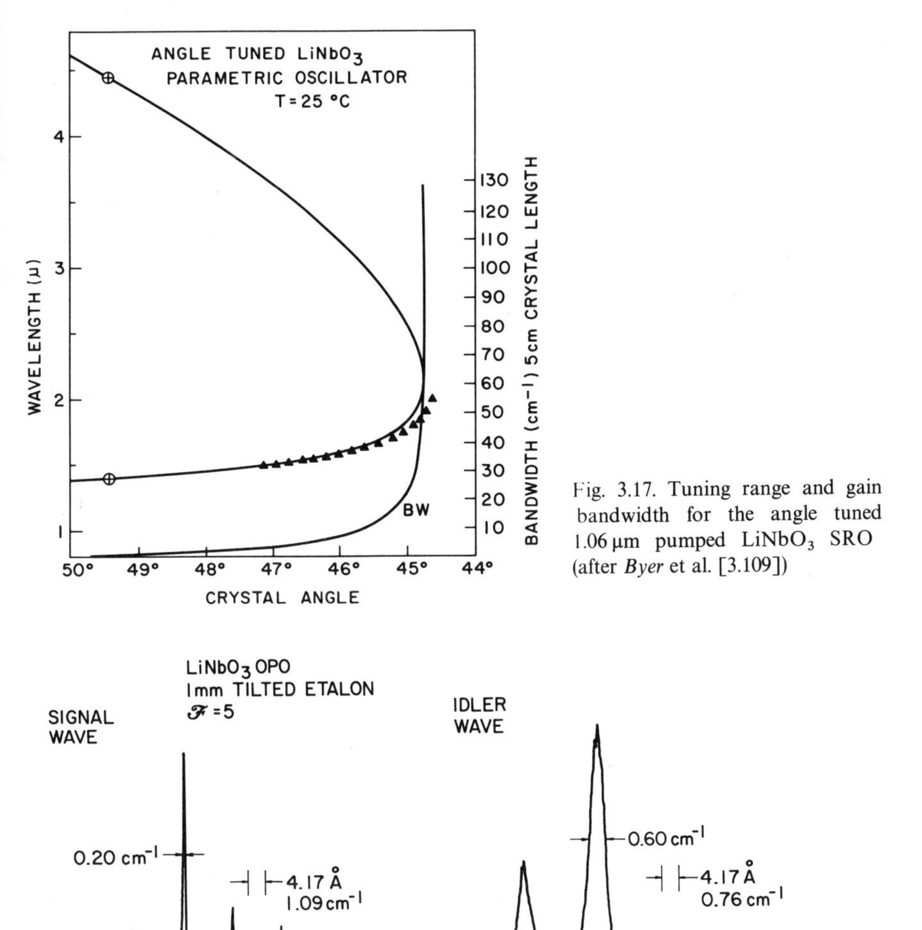

Fig. 3.17. Tuning range and gain bandwidth for the angle tuned 1.06 μm pumped $LiNbO_3$ SRO (after *Byer* et al. [3.109])

Fig. 3.18. $LiNbO_3$ SRO output spectrum at the resonant signal and non resonant idler wave with a 1 mm finesse of 5 tilted etalon (after *Byer* et al. [3.109])

losses to approximately 1% over the entire signal wavelength tuning range. The angle tuning required to cover the tuning range is small enough that a single $LiNbO_3$ crystal is used.

Figure 3.18 shows the $LiNbO_3$ SRO linewidth with a 1 mm tilted etalon inserted in the cavity. The resonant signal linewidth under each etalon transmission made is less than the resolution of the spectrometer while the nonresonant idler wave spectrum reproduces the Nd:YAG pump linewidth as expected. For operation on a single etalon mode a primary line narrowing element is needed to reduce the operating linewidth of the oscillator to near 1 cm^{-1} from the 10–130 cm^{-1} gain bandwidth.

A thin tilted etalon with a free spectral range of 50 cm^{-1} was first investigated as a primary line narrowing element. It operated as expected to reduce the oscillator linewidth to near 1 cm^{-1}. The insertion loss of the etalon is [3.152]

$$l = \frac{2}{(1-R)^2}\left(\frac{4\alpha t}{nw}\right)^2 \tag{3.69}$$

where R is the reflectivity, t the etalon thickness, α the tilt angle, n the index of refraction and w the oscillator spot size. As an example a 1 mm thick solid etalon with $R=60\%$ at $w=1.57$ mm has an insertion loss of 1.4% at $\alpha=2°$. Large spot sizes are therefore useful in reducing tilted etalon insertion loss.

The second primary line narrowing element tried was a two-element $LiNbO_3$ birefringent filter [3.153]. The tuning rate of the filter when operated at Brewster's angle and rotated through an angle A about an axis through the Brewster angle is

$$\delta = \frac{2\pi T}{\lambda \sin\theta_\beta}[n_0 - n_e(\gamma)] \tag{3.70}$$

where

$$\gamma = \tan^{-1}\left(\frac{n}{\cos A} - E\right)$$

with T the plate thickness, θ_B the Brewster angle and E the angle from the plate to the crystal optic axis. The two-element $LiNbO_3$ birefringent filter used plates of thickness $T_1 = 1.5$ mm and $T_2 = 3.0$ mm. It operated at a free spectral range of 48 cm^{-1} which could be tuned by a 6° rotation of the plates. The tuning rate is remarkably linear near 8 cm^{-1}/deg. As with the tilted etalon, rotation was accomplished by mounting the filter on a galvanometer which was electrically scanned over a $\pm 4°$ angular range with a resolution of 8 arc s. The birefringent filter reduced the $LiNbO_3$ oscillator's bandwidth to less than 4 cm^{-1}.

The third primary line narrowing element utilized was a grating operating in the Littrow mode. The resolution of the grating is

$$\frac{\lambda}{\Delta\lambda} = \frac{m}{2d\cos\varphi}\pi w_0$$

which reduces to an equivalent linewidth in cm^{-1} of

$$\Delta v = \frac{1}{\pi w_0 \tan\varphi}. \tag{3.71}$$

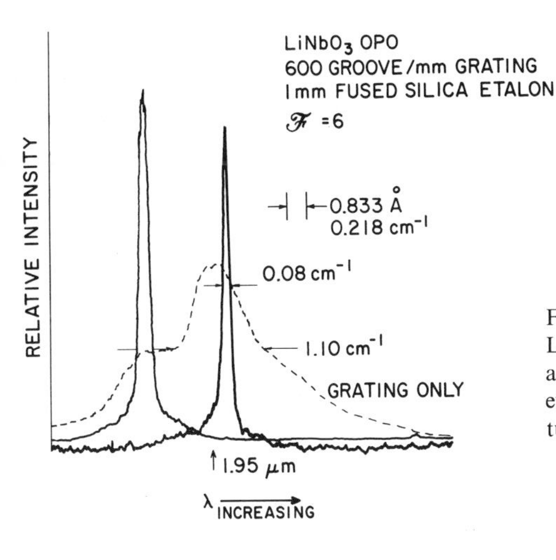

Fig. 3.19. The output spectrum of the $LiNbO_3$ SRO wiht a grating (dashed) and a grating plus a tilted etalon (solid). The etalong was set at two angles to illustrate tuning (after *Byer* et al. [3.109])

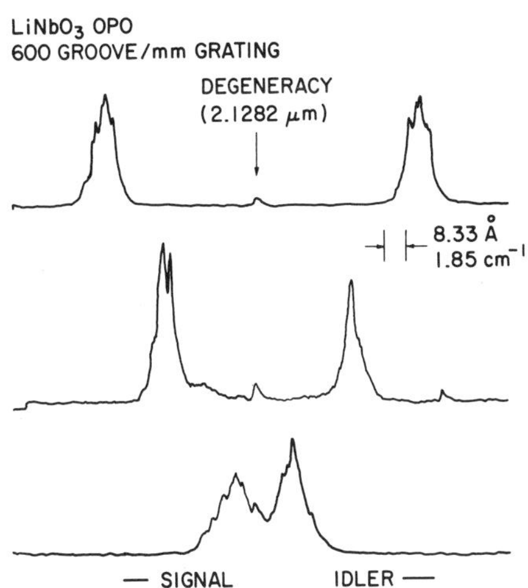

Fig. 3.20. Tuning of the $LiNbO_3$ SRO near degeneracy using the 600 lmm grating (after *Byer* et al. ([3.109])

For the 600 l/mm 1.8 μm blaze grating used $\omega \sim 20°$ and $\Delta v = 5.6\ cm^{-1}$ for $w_0 = 0.157$ cm. Again a larger spot size is advantageous for improving resolution. The 5.6 cm^{-1} resolution is broader than the expected oscillator linewidth due to multipass line narrowing by a factor of three. Figure 3.19 shows the measured output spectrum with the grating only (dashed) and with the additional tilted etalon operating at two angles to illustrate tuning. The linewidth of the etalon-narrowed output is less than 0.08 cm^{-1}. Somewhat surprisingly, the replica grating operated at over 100 MW/cm^2 incident intensity without damage. Due to the large TEM_{00} mode spot size of the parametric oscillator a beam expanding telescope is not needed prior to the grating. The grating angle was set by a stepper motor driven sin-drive stage with a resetability of ± 0.1 Å.

The operation of the parametric oscillator with a grating offers significant advantages in controlling the output spectrum. One of the advantages is accurate control of wavelength, especially near the degeneracy region. Figure 3.20 illustrates the $LiNbO_3$ SRO tuning near degeneracy with the grating. For these measurements the oscillator operated at 10 pps with a 20 ns pulse width 1.06 μm pump beam. Details of this oscillator, which forms a primary tuning element in a computer-controlled coherent spectrometer, are discussed by *Byer* et al. [3.109].

3.6 Conclusion

It is now possible to generate tunable coherent radiation over the entire infrared spectral range by second- and third-order nonlinear processes. Mixing and parametric oscillation in nonlinear crystals are the most widely used methods. However, the availability of high power tunable sources has led to increasing use of stimulated Raman down-shifting as a means of infrared generation. As alternate approach discussed in Section 3.4 is to mix input tunable radiation with induced coherent Raman vibrations to generated Raman shifted lower sidebands. Mixing from coherent Raman vibrations offers advantages over stimulated Raman scattering of high conversion efficiency at even low tunable input power levels. Experiments are underway to demonstrate the full potential of coherent Raman mixing.

The progress in the past ten years in infrared generation has come about because of the development of new nonlinear materials. The properties of the most useful nonlinear materials are given in Table 3.1. However, a more detailed discussion of material properties can be found in a number of review articles [3.3, 154–159]. Two books covering all aspects of nonlinear optics, including nonlinear materials, have also been published [3.160, 161].

Improvements and extensions in the generation of tunable infrared radiation will continue. For example, *Laubereau* et al. [3.162] and later *Kushida* et al. [3.163] have generated tunable picosecond pulses in the 1.4–4 μm range by parametric generation in $LiNbO_3$ and *Moore* and *Goldberg* [3.164] have also generated tunable picosecond pulses over a 1.13–5.6 μm range by mixing in $LiI\,0_3$. These experiments demonstrate increased capability for the generation and application of tunable infrared radiation.

Infrared generation in nonlinear materials offers the potential of generating widely tunable coherent radiation in a compact solid state system. Progress in constructing a tunable coherent spectrometer has been made recently. The device is based on a Nd:YAG pumped $LiNbO_3$ angle tuned parametric oscillator. The computer controlled parametric oscillator has a 1.4–4 μm tuning range and is the primary element in an extended wavelength computer controlled spectrometer. Extension of the $LiNbO_3$ oscillator's tuning range to 0.2600 μm is possible by second-harmonic generation in $LiNbO_3$ followed by

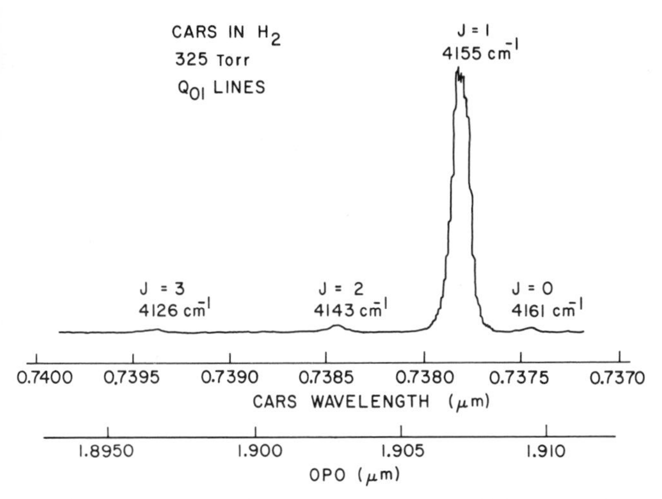

Fig. 3.21. Coherent anti-Stokes Raman spectrum of H_2 generated using the computer controlled $LiNbO_3$ SRO source

SHG in ADP. Extension to 25 μm is possible by mixing in $AgGaSe_2$ and GaSe or by coherent Raman mixing in H_2.

Figure 3.21 illustrates a coherent anti-Stokes Raman spectrum of H_2 [3.165] taken with the computer controlled SRO source. For this spectrum the 1.89–1.91 μm (ω_2) output of the parametric oscillator was mixed against 1.064 μm (ω_1) to generate the coherent anti-Stokes signal at $\omega_1+(\omega_1-\omega_2)$. The oscillator was step tuned over a 50 cm^{-1} interval in 0.25 cm^{-1} steps by a minicomputer which synchronously scanned the grating and the nonlinear crystal angle. The grating provides linewidth control to a resettability of $\pm 0.5\,\mathrm{cm}^{-1}$ at a $2\,\mathrm{cm}^{-1}$ bandwidth. With the insertion into the parametric oscillator of a tilted etalon the linewidth reduces to $0.1\,\mathrm{cm}^{-1}$. This tunable coherent spectrometer demonstrates the future potential for widely tunable coherent generation using nonlinear parametric and mixing methods.

Acknowledgements. The authors want to acknowledge support from ARO and ERDA and the Sloan Foundation (*R. L. Byer*). We also want to thank *Mary Farley* for preparation of the manuscript.

References

3.1 N. Bloembergen: *Nonlinear Optics* (Benjamin, New York 1965)
3.2 P. N. Butcher: *Nonlinear Optical Phenomena* (Ohio State University, Columbus, Ohio 1865)
3.3 S. K. Kurtz: In *Laser Handbook*, ed. by F. T. Arecchi, E. O. Schulz-DuBois (North Holland, Amsterdam 1972) p. 925
3.4 R. Bechmann, R. F. S. Hearmon, S. K. Kurtz: Landolt-Börnstein, New Series, Group 3, Vol. 2 (Springer, Berlin, Heidelberg, New York 1969) pp. 167–209

3.5 S.Singh: In *Handbook of Lasers*, ed. by R.J. Pressley (Chem. Rubber Co., Cleveland, Ohio 1971) p. 489
3.6 F.Zernike, J.E.Midwinter: *Applied Nonlinear Optics* (Academic Press, New York 1973)
3.7 P.S.Pershan: Phys. Rev. **130**, 919 (1963)
3.8 J.A.Armstrong, N.Bloembergen, J.Ducuing, P.S.Pershan: Phys. Rev. **127**, 1918 (1962)
3.9 G.D.Boyd, D.A.Kleinman: J. Appl. Phys. **39**, 3597 (1968)
3.10 R.C.Miller: Appl. Phys. Lett. **5**, 17 (1964)
3.11 B.Lax, J.Mavroides, D.Edwards: Phys. Rev. Lett. **8**, 166 (1962)
3.12 C.G.B.Garrett, Robinson,F.N.H.: IEEE J. QE-**2**, 328 (1966)
3.13 S.K.Kurtz, F.N.H.Robinson: Appl. Phys. Lett. **10**, 62 (1967)
3.14 D.A.Kleinman: Phys. Rev. **126**, 1977 (1962)
3.15 J.A.Giordmaine: Phys. Rev. Lett. **8**, 19 (1962)
3.16 P.D.Maker, R.W.Terhune, N.Nisenoff, C.M.Savage: Phys. Rev. Lett. **8**, 21 (1962)
3.17 S.A.Akhmanov, A.L.Kovrigin, R.V.Khoklov, O.N.Chunaev: Zh. Eksp. Teor. Fiz. **45**, 1336 (1963); Transl. Sov. Phys. JETP **18**, 919 (1964)
3.18 J.E.Bjorkholm: Appl. Phys. Lett. **13**, p. 36 (1968)
3.19 G.D.Boyd, A.Ashkin, J.M.Dziedzic, D.A.Kleinman: Phys. Rev. A. **137**, 1305 (1965)
3.20 J.E.Bjorkholm: Phys. Rev. **142**, 126 (1966)
3.21 D.A.Kleinman, A.Ashkin, G.D.Boyd: Phys. Rev. **145**, 338 (1966)
3.22 N.Bloembergen, P.S.Pershan: Phys. Rev. **128**, 606 (1962)
3.23 D.A.Kleinman: Phys. Rev. **128**, 1761 (1962)
3.24 S.E.Harris: Proc. IEEE **57**, 2096 (1969)
3.25 R.L.Byer: In *Treatise in Quantum Electronics*, vol. I, part B, ed. by H.Rabin, C.L.Tang (Academic Press, New York 1975) pp. 587–702
3.26 G.D.Boyd, A.Ashkin: Phys. Rev. **146**, 187 (1966)
3.27 S.A.Akhmanov, A.I.Kovrigin, V.A.Kolosov, A.S.Piskarkas, V.V.Fadeev, R.V.Khokhlov: JETP Lett. **3**, 241 (1966)
3.28 R.L.Byer: Parametric Oscillators. In *Laser Spectroscopy*, ed. by R.G.Brewer, A.Mooradian (Plenum Publishing Corp., New York 1973)
3.29 S.A.Akhmanov, A.P.Sukhorukov, R.V.Khohklov: Usp. Fiz. Nauk. **93**, 19 (1968); Transl. Sov. Phys. Usp. **10**, 609 (1968)
3.30 T.A.Rabson, H.J.Ruiz, R.L.Shah, F.K.Tittle: Appl. Phys. Lett. **21**, 129 (1972)
3.31 F.Zernike, P.R.Berman: Phys. Rev. Lett. **15**, 199 (1965)
3.32 S.Kielich: Opt. Elect. **2**, 125 (1970)
3.33 J.Warner: In *Treatise in Quantum Electronics*, vol. I, part B, ed. by H.Rabin, C.L.Tang (Academic Press, New York 1975) pp. 703–737
3.34 F.P.Schafer: *Topics in Applied Physics*, vol. 1. Dye Lasers (Springer, Berlin, Heidelberg, New York 1973)
3.35 C.F.Dewey, L.O.Hocker: Appl. Phys. Lett. **18**, 58 (1971)
3.36 D.W.Meltzer, L.S.Goldberg: Opt. Commun. **5**, 209 (1972)
3.37 D.C.Hanna, R.C.Smith, C.R.Stanley: Opt. Commun. **4**, 300 (1971)
3.38 D.C.Hanna, V.V.Rampel, R.C.Smith: Opt. Commun. **8**, 151 (1973)
3.39 G.B.Abdullaev, L.A.Kulevskii, A.M.Prokhorov, A.D.Savel'ev, E.Yu Salaev, V.V.Smirnov: JETP Lett. **16**, 90 (1972)
3.40 G.B.Abdullaev, K.R.Allakhverdiev, L.A.Kulevskii, A.M.Prokorov, Yu Salaev, A.D. Savel'ev, V.V.Smirnov: Sov. J. Quant. Electron **2**, 1228 (1975)
3.41 D.C.Hanna, V.V.Rampel, R.C.Smith: IEEE J. QE-**10**, 461 (1971)
3.42 H.Gerlach: Opt. Commun. **12**, 405 (1974)
3.43 H.Tashiro, T.Yajima: Opt. Commun. **12**, 129 (1974)
3.44 L.S.Goldberg: Appl. Optics, **14**, 653 (1975)
3.45 A.S.Pine: J. Opt. Soc. Am. **64**, 1683 (1974)
3.46 G.B.Abdulaev, L.A.Kulevskii, P.V.Nickles, A.M.Drokhorov, A.D.Savel'ev, E.Yu Salaev, V.V.Smirnov: Sov. J. Quant. Electron. **3**, 163 (1976)
3.47 R.L.Herbst, R.L.Byer: Appl. Phys. Lett. **19**, 527 (1971)
3.48 G.C.Bhar, D.C.Hanna, B.Luther-Davies, R.C.Smith: Opt. Commun. **6**, 323 (1972)

3.49 D.C.Hanna, B.Luther-Davies, R.C.Smith, R.Wyatt: Appl. Phys. Lett. **25**, 142 (1974)
3.50 R.L.Byer, M.M.Choy, R.L.Herbst, D.S.Chemla, R.S.Feigelson: Appl. Phys. Lett. **24**, 65 (1974)
3.51 J.R.Morris, Y.R.Shen: Opt. Commun. **3**, 81 (1971)
3.52 T.J.Bridges, A.R.Strnad: Appl. Phys. Lett. **20**, 382 (1972)
3.53 K.H.Yang, P.L.Richards, Y.R.Shen: Appl. Phys. Lett. **19**, 320 (1971)
3.54 K.H.Yang, J.R.Morris, P.L.Richards, Y.R.Shen: Appl. Phys. Lett. **23**, 669 (1973)
3.55 G.D.Boyd, T.J.Bridges, C.K.N.Patel, E.Beuhler: Appl. Phys. Lett. **21**, 553 (1972)
3.56 H.Kildal, J.C.Mikkelsen: Opt. Commun. **10**, 306 (1974)
3.57 C.K.N.Patel, E.D.Shaw: Phys. Rev. Lett. **24**, 451 (1970); see also Phys. Rev. B **3**, 1279 (1971)
3.58 V.T.Nguyen, T.J.Bridges: Phys. Rev. Lett. **29**, 359 (1972)
3.59 T.L.Brown, P.A.Wolff: Phys. Rev. Lett. **29**, 362 (1972)
3.60 Y.R.Shen: Appl. Phys. Lett. **23**, 516 (1973)
3.61 T.J.Bridges, V.T.Nguyen: Appl. Phys. Lett. **23**, 107 (1973)
3.62 N.Brignall, R.A.Wood, C.R.Pidgeon, B.S.Wherrett: Opt. Commun. **12**, 17 (1974)
3.63 B.Lax, R.L.Aggarwal, G.Favrot: Appl. Phys. Lett. **23**, 679 (1973)
3.64 R.L.Aggarwal, B.Lax, H.R.Fetterman, P.E.Tannenwald, B.J.Clifton: J. Appl. Phys. **45**, 3972 (1974)
3.65 N.Bloembergen, A.J.Sievers: Appl. Phys. Lett. **17**, 483 (1970)
3.66 S.Somekh, A.Yariv: Opt. Commun. **6**, 301 (1972)
3.67 Y.Yacoby, R.L.Aggarwal, B.Lax: Appl. Phys. **44**, 3180 (1973)
3.68 J.D.McMullen: Appl. Phys. **46**, 3076 (1975)
3.69 D.E.Thompson, J.D.McMullen, D.B.Anderson: Appl. Phys. Lett. **29**, 113 (1976)
3.70 E.J.Woodbury, W.K.Ng: Proc. IRE **50**, 2347 (1962)
3.71 G.Eckhardt, R.W.Hellworth, F.J.McClung, S.E.Schwarz, D.Weiner, E.J.Woodbury: Phys. Rev. Lett. **9**, 455 (1962)
3.72 Y.R.Shen, N.Bloembergen: Phys. Rev. **137**, A 1787 (1965)
3.73 N.Bloembergen, G.Bret, P.Lallemand, A.Pine, P.Simova: IEEE J. QE-**3**, 197 (1967)
3.74 N.Bloembergen: Am. J. Phys. **35**, 989 (1967)
3.75 W.Kaiser, M.Maier: Stimulated Rayleigh Brillouin and Raman Spectroscopy. In *Laser Handbook*, vol. II, ed by F.T.Arecchi, E.O.Schulz-Dubois (North Holland, Amsterdam 1972)
3.76 Y.R.Shen: Light Scattering in Solids, Vol. 8: In *Topics in Applied Physics*, ed. by M.Cardona (Springer, Berlin, Heidelberg, New York 1975) p. 278
3.77 P.O.Sorokin, J.J.Wynne, J.R.Lankard: Appl. Phys. Lett. **22**, 342 (1973)
3.78 R.H.Pantell, H.E.Puthoff: *Fundamental of Quantum Electronics* (John Wiley and Sons, New York 1969)
3.79 W.R.Fenner, H.A.Hyatt, J.M.Kellman, S.O.S.Porto: J. Opt. Soc. Am. **63**, 73 (1973)
3.80 D.Cotter, D.C.Hanna, R.Wyatt: Appl. Phys. **8**, 333 (1975)
3.81 W.Schmidt, W.Z.Appt: Naturforsch. **27**a, 1373 (1972)
3.82 W.Schmidt, W.Appt: Proc. 8th Intern. Quant. Elect. Conf., San Francisco, California (1974)
3.83 R.Frey, F.Pradera: Opt. Commun. **12**, 98 (1974)
3.84 M.Rokni, S.Yatsiv: Phys. Lett. **24**a, 277 (1967)
3.85 P.O.Sorokin, V.S.Shiren, J.R.Lankard, R.C.Hammond, T.G.Kazyaka: Appl. Phys. Lett. **10**, 44 (1967)
3.86 D.J.Bradley, G.M.Gale, P.D.Smith: J. Phys. B Atom. Molec. Phys. **4**, 1349 (1971)
3.87 D.Cotter, D.C.Hanna, P.A.Karkkainen, R.Wyatt: Opt. Commun. **15**, 143 (1975)
3.88 J.L.Carlsten, T.J.McIlrath: J. Phys. B Atom. Molec. Phys. **6**, L 80 (1973)
3.89 J.L.Carlsten, P.C.Dunn: Stimulated Stokes Emission with a Dye Laser: Intense Tunable Radiation in the Infrared. (to be published, Opt. Commun.)
3.90 J.A.Giordmaine, W.Kaiser: Phys. Rev. **144**, 676 (1966)
3.91 E.Garmire, F.Pandarese, C.H.Townes: Phys. Rev. Lett. **11**, 160 (1963)
3.92 J.A.Duardo, F.M.Johnson, L.J.Nugent: IEEE J. QE-**4**, 397 (1968)
3.93 S.E.Harris, R.L.Byer: Rept. 1918, Stanford University Microwave Laboratory
3.94 R.N.Fleming: Ph. D. Thesis, Stanford University (1976); Rept. 2521, Stanford University Microwave Laboratory

3.95 G.V.Venkin, G.M.Krochik, L.L.Kulyak, D.I.Maleav, G.Yu.Khronopulo: JETP Lett. **21** 105 (1975)
3.96 C.C.Wang, G.W.Racette: Appl. Phys. Lett. **6**, 169 (1965)
3.97 J.A.Giordmaine, R.C.Miller: Phys. Rev. Lett. **14**, 973 (1965)
3.98 S.A.Akhmanov, O.N.Chuneav, V.V.Fadeev, R.V.Khoklov, D.N.Klyshko, A.I.Kovrigin, A.S.Piskarskas: Parametric Generators of Light, presented at Symp. Mod. Opt., Polytechnic Institute of Brooklyn, Brooklyn, New York (1967)
3.99 A.G.Akhmanov, S.A.Akhmanov, R.V.Khokhlov, A.I.Kovrigin, A.S.Piskarskas, A.P. Sukhorukov: IEEE J. QE-**4**, 828 (1968)
3.100 J.M.Yarborough, G.A.Massey: Appl. Phys. Lett. **18**, 438 (171)
3.101 S.Sullivan, E.L.Thomas: Opt. Commun. **14**, 418 (1975)
3.102 J.A.Giordmaine, R.C.Miller: *Physics of Quantum Electronics* ed. by P.L.Kelley, B.Lax, P.E.Tannenwald (McGraw-Hill, New York 1966); also, Proc. Phys. Quant. Electron. Conf. San Juan, Puerto Rico (1965)
3.103 J.A.Giordmaine, R.C.Miller: Appl. Phys. Lett. **9**, 298 (1966)
3.104 J.E.Bjorkholm: Appl. Phys. Lett. **13**, 53 (1968)
3.105 J.Falk, J.E.Murray: Appl. Phys. Lett. **14**, 245 (1969)
3.106 E.O.Ammann, J.D.Foster, M.K.Oshman, J.M.Yarborough: Appl. Phys. Lett. **15**, 131 (1969)
3.107 R.W.Wallace, S.E.Harris: Laser Focus **7**, 42 (1970)
3.108 R.L.Herbst, R.N.Fleming, R.L.Byer: Appl. Phys. **25**, 520 (1974)
3.109 R.L.Byer, R.L.Herbst, R.N.Fleming: In *Laser Spectroscopy*, ed. by S.Haroche, J.C. Pebay-Peyroula, T.W.Hansch, S.E.Harris (Springer, Berlin 1975) pp. 207–225
3.110 B.C.Johnson, H.E. Puthoff, J.Soo Hoo, S.S.Sussman: Appl. Phys. Lett. **18**, 181 (1971)
3.111 L.S.Goldberg: Appl. Phys. Lett. **17**, 489 (1970)
3.112 A.L.Izrailenko, A.L.Kovrigin, P.V.Nikles: JETP Lett. **12**, 331 (1970)
3.113 A.J.Campillo, C.L.Tang: Appl. Phys. Lett. **19**, 36 (1971)
3.114 A.J.Campillo: IEEE J. QE-**8**, 809 (1972)
3.115 E.O.Ammann: IEEE J. QE-**11**, D 65 (1975)
3.116 R.L.Herbst, R.L.Byer: Appl. Phys. Lett. **21**, 189 (1972)
3.117 A.A.Davydov, L.A.Kulevskii, A.M.Prokhorov, A.D.Savel'ev, V.V.Smirnov: JETP Lett. **15**, 513 (1972)
3.118 J.A.Weiss, L.S.Goldberg: Appl. Phys. Lett. **24**, 389 (1974)
3.119 E.O.Ammann, J.M.Yarborough: Appl. Phys. Lett. **17**, 233 (1970)
3.120 D.C.Hanna, B.Luther-Davies, H.N.Rutt, R.C.Smith: Appl. Phys. Lett. **20**, 34 (1972)
3.121 D.C.Hanna, B.Luther-Davies, R.C.Smith: Appl. Phys. Lett. **22**, 440 (1973)
3.122 R.G.Smith: In *Laser Handbook*, ed by. F.T.Arecchi, E.O.Schulz-DuBois (North Holland, Amsterdam 1972) p. 837
3.123 R.G.Smith, J.E.Geusic, H.J.Levinstein, S.Singh, L.G.van Uitert: J. Appl. Phys. **39**, 4030 (1968)
3.124 R.G.Smith, J.E.Geusic, H.J.Levinstein, J.J.Rubin, S.Singh, L.G.van Uitert: Appl. Phys. Lett. **12**, 308 (1968)
3.125 R.L.Byer, M.K.Oshman, J.F.Young, S.E.Harris: Appl. Phys. Lett. **13**, 109 (1968)
3.126 C.Laurence, F.Tittel: J. Appl. Phys. **42**, 2137 (1971)
3.127 J.F.Weller, T.G.Giallorenzi, R.A.Andrews: J. Appl. Phys. **33**, 4650 (1972)
3.128 L.B.Kreuzer: Appl. Phys. Lett. **13**, 57 (1968)
3.129 J.E.Pearson, U.Ganiel, A.Yariv: IEEE J. QE-**8**, 433 (1972)
3.130 A.E.Siegman: Appl. Opt. **1**, 739 (1962)
3.131 J.E.Bjorkholm: IEEE J. QE-**5**, 293 (1969)
3.132 R.Fischer: Exp. Technik der Physik XIX, 193 (1971)
3.133 J.E.Bjorkholm: Appl. Phys. Lett. **13**, 399 (1968)
3.134 L.B.Kreuzer: Single and Multimode Oscillation of the Singly Resonant Optical Parametric Oscillator. Proc. Joint Conf. Lasers and Opto Electronics, University of Southampton, Southampton, England, (1969); see also Appl. Phys. Lett. **15**, 263 (1969)
3.135 J.E.Bjorkholm: IEEE J. QE-**7**, 109 (1971)
3.136 P.P.Bey, C.L.Tang: IEEE J. QE-**8**, 361 (1972)

3.137 L.B.Kreuzer: Appl. Phys. Lett. **13**, 57 (1968)
3.138 R.W.Wallace: Appl. Phys. Lett. **17**, 497 (1970)
3.139 A.I.Kovrigin, P.V.Nikles: JETP Lett. **13**, 313 (1971)
3.140 R.L.Byer, R.L.Herbst, R.S.Feigelson, W.L.Kway: Opt. Commun. **12**, 427 (1974)
3.141 R.L.Byer, S.E.Harris: Phys. Rev. **168**, 1064 (1968)
3.142 T.G.Giallorenzi, C.L.Tang: Phys. Rev. **166**, 225 (1968)
3.143 D.A.Kleinmann: Phys. Rev. **174**, 1027 (1968)
3.144 S.E.Harris, M.K.Oshman, R.L.Byer: Phys. Rev. Lett. **18**, 732 (1967)
3.145 D.Magde, H.Mahr: Phys. Rev. Lett. **18**, 905 (1967)
3.146 A.J.Campillo, C.L.Tang: Appl. Phys. Lett. **16**, 242 (1970)
3.147 A.Hordvik, H.R.Schlossberg, C.M.Stickley: Appl. Phys. Lett. **18**, 448 (1971)
3.148 R.W.Wallace: IEEE Conf. Laser Applications, Washington, D. C. (1971)
3.149 J.Pinard, J.E.Young: Opt. Commun. **4**, 425 (1972)
3.150 B.Damaschini: Compt. Rend. Acad. Sci. (Paris), **268**, 1169 (1969)
3.151 P.W.Smith: IEEE J. QE-**1**, 343 (1965)
3.152 W.R.Leeb: Appl. Phys. **6**, 267 (1975)
3.153 G.Holton, O.Teschke: IEEE J. QE-**10**, 577 (1974)
3.154 S.H.Wemple, M.DiDomenico, Jr.: In *Applied Solid State Science*, vol. 3, ed. by R.Wolfe (Academic Press, New York 1972)
3.155 J.Ducuing, C.Flytzanis: In *Optical Properties of Solids*, ed. by F.Abeles (North Holland, Amsterdam 1970)
3.156 I.S.Rez: Sov. Phys. Usp. **10**, 759 (1968)
3.157 K.F.Hulme: Rep. Prog. Phys. **36**, 497 (1973)
3.158 R.L.Byer: *Ann. Rev. Mater. Sci.* **4**, 147 (1974)
3.159 R.L.Byer, D.S.Chemla: *Nonlinear Materials* (to be published)
3.160 S.A.Akhmanov, R.V.Khokhlov: *Problems in Nonlinear Optics.* Moscow, Akad. Nauk. SSR, English Ed. (Gordon and Breach, New York 1973)
3.161 H.Rabin, C.L.Tang: *Quantum Electronics*, vols. I and II (Academic Press, New York 1975)
3.162 A.Leubereau, L.Greiter, W.Kaiser: Appl. Phys. Lett. **25**, 87 (1974)
3.163 T.Kushida, Y.Tanaka, M.Ojima, Y.Nakazahi: Japan J. Appl. Phys. **14**, 1097 (1975)
3.164 C.B.Moore, L.S.Goldberg: Opt. Commun. **16**, 21 (1976)
3.165 R.F.Begley, A.B.Harvey, R.L.Byer: Appl. Phys. Lett. **25**, 387 (1974)

4. Difference Frequency Mixing via Spin Nonlinearities in the Far-Infrared

V. T. Nguyen and T. J. Bridges

With 18 Figures

The lack of tunable, coherent, reasonably intense sources has severely hampered the use of far-infrared (FIR) radiation in scientific and practical applications. Recent development of the InSb spin-flip Raman (SFR) laser [4.1] magnetically tunable over a frequency band up to $150\,\mathrm{cm}^{-1}$ wide, led to a possible solution of this problem. By using difference mixing in nonlinear crystals between a fixed-frequency source and the tunable SFR laser, a tunable source covering the region between $10\,\mathrm{cm}^{-1}$ and $300\,\mathrm{cm}^{-1}$ could in principle be obtained with a linewidth as narrow as the original sources. This difference frequency mixing technique has already been well investigated by using two fixed-frequency sources [4.2]. Initially, the technique was not successful in producing a tunable source, because the primary tunable source has a somewhat lower power output than the fixed tuned laser. Recently however, a strong resonant optical nonlinearity due to the conduction electron spins has been reported in the generation of FIR in InSb [4.3]. This effect, which is directly related to the spin-flip Raman process, was subsequently used in combination with a SFR laser to produce FIR tunable over a range of $\approx 10\%$ near to $100\,\mathrm{cm}^{-1}$ [4.4, 5].

In the initial development of the InSb SFR laser (see Sect. 1.4 of this book) a Q-switched CO_2 molecular laser operating near 10.6 μm wavelength was used as the pump [4.1]. Peak powers of a few kilowatts, with pulse lengths of 200 ns gave tunable Stokes outputs up to a few hundred watts, tunable from 10.8 μm to 13 μm. It is this pulsed mode of operation that is of chief concern in the present chapter. Later, it was found possible to operate the InSb SFR laser under cw conditions [4.6] by pumping with ~ 5.3 μm radiation obtained from a CO molecular laser. This mode of oscillation is possible because of the large resonant enhancement of the Raman scattering cross section as the pump wavelength approaches the direct bandgap of the InSb at 5.3 μm (at 4 K). Oscillation threshold pump powers as low as 5–10 mW cw have been observed, and output power up to 1 Watt with conversion efficiencies $>50\%$. A general review of SFR lasers is given in [4.7].

In the first section of this chapter we shall describe the theoretical background and experiments concerned with the difference frequency mixing due to the spin-flip transition in InSb. In the subsequent section, we shall give details of experiments on the generation of pulsed tunable FIR using spin-flip Raman lasers, and include an analysis on the range of tunability and the linewidth of the generated FIR.

4.1 Difference Frequency Mixing via Spin-Flip Transitions

4.1.1 Theoretical Background

Optical nonlinearities of electrons due to spin motion constitute one of the most interesting phenomena in narrow-gap semiconductors. Electrons in such materials generally have large g-values, and large nonlinear spin-photon interaction. The latter effect was discovered by *Yafet* [4.8] in his theoretical analysis of the spin-flip Raman scattering process. The effective Raman interaction Hamiltonian of this process given by Yafet was complicated. A simplified form applicable to narrow-gap semiconductors was given by *Brown* and *Wolff* [4.9] as:

$$H_{\mathrm{R}} \approx \frac{i e^2}{4 m_{\mathrm{s}} c^2} \frac{\hbar\omega_1 E_{\mathrm{G}}}{E_{\mathrm{G}}^2 - (\hbar\omega_1)^2} \boldsymbol{\sigma} \cdot \boldsymbol{A}(\omega_1) \times \boldsymbol{A}^*(\omega_2) + cc\,. \tag{4.1}$$

Here E_{G} is the band gap, $m_{\mathrm{s}} = 2m/|g|$ is the spin-mass with g the electron g-value, $\boldsymbol{A}(\omega_1)$ and $\boldsymbol{A}(\omega_2)$ are vector potentials of two incident beams with frequencies ω_1 and ω_2.

Equation (4.1) bears a certain similarity to the standard Zeeman interaction;

$$H_{\mathrm{Z}} = (e\hbar/2mc)\, \boldsymbol{\sigma} \cdot \boldsymbol{H}\,. \tag{4.2}$$

In H_{Z}, the electron spin is coupled to a magnetic field; in (4.1) on the other hand, the spin couples to a nonlinear combination $\boldsymbol{A}(\omega_1) \times \boldsymbol{A}^*(\omega_2)$ of two electromagnetic fields. This consideration is crucial in the method for generating FIR at frequency $\omega_3 = \omega_1 - \omega_2$ by difference mixing in InSb of two radiations at frequencies ω_1 and ω_2. We see that the vector product $\boldsymbol{A}(\omega_1) \times \boldsymbol{A}(\omega_2)$ in (4.1) acts as an effective magnetic field which drives the spin system at $\omega_3 = \omega_1 - \omega_2$.

If the spin resonance frequency ω_{s} is tuned to the difference frequency ω_3, the spin system is driven on resonance. Moreover, because the electron spin-resonance linewidth is small, even modest fields can drive the spin precession to appreciable amplitude. Subsequently, this precessing magnetic moment radiates energy at FIR frequency ω_3. In effect, the spin system acts as a resonant, nonlinear medium for the difference mixing process. The FIR radiation is created by a magnetic dipole transition. Hence, the nonlinear medium need not be acentric for this process to be allowed. This is an important point since InSb lacks inversion symmetry, but the Hamiltonian of (4.1) is centrosymmetric.

The strength of the process described above was estimated by *Brown* and *Wolff* [4.9] considering the geometry illustrated in Fig. 4.1. An n-type InSb crystal is irradiated with two beams at frequencies ω_1 and ω_2 propagating along the y axis. The polarizations are orthogonal, with one parallel to the dc magnetic field along the z axis. The vector product $\boldsymbol{A}(\omega_1) \times \boldsymbol{A}(\omega_2)$ creates an

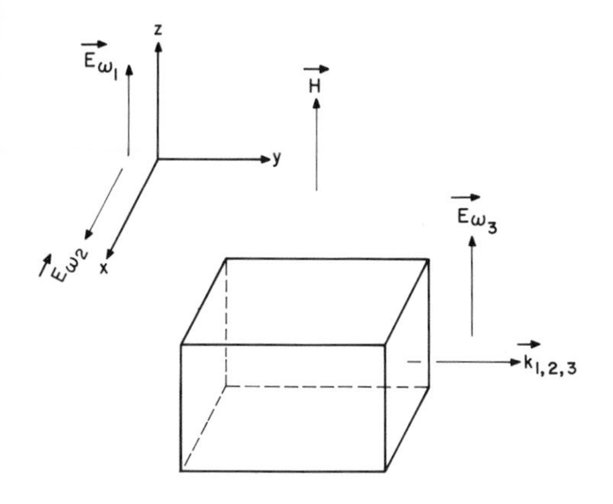

Fig. 4.1. Collinear difference frequency mixing in spin nonlinearities

effective magnetic field in the y direction which couples to the spins as indicated in (4.1). The subsequent motion is a standard problem in spin resonance. Consequently the spin magnetization along x can be written in the form [4.9]

$$M_x(\omega_3) = C(\omega_3)\, E_{\omega_1} E_{\omega_2} \exp[\mathrm{i}(\boldsymbol{k}_1 - \boldsymbol{k}_2)\, y] \tag{4.3}$$

where E_{ω_1} and E_{ω_2}, respectively, are electric field components of incident radiation, propagating with wave vectors $\boldsymbol{k}_1$ and $\boldsymbol{k}_2$ along the y axis; $C(\omega_3)$ is given by

$$C(\omega_3) = n_0 \mu^* \frac{\mathrm{e}^2}{m_s c^2} \frac{\hbar\omega_1 E_G}{E_G^2 - (\hbar\omega_1)^2} \frac{c^2}{\omega_1 \omega_2} [2\hbar(\omega_3 - \omega_s + \mathrm{i}\gamma/2)]^{-1} \tag{4.4}$$

where n_0 is the electron concentration, μ^* is the magnetic moment of electrons in InSb and γ is the linewidth (full width at half height). The magnetization described by (4.3) is the source of the FIR radiation at ω_3 propagating also along the y axis with wave vector $\boldsymbol{k}_3$. From Maxwell's equation, one can show that the electric component of the FIR is polarized along the z axis and the power output of FIR is given by

$$P_{\omega_3} = 32 \frac{\pi^2 (2d_{\mathrm{eff}}^{\mathrm{s}})^2}{c^3 n_1 n_2 n_3} \omega_3^2 \frac{P_{\omega_1} P_{\omega_2}}{w^2} T^3 I \tag{4.5}$$

where d_{eff} is the effective spin process nonlinear coefficient given by $d_{\mathrm{eff}}^{\mathrm{s}} = 1/2 n_3 C(\omega_3)$, P_{ω_1} and P_{ω_2} are the incident powers, $n_{1,2,3}$ are refractive indices at frequencies ω_1, ω_2, and ω_3, respectively, w is the radius of Gaussian incident beams inside the sample, T is the power transmission coefficient for each surface of crystal, and I is the coherence factor given in the case of a plane wave approximation by [4.10, 11].

$$I = \exp(-\alpha_3 l) \frac{1 + \exp(-\Delta\alpha l) - 2\exp(-\Delta\alpha l/2)\cos(\Delta k l)}{(\Delta k)^2 + (\Delta\alpha/2)^2}. \tag{4.6}$$

Here $\Delta \boldsymbol{k}=\boldsymbol{k}_1-\boldsymbol{k}_2-\boldsymbol{k}_3$ is the phase mismatch, $\Delta\alpha=\alpha_1+\alpha_2-\alpha_3$ with $\alpha_{1,2,3}$ being the absorption coefficients at three frequencies of interest and l is the length of crystal.

The treatment described above gives very good physical insight into the nonlinear spin-photon interaction. A more correct analysis of this problem using the theory of stimulated polariton scattering has been given by *Shen* [4.12]. It shows that the nonlinear coefficient d_{eff} of (4.5) derived from perturbation theory, as expected, is indentical to the one given by (4.4).

In the stimulated polariton process the electromagnetic fields E_{ω_1}, E_{ω_2} of incident radiations and E_{ω_3} of the generated FIR couple nonlinearly with one another. In addition, there is a material excitation at frequency ω_3 which couples directly with E_{ω_3} forming polaritons and nonlinearly with E_{ω_1} and E_{ω_2} through optical mixing. The material excitation, in our case, is an electronic excitation from the ground state to an excited state. This is the case of spin-flip transition in InSb.

Stimulated polariton scattering should then be described by a set of four coupled equations: three wave equations for E_{ω_1}, E_{ω_2}, and E_{ω_3} and one equation of motion for the spin-flip transition. The latter was derived by *Shen* from a general perturbation Hamiltonian $H'(\omega)$ given by:

$$H'(\omega)=-[\boldsymbol{P}\cdot+(c/\mathrm{i}\omega)\boldsymbol{M}\cdot\nabla X+\boldsymbol{Q}:\nabla+\ldots]\,\boldsymbol{E}(\omega)$$

where $\boldsymbol{P}$, $\boldsymbol{M}$, and $\boldsymbol{Q}$ are electric dipole, magnetic-dipole and electric-quadrupole operators, respectively.

Assuming that the depletion of the pump field E_{ω_1} is negligible and for the geometry illustrated in Fig. 4.1, the coupled equations for electromagnetic waves E_{ω_2} and E_{ω_3} are given by

$$\begin{aligned}&[\nabla^2+(\omega_3^2/c^2)\,\varepsilon(\omega_3)_{\text{eff}}]\,E_{\omega_3}=-(4\pi\omega_3^2/c^2)\,\chi_{\text{eff}}^{(2)}E_{\omega_1}E_{\omega_2}^*\\&[\nabla^2+(\omega_2^2/c^2)\,\varepsilon^*(\omega_2)_{\text{eff}}]\,E_{\omega_2}^*=-(4\pi\omega_2^2/c^2)\,\chi_{\text{eff}}^{(2)}E_{\omega_1}^*E_{\omega_3}\end{aligned}\tag{4.7}$$

where

$$\varepsilon(\omega_3)_{\text{eff}}=\varepsilon(\omega_3)-n_0\mu^{*2}\varepsilon(\omega_3)_{\text{eff}}\pi\Big/\left[\hbar\left(\omega_3-\omega_s+\mathrm{i}\frac{\gamma}{2}\right)\right]$$

$$\varepsilon(\omega_2)_{\text{eff}}=\varepsilon(\omega_2)+4\pi\chi_{\mathrm{R}}|E_{\omega_1}|^2$$

$$\chi_{\mathrm{R}}=-n_0\left(\frac{e^2}{2m_s\omega_1\omega_2}\right)^2\left(\frac{\hbar\omega_1E_{\mathrm{G}}}{E_{\mathrm{G}}^2-\hbar\omega_1}\right)^2\cdot\frac{1}{4\hbar\left(\omega_3-\omega_s+\mathrm{i}\dfrac{\gamma}{2}\right)}$$

$$\chi_{\text{eff}}^{(2)}=\chi^{(2)}+n_0\mu^*\frac{e^2}{m_s}\cdot\frac{\varepsilon(\omega_3)^{1/2}}{\omega_1\omega_2}\cdot\frac{\hbar\omega_1E_{\mathrm{G}}}{E_{\mathrm{G}}^2-(\hbar\omega_1)^2}\;\frac{1}{4\hbar\left(\omega_3-\omega_s+\mathrm{i}\dfrac{\gamma}{2}\right)},$$

n_0, μ^*, m_s, and γ were already defined above, $\varepsilon(\omega_2)$ and $\varepsilon(\omega_3)$ are the complex dielectric constant of frequencies ω_2 and ω_3, $\chi^{(2)}$ is the conventional nonlinear susceptibility which couples E_{ω_1}, E_{ω_2}, and E_{ω_3} directly, and χ_{R} may be called the spin-flip Raman susceptibility. Note also the second term of $\chi_{\text{eff}}^{(2)}$ is identical to the spin process nonlinear coefficient $d_{\text{eff}}^{\text{s}}$ given above.

Using the coupled (4.7), one can derive (4.5) for the FIR power output which includes the interference of two terms in $\chi_{\text{eff}}^{(2)}$. Equation (4.7) may also be used to describe the FIR generation in stimulated spin Raman scattering in which E_{ω_1} is the only pump field. In this case E_{ω_2} is the Stokes-shifted field at frequency ω_2 with $\omega_1 - \omega_2 = \omega_3 = \omega_s$. Assuming $n_0 = 2.2 \times 10^{15}\,\text{cm}^{-3}$, $\omega_1 = 1040\,\text{cm}^{-1}$, $\omega_s = 90\,\text{cm}^{-1}$, $|E_{\omega_1}|^2 = 10^3$ esu (2 Mw/cm^2) and phase-matching condition $\Delta k = 0$ being satisfied, *Shen* has predicted the ratio $P_{\omega_3}/P_{\omega_2} \approx 3 \times 10^{-5}$. So far, there has been no report on the observation of FIR radiation from a SFR laser pumped by conventional Q-switched CO_2 laser with output power of < 5 kW. For this level of pump power optimum tunable Stokes radiation at frequency ω_2 from SFR laser requires a higher electron concentration than that required for phase matching. The latter will be the subject of our discussion in the next section. However, recently, *E. D. Shaw*[1] observed FIR from a SFR laser using electron concentration in the range from $5\text{–}7 \times 10^{15}\,\text{cm}^{-3}$. The pump radiation was provided by a TEA CO_2 laser.

4.1.2 Phase-Matching Problem

The generation of appreciable amounts of FIR power is dependent on achieving the phase-matching condition of the three interacting waves. This condition requires in (4.6) that $\Delta \boldsymbol{k} = 0$. Additionally the nonlinear material should have small losses at the three frequencies of interest with ideally $\alpha_{1,2,3} = 0$. If these two conditions are obeyed then the coherence factor I achieves its maximum value equal to l^2. In practice the value of I is often considerably reduced by the presence of absorption. Also phase matching is not normally achieved because of dispersion of refractive index. Certain nonlinear materials which are birefringent can be phase matched by propagating at a particular direction in the crystal [4.13]. This scheme cannot be used in cubic materials such as InSb, which are nonlinear but have no birefringence. In the present application, phase matching ($\Delta k = 0$) can be achieved by using the negative dielectric effect of free electrons. The resulting free-electron loss however limits the value of I to $10^{-2}\text{–}10^{-3}$. This subject will be discussed in Subsection 4.1.3 for the case of collinear interaction, and in Subsection 4.1.4 for noncollinear interaction.

4.1.3 Collinear Interaction

In this scheme, as shown in Fig. 4.1, all incident and generated radiations propagate along the y axis. Phase matching then requires that

$$n_1 \omega_1 - n_2 \omega_2 = n_3 \omega_3 \tag{4.8}$$

[1] Private communication.

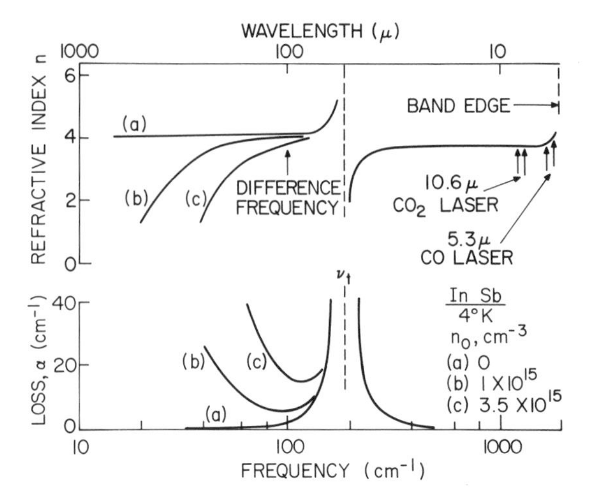

Fig. 4.2. Dispersion of refractive index of InSb at 4 K, including the effect of electron plasma, for three different electron concentrations. Magnetoplasma effects are neglected (see text)

where n_j ($j=1, 2, 3$) are real parts of the complex refractive index $n_j+i\kappa_j$ at frequency ω_j and by definition

$$(n_j+\mathrm{i}\kappa_j)^2=\varepsilon(\omega_j)$$

where $\varepsilon(\omega_j)$ is the dielectric constant at frequency ω_j. Since the value of n_3 is substantially affected by the presence of the electron plasma, it may be adjusted by correct choice of the electron concentration n_0. This effect may be used to satisfy (4.8) and thereby achieve phase matching. Since the magneto-plasma effects are negligible at the high frequencies ω_1 and ω_2 and zero at ω_3 because E_{ω_3} is parallel to the dc magnetic field, $\varepsilon(\omega_j)$ is independent of magnetic field and is given by [4.14]

$$\varepsilon(\omega_j)=\varepsilon^{\infty}\left[1+\frac{(\omega_{\mathrm{LO}}/\omega_{\mathrm{TO}})^2-1}{1-(\omega_j/\omega_{\mathrm{TO}})^2}-\frac{\omega_{\mathrm{p}}^2}{\omega_j(\omega_j+\mathrm{i}\nu_j)}\right] \tag{4.9}$$

where ε_∞ is the high-frequency dielectric constant, ω_{LO} and ω_{TO} are the longitudinal and transverse optical phonon frequencies, $\omega_{\mathrm{p}}=[4\pi n_0 e^2/m^*\varepsilon_\infty]^{1/2}$ is the plasma frequency with n_0 and m^* being, respectively, the electron concentration and the effective electron mass of the sample, and $\nu_j=1/\omega_j\tau$ where τ is the electron collisional relaxation time.

Typical behavior of n and α of InSb at 4 K, calculated from (4.9) as a function of wavelength (or frequency) is shown in Fig. 4.2 for different electron concentrations n_0. Also shown are the wavelengths of incident radiations around 10.6 μm or 5.3 μm corresponding to the output of the CO_2 and CO lasers, respectively. It is evident that 10.6 μm radiation from the CO_2 laser is a good choice for ω_1 and ω_2 because of the low dispersion of refractive indices at that point. Clearly a proper choice of n_0, and consequently of n and α, allows us, for 10.6 μm incident radiations, to obtain $\Delta k=0$ and comparatively small α and hence a usably large coherence factor I of (4.5). As an example, perfect phase matching $\Delta k=0$ occurs for $n_e=3.4\times10^{15}\,\mathrm{cm}^{-3}$ for $\omega_1=1045\,\mathrm{cm}^{-1}$, $\omega_2=944\,\mathrm{cm}^{-1}$ and difference frequency $\omega_3=101\,\mathrm{cm}^{-1}$. Because of the comparatively small SFR cross section in InSb at 10.6 μm however, operation is restricted to high peak power pulsed operation [4.1].

It will be seen from Fig. 4.2 that absorption coefficients α_1 and α_2 are always very small ($<1\,\mathrm{cm}^{-1}$) while α_3 in the region of $\omega_3=100\,\mathrm{cm}^{-1}$ is $\approx 2\,\mathrm{cm}^{-1}$ for $n_0=0$ and increases to above $7\,\mathrm{cm}^{-1}$ for $n_0>10^{15}\,\mathrm{cm}^{-3}$. Thus for the range of concentrations of interest $\Delta\alpha=-\alpha_3$ and for samples much thicker than $1/\alpha_3$ the exponential and periodic terms in (4.6) become negligible and $I\approx[(\Delta k)^2+1/4(\alpha_3)^2]^{-1}$. The phase matching is not critical provided that $\Delta k<1/2\alpha_3$. This advantage is obtained of course at the expense of mixing efficiency. For example, in an InSb crystal 1 cm long with an α_3 of $20\,\mathrm{cm}^{-1}$, the effective interaction length is ~ 0.05 cm and I is $10^{-2}\,\mathrm{cm}^2$ with a bandwidth to half-power points of $\Delta k=\pm 10\,\mathrm{cm}^{-1}$. On the other hand if $\alpha_3=\alpha_1=\alpha_2=0$, the interaction length is 1 cm and from (4.6) $I=1\,\mathrm{cm}^2$ and bandwidth $\Delta k=\pm 2.7\,\mathrm{cm}^{-1}$.

For incident radiation from the CO laser at 5.3 μm, close to the bandgap wavelength of InSb, the SFR cross section is greatly enhanced, a fact which has resulted in low-threshold cw operation of the SFR laser [4.6]. The same consideration might allow cw difference mixing, but it can be seen that the high dispersion of refractive index at 5.3 μm, as shown in Fig. 4.2, does not allow phase matching. Using values of refractive index given by *Brueck* and *Mooradian* [4.15] for incident radiation near to 5.3 μm one can show that, for $\omega_1=1900\,\mathrm{cm}^{-1}$, $\omega_2=1800\,\mathrm{cm}^{-1}$, $\omega_3=100\,\mathrm{cm}^{-1}$ and for the same range of electron concentration given in Fig. 4.2,

$$n_1\omega_1-n_2\omega_2>n_3\omega_3\,. \tag{4.10}$$

Consequently, the phase-matched generation of FIR by difference mixing of two radiations from CO laser at wavelength close to 5.3 μm is practically impossible even in the noncollinear scheme which will be discussed briefly in the following subsection.

4.1.4 Noncollinear Interaction

In this scheme, the wave vector triangle required by the phase matching condition $\Delta k=0$ is shown in Fig. 4.3 for different directions of incident and generated radiation inside the nonlinear crystal. As mentioned above, this scheme is still useful only for difference mixing in InSb with incident radiations $\gtrsim 10$ μm. It does however give us more freedom in the choice of electron concentration n_0 particularly allowing $n_0<10^{15}\,\mathrm{cm}^{-3}$, which would result in much lower absorption at the generated FIR frequencies. As an example,

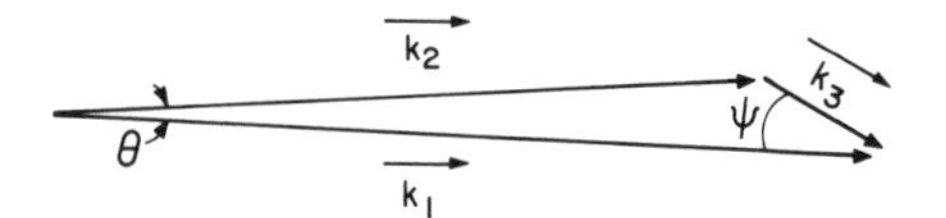

Fig. 4.3. Non-collinear difference frequency mixing

Shen [4.12] has calculated for $\omega_1 = 1040\,\mathrm{cm}^{-1}$, $\omega_3 = 90\,\mathrm{cm}^{-1}$, and $n_e = 2.2 \times 10^{15}\,\mathrm{cm}^{-3}$, $\Delta \boldsymbol{k} = 0$ could be obtained for $\theta = 1.7°$ and $\psi = 17.4°$. No experiment has been reported for noncollinear difference mixing in InSb.

4.1.5 Details of Experiments

In this subsection, we describe a difference frequency mixing experiment [4.3] performed in InSb for the observation of the electron spin nonlinearity discussed in Subsection 4.1.1. This nonlinearity depends on the shape of the conduction band, which in InSb is spherical (to a good approximation) but, as mentioned in Subsection 4.1.1, is not dependent on the crystal acentricities as in the conventional second-order nonlinearity. However, in InSb, the conventional nonlinear coefficient is present [4.16], and we take advantage of this to calibrate the effective electron-spin nonlinear coefficient d^s_{eff} of (4.5).

The schematic of the experimental setup is shown in Fig. 4.4. Two separate but synchronously Q-switched CO_2 lasers with a repetition rate of 250 Hz were used for the generation of 250 ns long linearly polarized pulses with peak powers of ≈ 1 kW. Fixed attenuators were used to reduce the power to desired levels. Gratings within the optical cavities allowed the selection of any one of many transitions near 9.6 μm (ω_1) and 10.6 μm (ω_2). In the present experiment the difference frequency ($\omega_3 = \omega_1 - \omega_2$) was fixed at $90.1\,\mathrm{cm}^{-1}$. A half-wave plate was used in order to make the polarization of the 9.6 μm beam orthogonal to that of the 10.6 μm beam. The two beams were combined collinearly in a beam splitter and focused onto the InSb sample mounted on a cold finger in a superconducting solenoid. The sample temperature was estimated to be 15 K. In a typical experiment the n-InSb sample ($n = 2.2 \times 10^{15}\,\mathrm{cm}^{-3}$) was cut and polished in a bar shape $3 \times 3 \times 4\,\mathrm{mm}^3$. A Ge:Ga detector [4.17] was used to detect the FIR output. Figure 4.5 shows the sample geometry. The propagation vectors were perpendicular to the magnetic field H and polarized so that the two orthogonal incident electric fields, E_{ω_1} at frequency ω_1 and E_{ω_2} at frequency ω_2 were parallel to the $\langle 1\bar{1}0 \rangle$ and $\langle 111 \rangle$ crystallographic axes, respectively, of the InSb sample. It should be pointed out that the geometry of the experimental setup shown in Fig. 4.5 is identical to one which has been used for the observation of the spin-flip Raman scattering process if we consider the radiation at frequency ω_1 as the pump and the radiation at frequency ω_2 as the *Stokes* frequency. This remark has important implications for the generation of tunable FIR which will be discussed in Section 4.2.

The geometry described above was chosen so that the electric field E_{ω_3} of the generated FIR both from the conventional nonlinearity and the electron

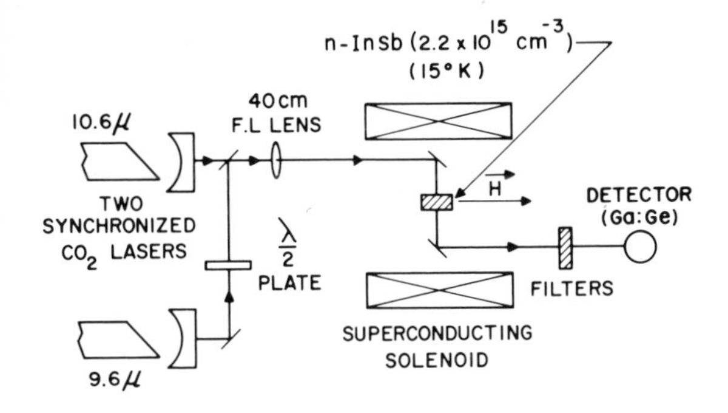

Fig. 4.4. Experimental arrangement for observing difference frequency mixing in spin nonlinearities in InSb

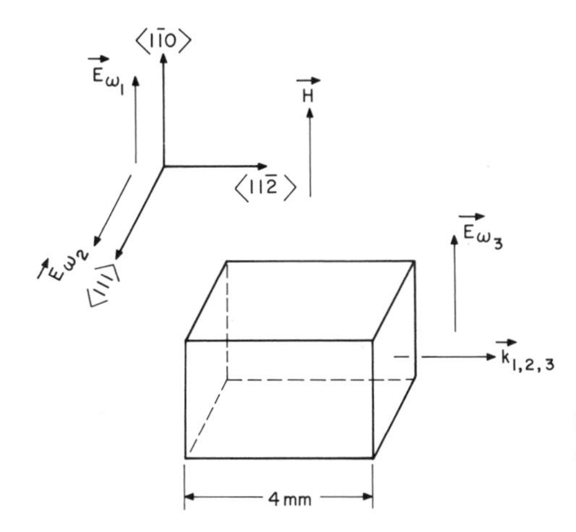

Fig. 4.5. Sample geometry for experiment of Fig. 4.4

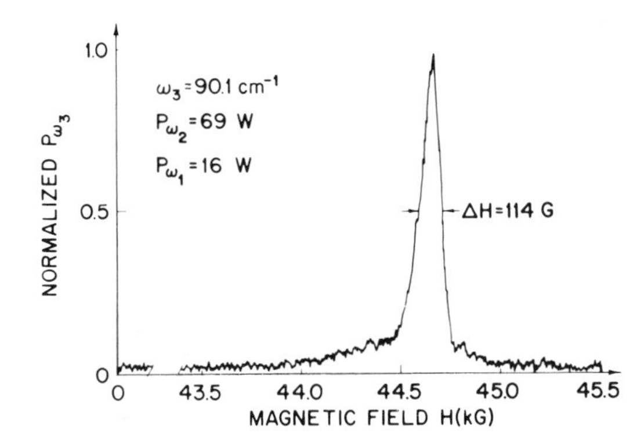

Fig. 4.6. Experimental result of experiment of Fig. 4.4

spin nonlinearity was parallel to the dc magnetic field H [4.3]. For this geometry, from (4.9), the contribution of free electrons to the complex refractive index at frequencies ω_1, ω_2 and particularly at FIR frequency ω_3 is independent of the magnetic field strength and identical for both types of nonlinearity. Therefore, the coherence factor I of (4.5) remains essentially unchanged as the magnetic field H varies, and the power output $P_{\omega_3}^{c,s}$ both from conventional nonlinearity and from electron-spin nonlinearity is given by (4.5). To obtain $P^{c}_{\omega 3}$, the nonlinear coefficient d^{s}_{eff} is replaced with $d^{c}_{eff} = d_{14}/\sqrt{3}$ [4.3].

Results of an experiment with input powers $P_{\omega_1} = 69$ W and $P_{\omega_2} = 16$ W are shown in Fig. 4.6 for normalized P_{ω_3} as a function of the dc magnetic field H. A sharp peak of the FIR power output was observed at $H = 44.63$ kG where $\omega_1 - \omega_2 = \omega_3 = \omega_s$. Away from resonance, the signal is due to the conventional nonlinearity. The ratio $Q = P^{s}_{\omega 3}/P^{c}_{\omega_3}$ at resonance is found to be 88 for a linewidth of 114 G (≈ 0.2 cm^{-1}). The power output at the resonance peak is ≈ 0.2 μW. From (4.5) it is found that $d^{s}_{eff} = 5.4\, d_{14}$. Taking d_{14} as 5×10^{-7} esu [4.11], $d^{s}_{eff} = 2.7 \times 10^{-6}$ esu in reasonable agreement with $d^{s}_{eff} = 4 \times 10^{-6}$ esu calculated from $d^{s}_{eff} = 1/2 n_3 C(\omega_3)$ with $C(\omega_3)$ given by (4.4). The measured linewidth of 114 G is also in good agreement with the measured value of *McCombe* and *Wagner* [4.18] for the same range of electron concentration, and a sample temperature of 15 K. The asymmetry of the curve in Fig. 4.6 was well explained by *Shen* [4.12]. It arises from the interference of the conventional nonlinearity and of the electron-spin nonlinearity. Before going to the next section for the discussion of the generation of tunable FIR, we would like to point out that the experimental method described in this section is an excellent tool for studying the line shape and other physics of the spin-resonance process.

4.2 Generation of Tunable Far-Infrared Using Spin-Flip Raman Lasers

In the mixing experiment described in Subsection 4.1.5, frequencies ω_1, ω_2, and ω_3 were fixed. The power output of FIR at frequency ω_3 peaks strongly when ω_3 is equal to the spin resonance frequency ω_s. It was also pointed out in Subsection 4.1.5 that the experimental arrangement is identical for the observation of the spin-flip Raman scattering mechanism and of the resonant spin nonlinearity process. It is clear that the generation of tunable FIR could be obtained by replacing the two frequencies ω_1 and ω_2 by the pump and tunable *Stokes* radiation from a 10.6 μm pumped spin-flip Raman laser [4.1] which automatically differ by the spin resonance frequency ω_s. Consequently, the linewidth and tunability of the generated FIR depend on the tuning characteristics and the bandwidth of the 10.6 μm pumped SFR laser output radiation. These two important parameters will be discussed in the following subsection.

4.2.1 Linewidth and Tunability of 10.6 μm Pumped SFR Laser

The mechanism and properties of the SFR laser have been reviewed by a great number of papers [4.7, 19] and also in Section 1.4 of this book. We discuss in this section only the bandwidth and tuning characteristics of 10.6 μm pumped SFR laser output radiation.

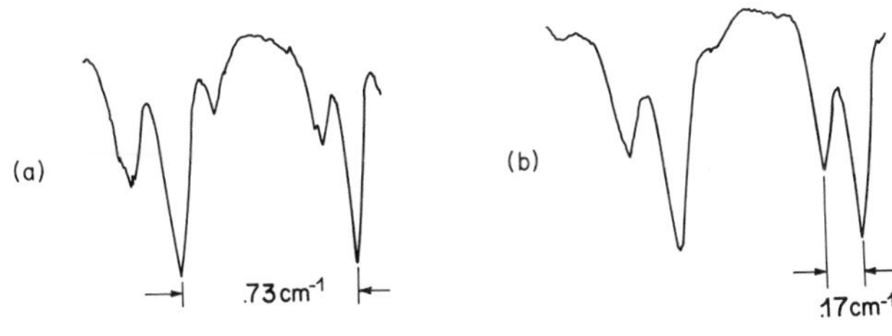

Fig. 4.7a and b. Fabry-Perot scanning interferometer spectrum of output from pulsed, 10.6 μm pumped SFR laser, showing (a) three and (b) two longitudinal modes. The free spectral range (0.73 cm^{-1}) provides a calibration

Early work on the 10.6 μm pumped SFR laser [4.1] showed that the SFR radiation was continuously tunable with a bandwidth $<0.03\ cm^{-1}$. However, an analysis [4.20] with a scanning Fabry-Perot interferometer of the output radiation from a SFR laser cavity having input and output faces parallel to within 10^{-4} radian has shown that the SFR laser oscillation occurred on two or three longitudinal modes at one time or possibly jumped from one mode to another in consecutive pulses [4.7]. The typical output radiation spectrum is shown in Fig. 4.7 (a, b) for two values of magnetic field. The spacings and the number of modes were in agreement with those calculated from the cavity length and consistent with a Raman gain linewidth of about 0.5 cm^{-1}. It is shown also in Fig. 4.7 (a, b) that, as the gain curve was tuned by scanning the magnetic field, the positions of the modes did not change continuously. Only the relative intensity of the output in each mode varied. Some pulling of individual modes ($\approx 0.005\ cm^{-1}\ G^{-1}$) was observed as the gain line was tuned past each mode. This amount is in rough agreement with that measured by *Brueck* and *Mooradian* [4.21] for a cw SFR laser with approximately the same cavity and material characteristics. The results of this analysis were confirmed recently by *Ganley* et al. [4.22] who, by reducing the pump power, were able to obtain the SFR laser operating on a single longitudinal mode with a linewidth of 0.013 cm^{-1}. As discussed in the next section, the performance cited above of the 10.6 μm pumped SFR will severely reduce the usefulness of FIR as a spectroscopic tool. One way to resolve the problem of the presence of multiple longitudinal modes of the SFR laser is to destroy the resonance of the SFR laser cavity and achieve a nonresonant superradiant spin-flip Raman laser (NRSSFR). Superradiance has been previously discussed in connection with measurements of linewidth at 10.6 μm [4.1] and 5.3 μm [4.23]. An experiment was performed with a NRSSFR laser as shown in Fig. 4.8. The shape and dimensions of the crystal are given, together with the polarization of the electric field of the pump radiation. The optical path was folded to improve uniformity of magnetic field over the active region. Making the polished input and output surfaces nonparallel by 5° effectively prevented resonance of the sample. The results of linewidth measurements using a scanning Fabry-Perot interferometer are shown in Fig. 4.9 for the NRSSFR, together with the

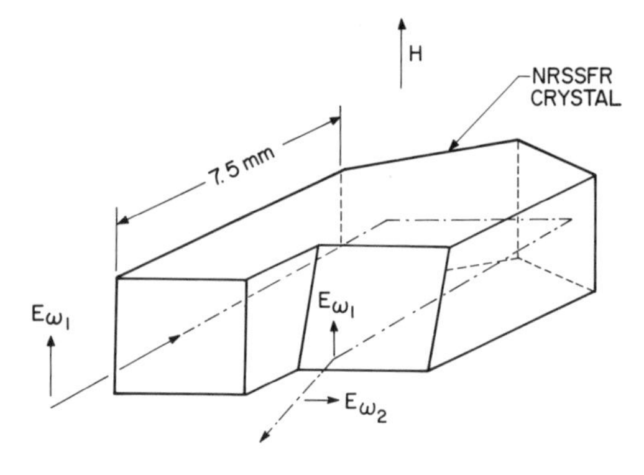

Fig. 4.8. Nonresonant superradiant spin-flip Raman (NRSSFR) laser showing folded optical path, and nonparallel input and output faces

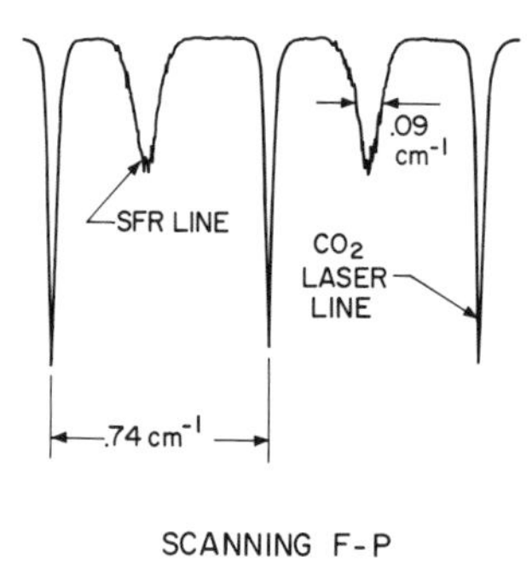

Fig. 4.9. Fabry-Perot scanning interferometer spectrum of the NRSSFR laser of Fig. 4.8. The tunable output is shown and also the CO_2 laser pump. The free spectral range ($0.74\,cm^{-1}$) provides a calibration

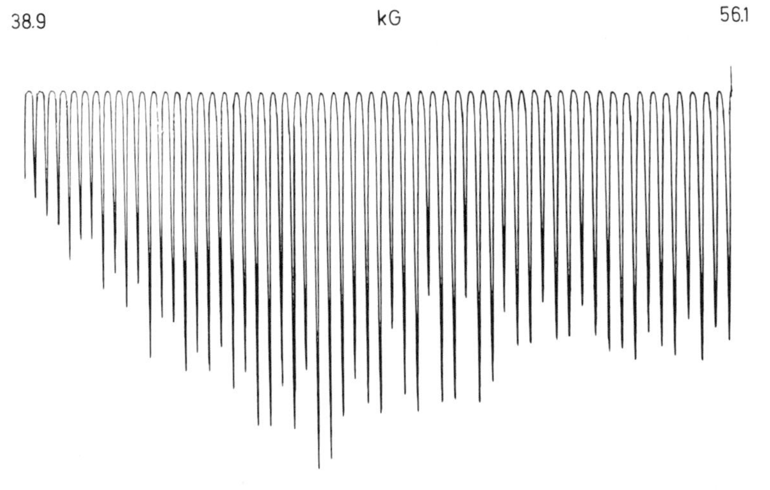

Fig. 4.10. Response of fixed Fabry-Perot interferometer to NRSSFR output when tuned from $867\,cm^{-1}$ (38.9 kG) to $837\,cm^{-1}$ (56.1 kG). The tuning is continuous with a rate of $1.8\,cm^{-1}/kG$

pump line of the CO_2 laser which indicates the resolving power of the interferometer. The NRSSFR linewidth of $0.09\,\mathrm{cm}^{-1}$ is in agreement with the theoretical estimate of linewidth given by *Patel* and *Shaw* [4.1] considering the spontaneous linewidth of an InSb sample with electron concentration of $10^{16}\,\mathrm{cm}^{-3}$ and the pump intensity of about $10^5\,\mathrm{W/cm}^2$. In Fig. 4.10, we show the result of the analysis of the output of the NRSSFR as a function of the magnetic field using a Fabry-Perot interferometer with fixed spacing. The regular pattern of the spectrum of Fig. 4.10 for the magnetic field from 38.9 kG to 56 kG implies that, within the resolution of our interferometer, the tuning of NRSSFR radiation is continuous with a tuning rate of $\approx 1.8\,\mathrm{cm}^{-1}/\mathrm{kG}$.

4.2.2 Experiments on Generation of Tunable Far-Infrared

In this subsection, we describe experiments on generation of tunable FIR by difference mixing between the pump frequency and the magnetically tunable Stokes frequency of the SFR laser. As pointed out in Subsection 4.1.1, in principle the generation of Stokes radiation and the difference mixing between the pump and the Stokes frequency could be combined in the same crystal, and so a SFR laser could itself be a generator of tunable FIR. However, optimum tunable Stokes radiation from a 10.6 μm pumped SFR laser requires a higher electron concentration that that required for phase matching of the mixing process. Consequently separate and adjacent crystals of InSb of different doping concentration for the SFR laser and for the mixer crystal have been used [4.4, 5] in the geometry of Fig. 4.11. The primary pump source in [4.4] was a *Q*-switched CO_2 laser and in [4.5] a TEA CO_2 laser. Results of both experiments are similar.

The experiment of [4.4] will now be described in detail. The crystals (Fig. 4.11) which had polished faces were mounted on a cold finger at an estimated temperature of 15 K in a superconducting magnet capable of fields up to 50 kG. The experimental arrangement of the generation of tunable FIR is shown in Fig. 4.12. Pump power at a frequency of $944.2\,\mathrm{cm}^{-1}$ was provided by a single-line single mode CO_2 laser, *Q*-switched by a rotating mirror at 120 pps. The power (5 kW peak, 200 ns pulse length) was focused into the SFR laser crystal ($n_0 = 1 \times 10^{16}\,\mathrm{cm}^{-3}$) with a 100 cm focal length lens giving a beam diameter in the crystal of 2 mm. As shown in Fig. 4.11, the geometry was such that $E_{\omega_1} \| H, \boldsymbol{k}_1 \perp H$ where E_{ω_1}, $\boldsymbol{k}_1$ and H were already defined in Subsection 4.1.1.

Oscillation of the SFR laser was obtained collinearly with the pump beam. The resulting Stokes radiation at frequency ω_2 ($P_{\omega_2} \approx 10$ W, $E_{\omega_2} \perp H$) and the transmitted part (2 kW) of the pump power leave the laser crystal together in the forward direction and enter the mixing crystal ($n_0 = 3.9 \times 10^{15}\,\mathrm{cm}^{-3}$). As explained in Subsection 4.2.1, the SFR oscillation occurred on two or three longitudinal modes which effectively spread the linewidth of the output to $\approx 0.25\,\mathrm{cm}^{-1}$. In order to ensure that the difference frequency mixing resulting only from the free-electron spins, the crystal was mounted so that E_{ω_1}, E_{ω_2}

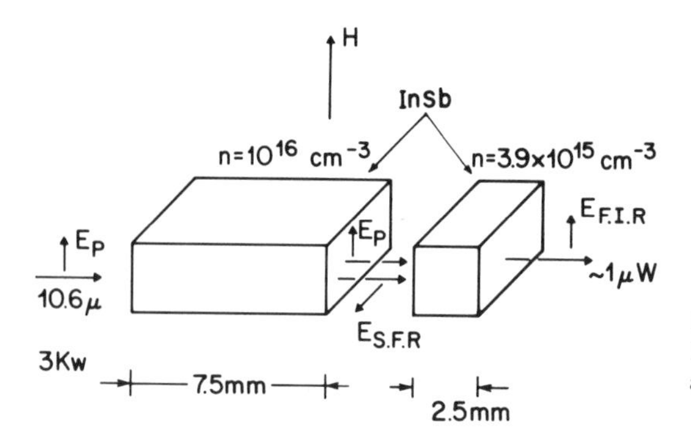

Fig. 4.11. Continuously tunable FIR source using InSb SFR laser and mixing crystal

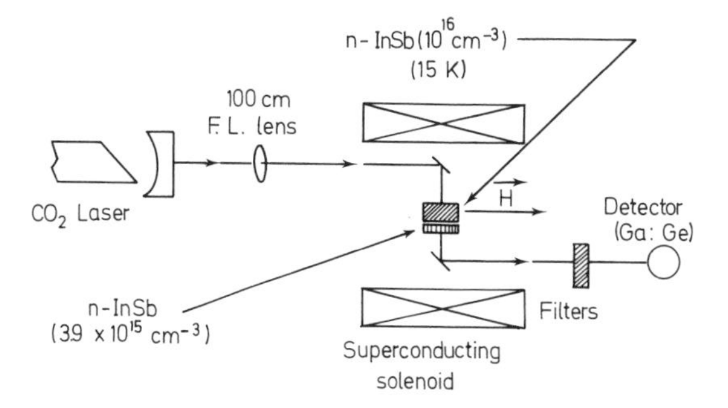

Fig. 4.12. Experimental arrangement for experiment of Fig. 4.11

were parallel, respectively, to the $\langle 100 \rangle$ and $\langle 010 \rangle$ crystallographic axes of the InSb mixing crystal [4.3]. The FIR difference frequency power P_{ω_3} leaving the mixer was observed with a Ge:Ga detector [4.17]. A crystalline quartz filter rejected all residual power near 10 μm. In agreement with the theory and the mixing experiment described in Subsection 4.1.3, the FIR output was polarized with the electric field vector parallel to H. A boxear integrator with an integration time of 3 s was used to process the signal. Figure 4.13 shows a scan of FIR output power vs. wave number, produced by slowly scanning the magnetic field and applying the detector output to a pen recorder. Maximum signal-to-noise ratio was 500. The observed signal fluctuations were almost entirely due to pulse-to-pulse variations in the pump laser power.

The electron concentration of $3.9 \times 10^{15}\,\mathrm{cm}^{-3}$ of the mixing crystal was chosen for the perfect phase-matching condition $\Delta k = 0$ already mentioned in Subsection 4.1.2 when $\omega_1 = 1045\,\mathrm{cm}^{-1}$, $\omega_2 = 944\,\mathrm{cm}^{-1}$ and $\omega_3 = 101\,\mathrm{cm}^{-1}$. It was also mentioned in Subsection 4.1.2 that as well as phase matching, the free carrier absorption loss at ω_3 plays a significant part in the difference mixing process. Taking these factors into account, the FIR power generated can be estimated from (4.5). Since many of the parameters are approximately constant over the range of the magnetic field variation from 45 to 50 kG, the FIR output P_{ω_3} becomes proportional to $\omega_3^2 I$ where I is given by (4.6). The computed value of $\omega_3^2 I$ as a function of ω_3 is shown by a dotted line normalized

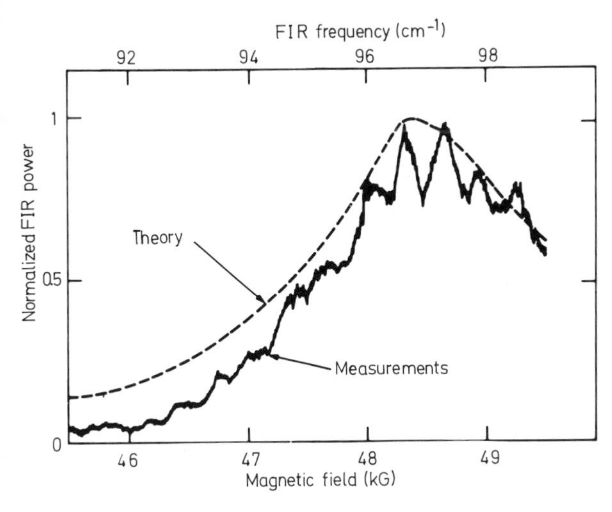

Fig. 4.13. Spectral scan from source shown in Figs. 4.11, 12

at the peak power in Fig. 4.13. There is approximate agreement between the theoretical and experimental curves. Periodic fluctuations in the experimental curve have a free spectral range corresponding to Fabry-Perot resonances in the mixing crystal. In estimating an absolute output power from (4.5), we used $P_{\omega_1}=2$ kW, $P_{\omega_2}=10$ W and $\omega_3=98\,\mathrm{cm}^{-1}$, obtaining a value of 6 μW peak power for P_{ω_3}. The experimental value of P_{ω_3} is 2 μW. This discrepancy is due mostly, for the experimental arrangement shown in Fig. 4.11, to the problem of spin resonance matching between the oscillator and the mixer crystal [4.5, 24]. The mixing resonance condition is automatically obtained only if the SFR oscillator and the mixing crystal are located in the same magnetic field and have the same g-factor. This condition may be expressed by

$$\omega_3=\omega_1-\omega_2=\mu_B gH=\omega_s \tag{4.11}$$

where μ_B is the Bohr magneton. In InSb, the g-factor and hence the spin resonance frequency ω_s depend on the electron concentration and the magnetic field mostly through the nonparabolicity of the conduction band [4.24]. The effective spin resonance or spin-flip frequency ω_s is given by

$$\omega_s=\omega_s(0)\left(1-\frac{2E_F}{3E_g}\right) \tag{4.12}$$

where $\omega_s(0)$ is the concentration independent spin resonance frequency and E_F is the Fermi energy. For an InSb crystal with an electron concentration $n_0>10^{15}\,\mathrm{cm}^{-3}$ at temperatures <20 K and for a magnetic field $H>20$ kG, E_F is given by [4.25]

$$E_F(\mathrm{cm}^{-1})=3.711\times10^{-23}\frac{n_0^2}{H^2} \tag{4.13}$$

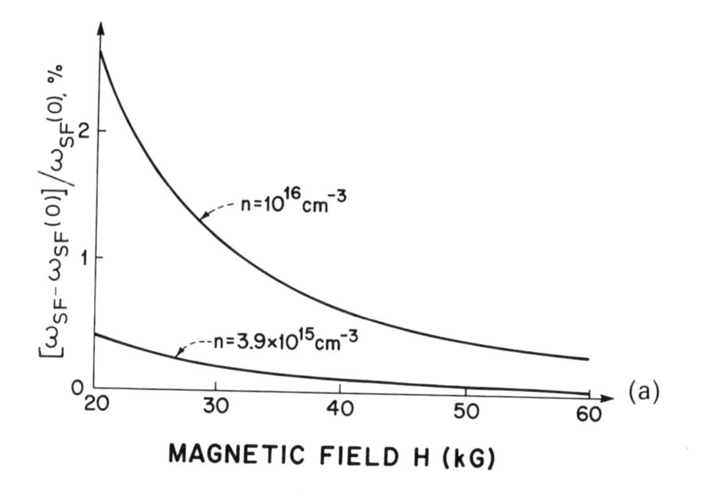

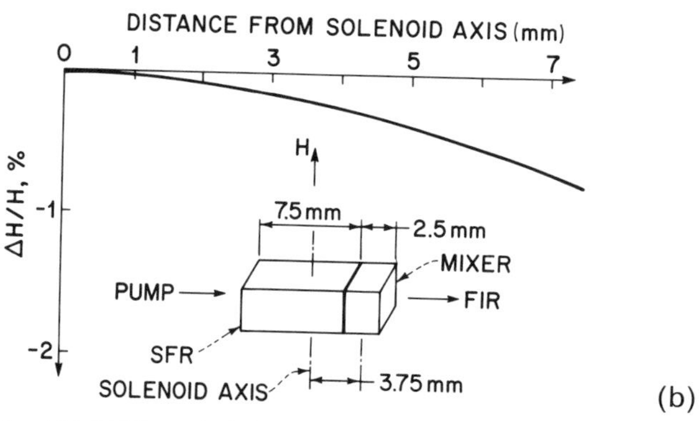

Fig. 4.14. (a) Theoretical deviation of the spin-flip frequency for two electron concentrations as a function of magnetic field. (b) Measured change of magnetic field of superconducting magnet as a function of distance from magnet axis. Inset shows physical arrangement of SFR laser ($n=1\times10^{16}$ cm^{-3}) and mixer crystal ($n=3.9\times10^{15}$ cm^{-3}) in the experiment of Figs. 4.11, 12

where n_0 is the electron concentration per cm^3 and H is the magnetic field in gauss. The curves of Fig. 4.14(a) show the percentage deviation $[\omega_s-\omega_s(0)]/\omega_s(0)$ as a function of the magnetic field H for two electron concentrations used in the FIR generation experiment described above. In Fig. 4.14(b) is shown the percentage variation of the magnetic field as a function of distance from the axis in the superconducting solenoid. The generated FIR power output was optimum at $H=48$ kG, at which point there is a difference between the spin resonance frequency of oscillator and mixer of 0.4% corresponding to 0.38 cm^{-1}. Since the spin resonance linewidth of the mixer was only 0.28 cm^{-1} [4.3], this would put the mixer completely off resonance. The error was partly compensated by placing the mixing crystal off-axis in the solenoid by 3.75 mm at a position of reduced magnetic field [see inset of Fig. 4.14(b)]. The net

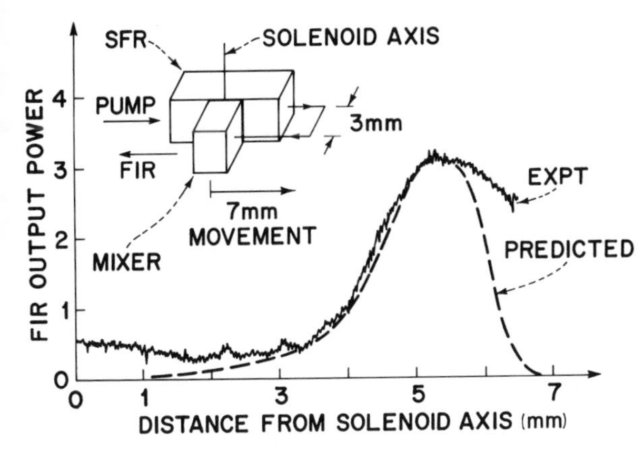

Fig. 4.15. Experimental FIR output (in arbitrary units), as a function of mixer crystal position, compared with theoretically predicted result. Inset shows the physical arrangement, the crystals being the same ones as shown in Figs. 4.11, 12, 14

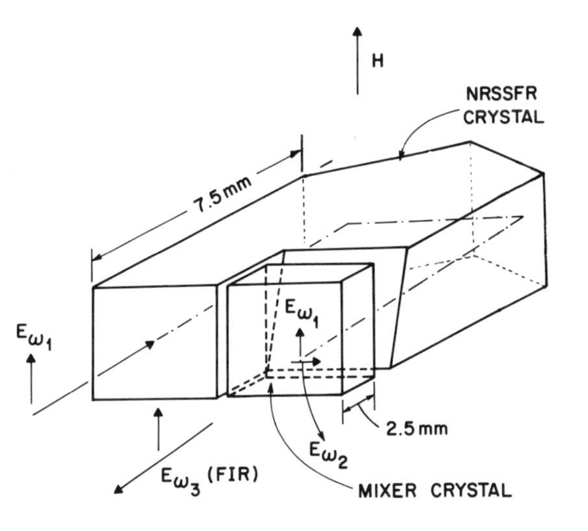

Fig. 4.16. Addition of mixing crystal to the NRSSFR laser of Fig. 4.8, to obtain true continuously tunable FIR

deviation was then 0.2% or 0.19 cm^{-1} resulting in a reduction of ~ 3 in FIR output from the maximum possible at the optimum value of field. This explains the discrepancy mentioned earlier between the predicted and observed FIR output power. To optimize the spin resonance matching, the experiment was repeated with an assembly in which the mixing crystal could be moved in the magnetic field. The arrangement is shown in the inset of Fig. 4.15. The field at the mixer could be varied continuously from 100% to 94% of the value at the oscillator by a movement of 6.5 mm. The experimental change of FIR power with movement shown in Fig. 4.15 indicates an optimum position of the mixer crystal 5.3 mm from the solenoid axis. Also shown in Fig. 4.15 is a predicted output curve calculated from data of Fig. 4.14(a, b) and from the linewidth of the mixing crystal (distances measured from center of crystals) with the amplitude the only fitted parameter. The optimum position of the crystal is

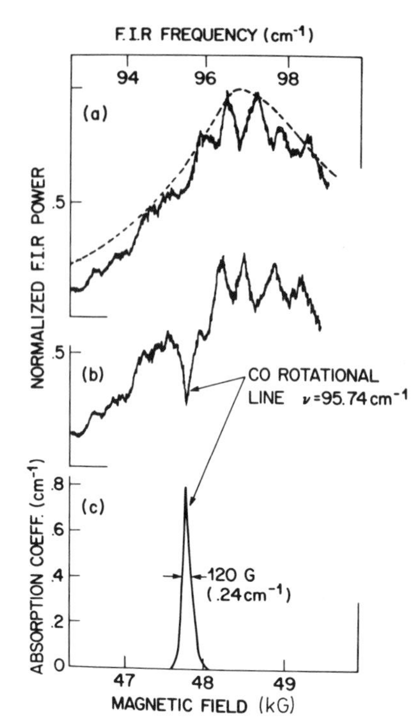

Fig. 4.17. (a) Tunable FIR output as a function of frequency compared with theoretical curve (dotted). (b) Absorption line of CO. (c) Apparent linewidth of CO absorption

reasonably well predicted, but the peak is broader than expected. This broadening may be explained by the multimode nature of the output of the SFR already mentioned in Subsection 4.2.1. The presence of two or three simultaneous longitudinal modes spaced by the 0.18 cm^{-1} free spectral range of the SFR cavity can substantially broaden the observed line. The peak power measured in this experiment at the optimum point was 20 μW, a factor of 10 greater than that of the previous experiment. About a third of this improvement is attributed to better spin resonance matching, and the remainder to increased pump power and improved SFR oscillator performance.

The nonresonant superradiant spin-flip Raman laser has also been used to generate FIR [4.20] which has true continuous tunability. The position of the mixer crystal is shown in Fig. 4.16. Spectral characteristics of this arrangement are discussed in the next subsection.

4.2.3 Linewidth and Tunability of Far-Infrared

The linewidth and the tunability of the generated FIR, which results from the difference frequency mixing between the pump and tunable Stokes radiation from a SFR laser, are determined by the linewidth and the tuning characteristics of the SFR laser. In Subsection 4.2.1, we have discussed the multi-longitudinal mode operation and the "jump" tuning of the 10.6 μm pumped

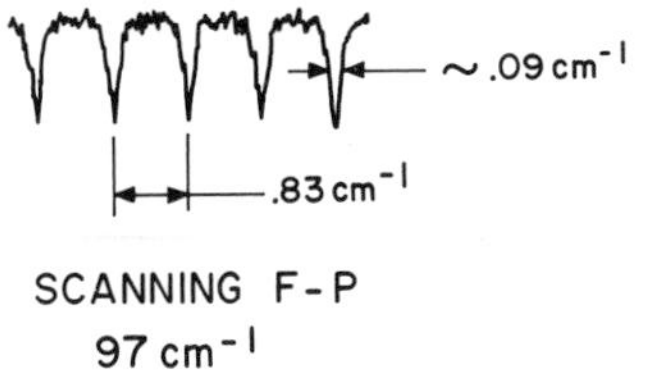

Fig. 4.18. Spectrum of continuously tunable FIR source of Fig. 4.16 taken with a Fabry-Perot interferometer with fixed metal mesh mirrors. The free spectral range of $0.83\ \mathrm{cm}^{-1}$ provides a calibration and the linewidth of the FIR ($0.09\ \mathrm{cm}^{-1}$) is the same (see Fig. 4.9) as that of the tunable output of the primary source of Fig. 4.8

SFR laser. These characteristics limit the usefulness of the generated tunable FIR as a spectroscopic tool as shown in a simple spectroscopic experiment. In this experiment, the pure rotational absorption lines of CO gas were observed. These lines are simple and widely spaced and so provide an uncomplicated test of the FIR source. In Fig. 4.17(a) we show the FIR output as a function of the magnetic field and of the FIR frequency at zero pressure of CO gas. The small pulling of the longitudinal modes mentioned previously (Subsect. 4.2.1) can be neglected on this scale. In Fig. 4.17(b) the $v=0$, $J=24$ absorption [4.26] at $95.74\ \mathrm{cm}^{-1}$ is shown, and in Fig. 4.17(c) the measurement is converted to absorption factor. The measured linewidth of $0.24\ \mathrm{cm}^{-1}$ was found to be independent of pressure up to 400 Torr. Since the true pressure broadened linewidth (full width at half height) is $0.05\ \mathrm{cm}^{-1}$ at 400 Torr [4.27], the measured linewidth is due to the generated FIR. The reason for this behavior is the presence, already mentioned in Subsection 4.2.1, of two or three longitudinal modes with spacing of $0.18\ \mathrm{cm}^{-1}$ in the $10.6\ \mu\mathrm{m}$ pumped SFR laser radiation.

We have discussed also in Subsection 4.2.1 the characteristics of the nonresonant superradiant spin-flip Raman laser used for the generation of the true continuously tunable FIR. The experimental arrangement is shown in Fig. 4.16. The linewidth and the continuous tuning of the NRSSFR laser were shown in Figs. 4.9 and 4.10. As expected, the linewidth of $0.09\ \mathrm{cm}^{-1}$ was found in the analysis of the FIR shown in Fig. 4.18 using a scanning FIR Fabry-Perot interferometer with metal mesh mirrors.

4.3 Concluding Remarks

In this chapter, we have shown how magnetically tunable FIR radiation at the spin resonance frequency in InSb can be obtained by a combination of splin-flip Raman laser action and subsequent nonlinear difference mixing in the electron spins. At present, phase-matching requirements have necessitated separation of the functions into two separate crystals, but in principle the spin-flip Raman laser itself could be a generator of FIR.

Experiments have been confined to pulsed operation using a CO_2 laser at $10.6\ \mu\mathrm{m}$ as a pump. However, in the same way that cw SFR laser action is obtained by pumping near the InSb bandgap at $5.3\ \mu\mathrm{m}$, it should be possible to generate cw FIR output. To do this would require some way of phase matching in the presence of the large dispersion of refractive index near the bandgap.

The combination of narrow linewidth ($<0.1\ \mathrm{cm}^{-1}$) and continuous or near continuous tunability offered by this InSb spin resonance source is unique for this frequency range and will be invaluable in the further investigation of the far-infrared region.

References

4.1 C.K.N. Patel, E.D. Shaw: Phys. Rev. **B 3**, 1279 (1971)
4.2 G.D. Boyd, T.J. Bridges, E. Buehler, C.K.N. Patel: Appl. Phys. Lett. **21**, 553 (1972), and references therein
4.3 V.T. Nguyen, T.J. Bridges: Phys. Rev. Lett. **29**, 359 (1972)
4.4 T.J. Bridges, V.T. Nguyen: Appl. Phys. Lett. **23**, 107 (1973)
4.5 N. Brignall, R.A. Wood, C.R. Pidgeon, B.S. Wherrett: Opt. Commun. **12**, 17 (1974)
4.6 A. Mooradian, S.R.J. Brueck, F.A. Blum: Appl. Phys. Lett. **17**, 481 (1970)
4.7 M.J. Colles, C.R. Pidgeon: Rept. Progr. Phys. **38**, 387 (1975)
4.8 Y. Yafet: Phys. Rev. **152**, 858 (1966)
4.9 T.L. Brown, P.A. Wolff: Phys. Rev. Lett. **29**, 362 (1972)
4.10 V.T. Nguyen, C.K.N. Patel: Phys. Rev. Lett. **22**, 463 (1969)
4.11 C.K.N. Patel, V.T. Nguyen: Appl. Phys. Lett. **15**, 189 (1969)
4.12 Y.R. Shen: Appl. Phys. Lett. **23**, 516 (1973)
4.13 See for example A. Yariv: *Quantum Electronics* (John Wiley and Sons, New York, London, Sydney 1967) pp. 352—356
4.14 E.D. Palik, G.B. Wright: In *Semiconductors and Semimetals*, Vol. 3, ed. by R.K. Willardson, A.C. Beer (Academic Press, New York, London 1967) pp. 421—458
4.15 S.R. Brueck, A. Mooradian: IEEE J. Quant. Electron. QE-**10**, 634 (1974)
4.16 P.S. Pershan: In *Semiconductors and Semimetals*, Vol. 2, ed. by R.K. Willardson, A.C. Beer (Acedemic Press, New York, London 1966) p. 283
4.17 W.J. Moore, H. Shenker: Infrared Phys. **5**, 99 (1965)
4.18 B.D. McCombe, R.J. Wagner: Phys. Rev. **B 4**, 1285 (1971)
4.19 C.K.N. Patel: In *Proc. Symp. Laser Physics and Applications, Esfahan*, Sept. 1971 (John Wiley and Sons, New York, London, Sydney 1973) pp. 689—721;
C.K.N. Patel: In *Coherence and Quantum Optics*, ed. by L. Mandel, E. Wolf (Plenum Press, New York, London 1973) pp. 567—593
4.20 V.T. Nguyen, T.J. Bridges: In *Laser Spectroscopy*, ed. by R.G. Brewer, A. Mooradian (Plenum Press, New York, London 1974) pp. 513—521
4.21 S.R.J. Brueck, A. Mooradian: Appl. Phys. Lett. **18**, 229 (1971)
4.22 J.T. Ganley, F.B. Harrison, W.T. Leland: J. Appl. Phys. **45**, 4980 (1974)
4.23 R.A. Wood, R.B. Dennis, J.W. Smith: Opt. Commun. **4**, 383 (1972)
4.24 V.T. Nguyen, T.J. Bridges: Appl. Phys. Lett. **26**, 452 (1975)
4.25 L.M. Roth, R.N. Argyres: In *Semiconductors and Semimetals*, Vol. 1, ed. by R.K. Willardson, A.C. Beer (Academic Press, New York, London 1966) p. 173
4.26 K.D. Möller, W.G. Rothschild: *Far-Infrared Spectroscopy* (Wiley-Interscience, New York, London, Sydney 1971) p. 51
4.27 D.A. Draegert, D. Williams: J. Opt. Soc. Am. **58**, 1399 (1968)

5. Optical Mixing in Atomic Vapors

J. J. Wynne and P. P. Sorokin

With 33 Figures

Vapors play a very different historical role in the field of nonlinear optics from the roles they have played in fields such as optical spectroscopy. Although vapors are much simpler in structure than solids, it has been solid media that have garnered all of the attention of experimental nonlinear optics from the advent of lasers, in 1960, until relatively recently. The main reason for this apparent anomaly can be found in a very simple symmetry argument. Quite naturally the first type of optical nonlinearity studied was a quadratic nonlinearity which resulted in second-harmonic generation of light [5.1]. It is well known that media possessing bulk inversion symmetry (i.e., centrosymmetric media) cannot have macroscopic quadratic optical nonlinearities (in the dipole approximation which we shall assume to hold for all of the phenomena to be discussed in this chapter).

Thus quadratic nonlinearities are nonvanishing only in acentric crystals, and these have been the subject of intensive study since the dawn of nonlinear optics in 1962. Nonlinear optics in centrosymmetric media were also studied quite early [5.2]. This requires a higher-order nonlinearity. The cubic or third-order nonlinearity is nonvanishing in all media. The study of nonlinear optics in vapors began with measurements of third-harmonic generation (THG) from atomic vapors irradiated with a high-power ruby laser beam [5.3, 4]. Vapor densities are very small compared to the high densities found in solids. The nonlinearities in the vapors were found to be small relative to those in solids, directly reflecting the relative densities. It is only with the growing availability and widespread use of tunable lasers [5.5] that vapors have "come into their own" as widely studied, physically interesting and "practical" nonlinear media. We shall now proceed to describe how to use nonlinear parametric mixing processes in vapors to extend greatly the tuning range of "practical" tunable laser sources.

The subject matter of this chapter is the generation of light by nonlinear optical mixing in atomic vapors. In Section 5.1 we describe some experimental apparatus and techniques. The vapors respond nonlinearly to the application of strong electric fields from laser beams. The response of interest is cubic. The induced nonlinear polarization corresponds to a phased array of electric dipoles which are capable of radiating. This polarization contains Fourier components which are the sum and difference combinations of the frequencies of the applied light fields, so that the radiated light will contain Fourier components that were not part of the input light. In Section 5.2 we shall review the theory of the cubic nonlinear susceptibility. Section 5.3 deals with infrared (IR) generation via stimulated Raman scattering and difference mixing. One advantage of working with atomic vapors is that they are completely transparent in the infrared since they have no vibrational or rotational modes to absorb the generated IR light. The optical nonlinearities are quite large due to resonance enhancement, and high conversion efficiencies are theoretically possible. However, there are other nonlinear effects which must be considered as the input lasers become more powerful, and the net result is an effective saturation of the nonlinearity, limiting the conversion efficiency. The spectral width of the generated infrared will be seen to be limited by the spectral width of the input lasers. An extension of this technique to sum mixing allows us to generate tunable vacuum ultraviolet (VUV) light with output intensities high enough to permit many spectroscopic applications. Ultraviolet generation is the subject of Section 5.4.

5.1 Experimental Techniques

All of the experiments carried out by us and described in this chapter have made use of nitrogen-laser-pumped dye lasers. A typical experimental configuration for VUV generation is shown in Fig. 5.1. A single nitrogen laser simultaneously pumps two independently tunable dye lasers. We have used both a 100 kW (Avco C 950) and a 1 MW (Molectron UV-1000) nitrogen laser. Depending on the dye used, the dye lasers vary in efficiency (for conversion of 3371 Å nitrogen laser light to dye laser output light) from ~5–20% throughout the visible. The nitrogen lasers have pulse widths of ~10 ns and repetition rates to 100 pps. The dye lasers are designed along the lines described by *Hänsch* [5.6] and incorporate transverse pumping, internal beam expansion, and a diffraction grating as one end reflector. A schematic diagram of our dye laser design is presented in Fig. 5.2. The nitrogen laser beam is shaped and focused onto the dye cell by a combination of a spherical and cylindrical lens. The dye cell is kept short (~1 cm) in the direction of lasing to minimize the problem of superradiant emission [5.6]. The spherical lens is set to match the long dimension of the nitrogen laser beam with the length of the dye cell. The cylindrical lens then shrinks the transverse dimension of the pump beam to produce a focal "line" which efficiently pumps the dye for the geometry being used. It is possible to "overpump" many dyes with the powerful 1 MW nitrogen laser, so, in practice, the pump beam is not focussed too tightly in the transverse dimension. The width of the focus is adjusted empirically to give optimum output power. The width of the focused line is typically ~0.3 mm. The gain in the strongly pumped region of the dye can be several orders of magnitude per pass, so that the laser may oscillate even with very small feedback. Therefore, it is necessary to prevent the windows of the dye cell from feeding back the Fresnel reflection. This is accomplished by tilting the windows at an angle of ~5° as shown in Fig. 5.2. The dye cells are constructed with inlet

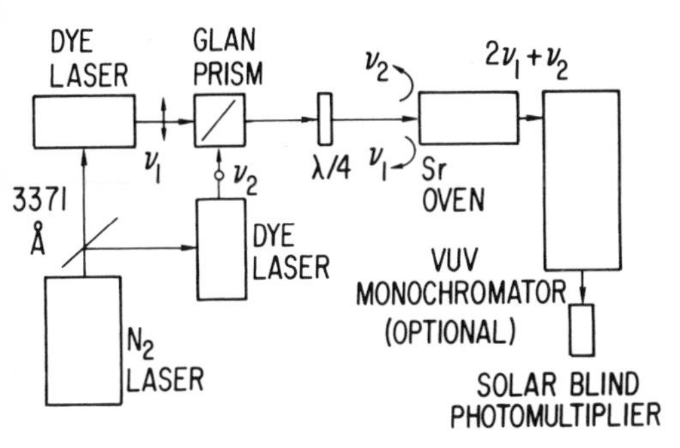

Fig. 5.1. Method of combining beams from two dye lasers, simultaneously pumped by a single nitrogen laser, to produce a single-frequency, tunable, coherent vacuum ultraviolet output (see Subsect. 5.4.3)

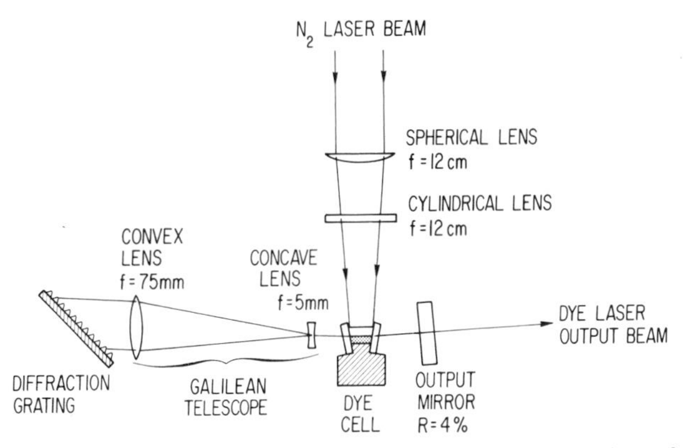

Fig. 5.2. Schematic diagram of tunable dye laser pumped by beam from nitrogen laser

and outlet pipes to provide dye flow transverse to the pumping and lasing directions. The dye region is represented by the dotted area of the dye cell in Fig. 5.2. The intracavity Galilean telescope expands the beam emerging from the dye cell by a factor ~15 so that it fills a sizable area of the diffraction grating, providing for improved resolution and, thereby, spectral narrowing. The spectral width of the output is dependent on the nature of the grating. We often use an echelle grating in the 9-th order with 316 lines/mm and blazed at 63°. In this manner output linewidths of ~0.1 Å are obtained.

Nitrogen-laser-pumped dye lasers are very easy to construct and align. The high gain and rapid repetition rate permit optimization of alignment by simple adjustments while the laser is running and the output power is being monitored. If narrower linewidths are desired, intracavity etalons may be used, but only at the expense of output power. For the experiments we have carried out, narrower linewidths are not required but high output power is desirable. Further information on nitrogen-laser-pumped dye lasers may be found in [5.7].

The only other feature of our experimental setups which is "unconventional" is the use of heat-pipe ovens to contain the atomic vapors. Such ovens were

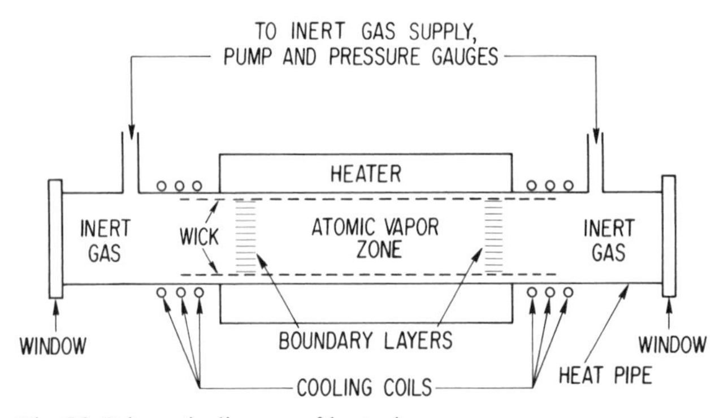

Fig. 5.3. Schematic diagram of heat-pipe oven

first developed by *Vidal* and *Cooper* [5.8] for spectroscopic applications. They provide a means of producing a well-defined, homogenous vapor region at temperatures which are ordinarily difficult to work with for optical transmission applications. A diagramatic representation of the oven is shown in Fig. 5.3. In the heat-pipe oven, heat is applied to a region of pipe where the atoms of interest are present in liquid form. The liquid evaporates, diffuses toward a colder region of pipe, and condenses. However, a path is provided for returning the liquid to the heated region by the presence of a "wick". The wick causes the liquid metal to flow against the thermal gradient by capillary action. The net effect is that the heat-pipe oven acts as a diffusion pump with the atomic vapor carrying any impurities out of the central vapor region. A well-defined boundary layer is established between the atomic vapor of interest and the inert gas which is used to establish the operating pressure (and temperature). The region between the two boundary layers consists of pure, homogenous, hot atomic vapor; the region between the boundary layers and the windows is pure, cold, inert gas.

To construct the heat-pipe ovens, we have used pipes made of #304 stainless steel with wicks made of 100 mesh, 4.5 mil stainless wire screen, for use with alkali metals. Pipes and wicks made of nickel were used for alkaline earth materials. Tantalum makes an excellent wick material for the alkaline earths, and for some rare earths. Pipe dimensions are not critical. The windows should be several cm from the cooling coils to minimize vapor condensation on the windows. The pipe should be long compared to its diameter. We have used pipes 1 meter in length with a 3 cm inner diameter. We have also used pipes 15 cm in length with a 1 cm diameter. The only difficulties occur when the melting point of the atomic system being studied is close to the operating temperature. Then the atomic vapor has a tendency to solidify and to build up at the location of condensation to the point where the pipe becomes optically blocked. This problem occurred for the materials Sr and Yb used in VUV generation experiments and necessitated scraping of the inside of the pipes before operation could be resumed.

To produce mixed vapor systems, we employed the concentric heat-pipe oven design of *Vidal* and *Hessel* [5.9]. One pipe containing the mixture (e.g., Na and K) was placed inside of another containing just one of the components (e.g., Na) of the center pipe. A single external heater warmed the outer pipe which, when operating in the heat-pipe mode, produced a constant temperature bath in which the inner pipe operated. By controlling the inert gas pressures of both pipes independently, controllable mixtures of the two atomic vapors were produced in the inner pipe. The two pipes had separate pressure and vacuum controls. The reader is referred to [5.9] and [5.10] for more details about the concentric heat-pipe operation and design.

Finally, we mention a simple technique which allows us to change rapidly from one atomic system to another when the solid form is not very reactive with air at room temperature. We use open-ended tantalum cassettes of small enough diameter to fit inside of the main heat pipe. We can simply take the cassette (with solid matter condensed on the inside) out of the main pipe and substitute another cassette with another material. For the alkaline earths and the rare earths Eu and Yb, we found this technique to be very advantageous. When using the cassettes, the heat pipes were not operated in the heat-pipe mode. Instead, a pressure of inert gas higher than the vapor pressure of the atomic system of interest was added to the pipe. This inert gas, plus small ($\sim$6 mm) apertures at the ends of the cassettes, reduced the rate of diffusion of the atomic vapor to the colder regions of the pipe. In practice, for vapor pressures of several torr, continuous operation could be maintained for days before it was necessary to scrape out the cassettes to reopen the optical aperture.

5.2 Theory of the Optical Nonlinearity in Vapors

Before discussing specific experiments, a general review of the theory relevant to the experiments is appropriate. The nonlinear response of a material system to applied light fields is described by the macroscopic nonlinear susceptibility tensor. This relates the macroscopic induced nonlinear polarization to the applied fields. Of course the material response is frequency dependent, and the most useful way to proceed is to Fourier analyze the time-dependent nonlinear polarization, producing expressions for the nonlinear susceptibilities which are functions of frequency. Following the approach used by many authors including *Bloembergen* [5.11], we may write

$$P_\alpha^{\mathrm{NL}}(\omega_4)=D\chi^{(3)}_{\alpha\beta\gamma\delta}(-\omega_4,\omega_1,\omega_2,\omega_3)\,E_\beta(\omega_1)\,E_\gamma(\omega_2)\,E_\delta(\omega_3)\,, \tag{5.1}$$

where $\omega_4=\omega_1+\omega_2+\omega_3$. P_α^{NL} is the α-th cartesian coordinate of the macroscopic cubic nonlinear polarization induced in a medium by the application of the electric fields E. The field and polarization Fourier components are related to the time-dependent field and polarization by the convention

$$E(t)=\sum_{\omega} E(\omega)/2\,\mathrm{e}^{-\mathrm{i}\omega t} \tag{5.2a}$$

and

$$P(t)=\sum_{\omega} P(\omega)/2\,\mathrm{e}^{-\mathrm{i}\omega t} \tag{5.2b}$$

where $E(-\omega)=E^*(\omega)$ and $P(-\omega)=P^*(\omega)$. D is a degeneracy factor which accounts for all permutations of the field components when the frequencies are not all identical. These equations provide a definition for the bulk cubic nonlinearity $\chi^{(3)}$ and are consistent with *Maker* and *Terhune* [5.12]. Note that $\chi^{(3)}$ as defined here is a factor 4 smaller than that defined by *Bloembergen* [5.11], and *Miles* and *Harris* [5.13]. There are general procedures for calculating the nonlinear susceptibilities [5.11]. These susceptibilities, calculated from time-dependent perturbation theory, consist of many different terms. Due to resonance enhancement, some of these terms will be much larger than the others, and it is possible to simplify greatly the expressions for the susceptibilities.

Several examples will serve to clarify what has just been said. Consider the simple case of THG from atoms having the simplified energy-level diagram depicted in Fig. 5.4. There are four electronic energy levels of interest, levels g (the only occupied level) and j', having the same parity which is opposite to that of levels j and j''. The nonvanishing dipole matrix elements are, as usual, between levels of opposite parity. Of the many possible terms which contribute to $\chi^{(3)}$, the only one of importance corresponds to the left-to-right order of the vertical arrows drawn in Fig. 5.4. This term may be simply written as

$$\begin{aligned}&\chi^{(3)}_{\alpha\beta\gamma\delta}(-3\omega,\omega,\omega,\omega)\\&\simeq\frac{N}{4\hbar^3}\frac{\mu^{\alpha}_{gj''}\mu^{\beta}_{j''j'}\mu^{\gamma}_{j'j}\mu^{\delta}_{jg}}{(\omega_{j''g}-3\omega-\mathrm{i}\Gamma_{j''g})(\omega_{j'g}-2\omega-\mathrm{i}\Gamma_{j'g})(\omega_{jg}-\omega-\mathrm{i}\Gamma_{jg})}\end{aligned} \tag{5.3}$$

or, in the case of frequency offsets large compared to Γ's,

$$\chi^{(3)}_{\alpha\beta\gamma\delta}(-3\omega,\omega,\omega,\omega)\simeq\frac{N}{4}\frac{\mu^{\alpha}_{gj''}\mu^{\beta}_{j''j'}\mu^{\gamma}_{j'j}\mu^{\delta}_{jg}}{(\Delta E_3)(\Delta E_2)(\Delta E_1)}. \tag{5.4}$$

In these expressions, μ^{ε}_{ab} is the ε'-th cartesian coordinate of the electric dipole matrix element between states a and b, N is the volume density of atoms, $\hbar\omega_{ab}$ is energy difference between states a and b, and Γ_{ab} is the damping constant which reflects the combined linewidth of the laser beam at frequency ω and the material linewidth due to various broadening mechanisms. Other terms which contribute to $\chi^{(3)}$ will have much larger energy denominators and thus be much less important. The lesson of this example is that when resonance

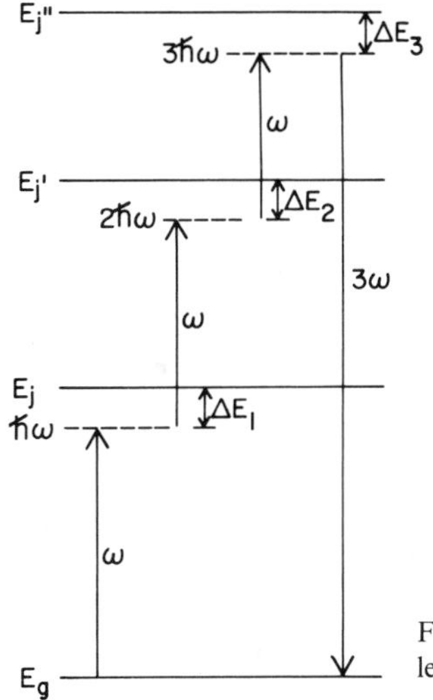

Fig. 5.4. Third-harmonic generation process from simple four-level system

is approached, the terms which display this resonance become the only important contributions to $\chi^{(3)}$.

Another example, which involves a difference frequency process, is shown in Fig. 5.5. Here three beams at frequencies ω_1, ω_2, and ω_3 are incident on an atomic medium where they produce a nonlinear polarization at $\omega_4=\omega_1-\omega_2-\omega_3$. The susceptibility describing this process is dominated by the following term:

$$\chi^{(3)}_{\alpha\beta\gamma\delta}(-\omega_4,\omega_1,-\omega_2,-\omega_3)\simeq\frac{N}{D4\hbar^3}\cdot\frac{\mu^{\alpha}_{gj''}\mu^{\beta}_{j''j'}\mu^{\gamma}_{j'j}\mu^{\delta}_{jg}}{(\omega_{j''g}-\omega_4-\mathrm{i}\Gamma_{j''g})(\omega_{j'g}-\omega_1+\omega_2-\mathrm{i}\Gamma_{j'g})(\omega_{jg}-\omega_1-\mathrm{i}\Gamma_{jg})} \tag{5.5}$$

or

$$\chi^{(3)}_{\alpha\beta\gamma\delta}(-\omega_4,\omega_1,-\omega_2,-\omega_3)\simeq\frac{N}{D4}\frac{\mu^{\alpha}_{gj''}\mu^{\beta}_{j''j'}\mu^{\gamma}_{j'j}\mu^{\delta}_{jg}}{(\Delta E_3)(\Delta E_2)(\Delta E_1)}\,. \tag{5.6}$$

Again this term is resonantly enhanced by making the energy denominators small.

The nonlinear susceptibilities tell us the magnitude of the various Fourier components of the nonlinear polarizations. To determine the intensities at the new frequencies which will emerge from the nonlinear medium, we must understand how the nonlinear polarizations lead to radiation. We are guided again by *Bloembergen* [5.11] who shows how to use the nonlinear polarizations as source terms in the wave equations. One wishes to deal with as few Fourier

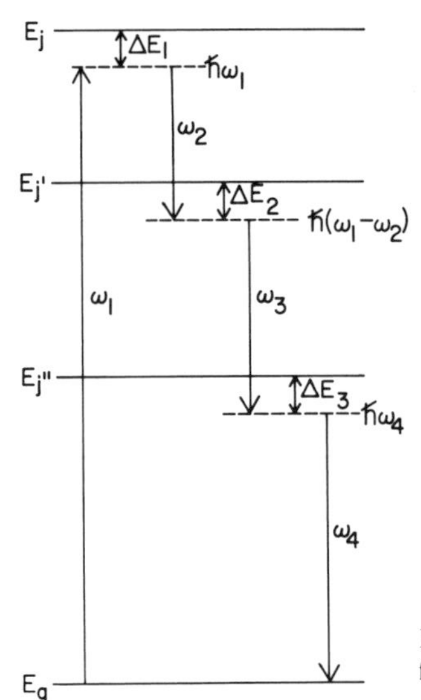

Fig. 5.5. Four-wave difference mixing process from simple four-level system

components as possible, so it is important to know to which other waves the wave of interest is coupled. There are usually good physical grounds for selecting a minimal number of important waves, thereby making the theoretical problem tractable.

Next, it is important to treat the problem of the growth of waves at new Fourier components by the method of coupled waves rather than just solving rate equations or finding transition probabilities. Those approaches are valid only when the phases of the waves are unimportant. An example where the phases are important is in the case of third-harmonic generation. An example where the relative phases are not important and where rate equations will give the correct answer is stimulated Raman scattering where the Stokes wave grows at the expense of the input laser with no dependence on the relative phases. Intimately related to the question of relative phases is the concept of phase matching. In those cases where relative phases are important, there is a driving nonlinear polarization propagating with a certain phase velocity which is tied to the phase velocities of the driving waves. The wave radiated by the nonlinear polarization wants to propagate freely with its own phase velocity. In general the two phase velocities are not equal and after a certain distance, called the coherence length, the waves get out of phase. The longer the coherence length, the greater the in-phase interaction length and the larger the generated signal. Those cases where relative phase is unimportant correspond to the nonlinear polarization automatically having the same phase velocity as the generated wave.

Lastly, one must be concerned about loss mechanisms for the various waves. Of greatest importance are linear absorptions of the light waves. Any direct absorption of the input laser beams reduces the resulting nonlinear polarization and thereby decreases the amount of generated radiation. In addition, direct linear absorption at the newly radiated frequencies will drastically decrease the output intensities. There is some tradeoff with the resonance enhancement of the nonlinear susceptibilities when one of the fundamentals or the generated signal is near resonance. Detailed analysis is required to see if one gains or loses in signal strength when the single-photon resonances are approached. In general, if the absorption is relatively weak, but if the nonlinearity is strongly resonantly enhanced by the proximity to this weak transition, then it pays to tune near to resonance. As for the problem of phase matching, proximity to a single-photon resonance introduces strong dispersive effects with concomitant effects on the coherence lengths for phase matching. While the problems of dispersion and absorption often make it undesirable to tune too close to single-photon resonant transitions, the nonlinearity may be resonantly enhanced without having to worry about the associated absorptive and dispersive processes by tuning near a two-photon resonance. *Maker* and *Terhune* [5.12] were the first authors to discuss resonant enhancement due to two-photon processes. As the reader shall see, the feature of two-photon resonance enhancement is common to all the non-linear generation processes we shall discuss in the balance of this chapter.

5.3 Infrared Generation

A difference-frequency mixing process of the type depicted in Fig. 5.5 provides a method of generating long-wavelength, tunable infrared radiation. With frequencies ω_1 and ω_2 chosen so as to minimize ΔE_2, an infrared output at ω_4 will tune in response to tuning ω_3. As ω_3 is tuned to approach $(E_{j'}-E_g)/\hbar=\omega_{j'g}$, ω_4 approaches 0 so that the infrared output tunes into the far infrared.

One method of automatically satisfying the condition of small ΔE_2 is to have ω_2 generated by stimulated Raman scattering. When the intensity of an input laser beam of frequency ω_1 is high enough, coherent light will be emitted at frequency $\omega_2=\omega_1-\omega_0$, where ω_0 is the natural frequency of a Raman active excitation. This effect has been thoroughly studied and good reviews are available in the literature [5.14]. In atomic vapors the only modes of excitation are electronic, and Raman scattering occurs with the atoms making a transition from an occupied state to an unoccupied state of the same parity. The first observations of stimulated electronic Raman scattering (SERS) were reported by *Sorokin* et al. [5.15] and *Rokni* and *Yatsiv* [5.16].

We have studied SERS in alkali metal vapors using a tunable dye laser as the input light. Threshold for SERS is exceeded over a wide tuning range so that the Raman scattered light at frequency ω_2 constitutes a tunable infrared source.

Infrared generation, using SERS both as the direct source of the infrared and as a part of various four-wave mixing processes, will be discussed in this section.

5.3.1 Infrared Generation by Stimulated Electronic Raman Scattering

The process of Raman scattering is one in which an input photon is inelastically scattered from a material excitation with a Stokes photon emitted. Normal Raman scattering is incoherent, with each scattering site acting independently of the others. The total scattered light is proportional to the volume density of scattering centers. The elementary process of Stokes emission is depicted in Fig. 5.6. In the case of electronic Raman scattering, the energy levels in Fig. 5.6 correspond to discrete electronic states of the material system, and the Raman active excitation is the electronic excitation from level g to level f. The stimulated process becomes important when the rate of Raman scattering is such that the number of Stokes photons exceeds the number of vacuum electromagnetic modes in the frequency interval of the scattered light. Then the scattering sites emit Stokes photons in phase with the photons already present. The theoretical framework built around coupled waves and nonlinear polarizations serves to describe the growth of the coherent Stokes wave. This problem has been treated in detail by *Shen* and *Bloembergen* [5.17]. We will adopt this approach to describe SERS, which can alternatively be described by the rate equation approach.

In our studies the input laser is nearly resonant with an allowed transition from the ground state. This resonantly enhances the nonlinear susceptibility $\chi^{(3)}(-\omega_S, \omega_L, -\omega_L, \omega_S) = \chi^{(3)}(\omega_S)$ responsible for the nonlinear polarization at the Stokes frequency. The growth of the stimulated Stokes wave can be described entirely in terms of the process depicted in Fig. 5.7. The resonant enhancement comes from having ΔE_2 be identically zero (where we are treating only the on-resonance or maximum gain case), and from the reduction of ΔE_1 and ΔE_3 when ω_L is close to resonance with the allowed transitions from the ground state, g, to levels j and j''. Allowing for several closely spaced levels around E_j and $E_{j''}$, explicit calculation [Ref. 5.11, pp. 37–44] gives

$$\begin{aligned}\chi^{(3)}(\omega_S) &= \frac{-\mathrm{i}Ne^4}{24\hbar^3\Gamma_{j'g}} \sum_{j,j''} \frac{x_{gj''}x_{j''j'}x_{j'j}x_{jg}}{(\omega_{j''g}-\omega_L+\mathrm{i}\Gamma_{j''g})(\omega_{jg}-\omega_L-\mathrm{i}\Gamma_{jg})} \\ &= \frac{-\mathrm{i}Ne^4}{24\hbar^3\Gamma_{j'g}} \left| \sum_j \frac{x_{gj}x_{jj'}}{\omega_{jg}-\omega_L+\mathrm{i}\Gamma_{jg}} \right|^2 \end{aligned} \quad (5.7)$$

when all waves are linearly polarized in the x direction. The x_{ab} are the matrix elements of x. The factor D is equal to 6 in this case. As we shall see when numerical values are calculated for specific atomic systems with $N \sim 10^{17}$/cc, $\chi^{(3)}(\omega_S)$ may easily exceed that of a very nonlinear solid. Equation (5.7) shows that $\chi^{(3)}(\omega_S)$ is proportional to the product of matrix elements x_{ab}, and inversely proportional to the linewidth $\Gamma_{j'g}$. This linewidth is an effective linewidth related to the material linewidth for the Raman transition from g to j' and the laser linewidth. (We assume Lorentzian lineshapes throughout). In the absence of various active broadening mechanisms such as multiphoton ionization

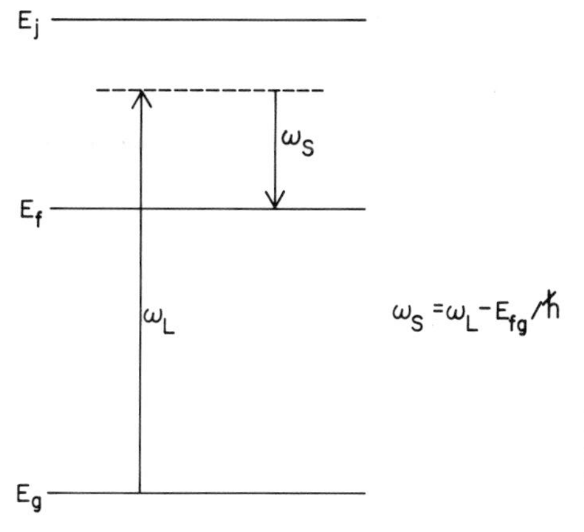

Fig. 5.6. Elementary Raman-Stokes process seen as inelastic scattering from simple three-level system

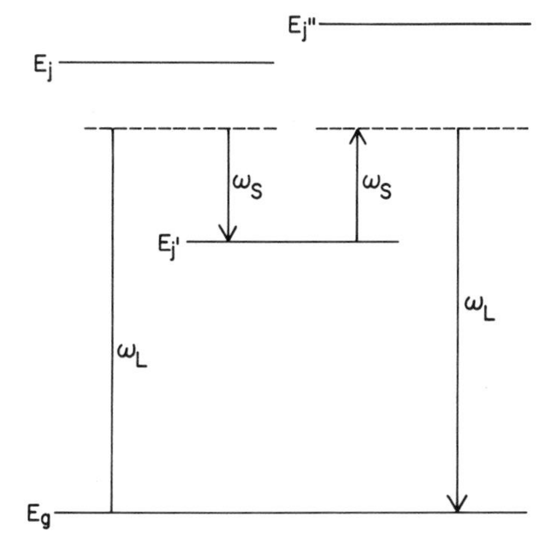

Fig. 5.7. Coherent Raman-Stokes generation process seen as four-wave interaction

(see Sect. 5.4), the material linewidth is usually small compared to the laser linewidth ($\sim 0.1\ \mathrm{cm}^{-1}$ or greater) so that all the Γ's can be taken as equal to the laser linewidths. Note that, except right near the one-photon resonances, Γ_{jg} and $\Gamma_{j''g}$ may be neglected. Note also that the nonresonant contributions to $\chi^{(3)}$ are negligible and are ignored here and throughout this chapter.

From $\chi^{(3)}(\omega_S)$ one can determine the nonlinear polarization at the Stokes frequency and then, by solving the driven wave equation, derive the gain of the Stokes wave. The output power at ω_S is expected to increase as the gain increases, but the detailed relationship between Stokes power and gain has not been treated theoretically with much success. One can construct a simple model, and we shall do this below, but the results of the model differ significantly from experiment. We shall first discuss our experimental results.

Experimental Results

The alkali metal vapors are the material systems we have studied. The energy-level diagram for K atoms is shown in Fig. 5.8. With an input laser tuned near the $5p$ resonance lines (4044, 4047 Å) there is strong resonance enhancement of the SERS process, with the K atoms making a transition from the ground $4s$ state, to the $5s$ state, the $5p$ states acting as the near-resonant intermediate states. The Stokes shift for this transition is 21026.8 cm^{-1}, and the Stokes wave falls in the infrared near 3700 cm^{-1} or 2.7 μ. With an input beam from a nitrogen-laser-pumped dye laser, we have observed SERS over a tuning range of $>1000\,cm^{-1}$. The dye laser utilized a solution of pyridine derivative dye called CSA-35 [5.18], and produced a multimode beam of 80 kW at ~4040 Å with a linewidth of ~0.3 cm^{-1}. The dye output power fell to half its peak value at ~4000 Å and ~4375 Å. The beam was focused into a heat-pipe oven [5.8] containing K vapor at ~10 torr. The focusing lens had a focal length of 35 cm. The focused beam was ~200 μ diameter at its waist and doubled in area in 2.5 cm. Utilizing this dye laser, SERS was observed in the forward direction as the laser was tuned from 3973 Å to 4149 Å, a 1087 cm^{-1} tuning range, with the Stokes tuning from 2.40 μ (4162 cm^{-1}) to 3.25 μ (3075 cm^{-1}). The Stokes wave was detected with an InSb photoconductive detector, and showed a maximum output when the laser was tuned to 4032 Å. Using a lower power dye laser (~5 kW) we have measured the relative Stokes power as a function of $\nu_L = \omega_L/2\pi c$[1], and the results are plotted in Fig. 5.9. Earlier results by *Sorokin* et al [5.19] were taken with even lower power (~1 kW), with the Stokes power being detected by a PbS photodetector with rapidly changing responsivity near 2.8 μ. This detector weighted the shorter wavelengths too heavily. From Fig. 5.9 one sees that the Stokes power increases as one approaches resonance from either side, until a peak is reached, whereupon the output falls dramatically as one tunes still closer to the resonance lines. Deep minima occur exactly at resonance. There is a local maximum between the resonance lines. The Stokes output tunes over a wider range at the higher pressure, as is to be expected since $\chi^{(3)}(\omega_S)$ is proportional to N. The appearance of the structure between 24625 cm^{-1} and 24700 cm^{-1} is due to CO_2 and H_2O absorption in the atmosphere. The path length through air between the detector and the Raman source was ~1 m.

The same process of SERS can be studied from analogous energy levels in Rb vapor. With an input dye-laser beam tuned near the $6p$ response lines (4202 Å, 4215 Å), the atoms undergo SERS, making transitions from the ground $5s$ state to the $6s$ state with a Stokes shift of 20133.6 cm^{-1}. As in the case of K, the Stokes wave falls in the infrared near 3700 cm^{-1}. We have carried out a study of the Stokes gain and the Stokes linewidth in this system. In addition we have made an absolute measurement of the Stokes output power and can thereby determine the conversion efficiency of the Stokes generation

[1] ν in cm^{-1} will be used instead of ω wherever quantitative values or graphical representations of dispersion are used.

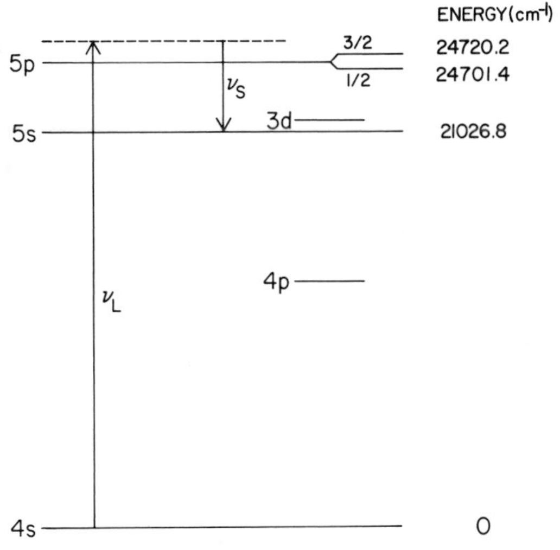

Fig. 5.8. Energy level diagram for K atoms showing Raman process from 4*s* ground state to 5*s* excited state, resonantly enhanced by 5*p* states

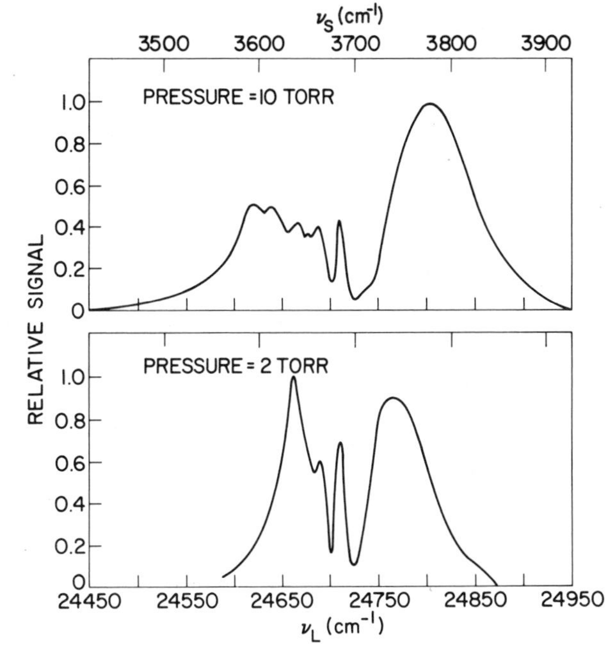

Fig. 5.9. Stokes generation as a function of frequency for two different pressures of K vapor

process. For these experiments the dye laser power was sufficiently high so that tight focusing was not necessary to exceed the threshold for SERS. Almost all the data were taken using a collimated laser beam ~2 mm in diameter.

Using an oven with a 60 cm long zone of Rb vapor at ~7 torr pressure, and a laser input power of ~40 kW (or 1.3 MW/cm^2), we measured the relative Stokes output, with an InSb photovoltaic detector. The data are plotted as a

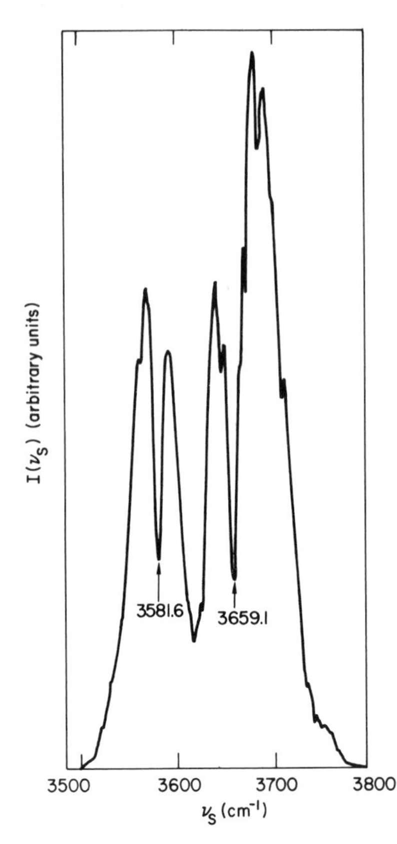

Fig. 5.10. Stokes generation as a function of frequency from Rb vapor at ~ 7 torr pressure. The dips at 3581.6 and 3659.1 cm^{-1} correspond to the laser being at resonance with the $6p_{1/2}$ and $6p_{3/2}$ lines, respectively

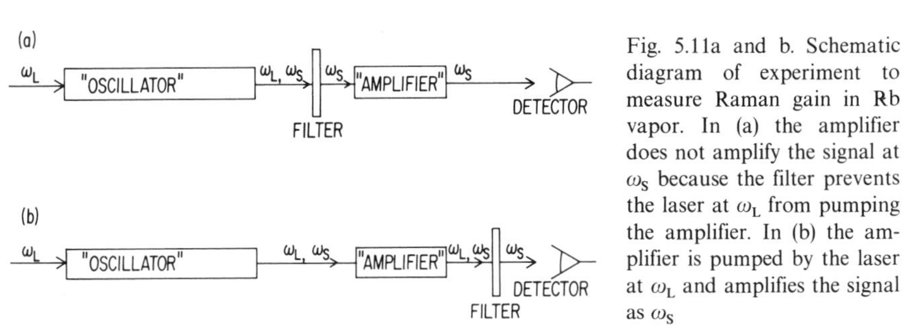

Fig. 5.11a and b. Schematic diagram of experiment to measure Raman gain in Rb vapor. In (a) the amplifier does not amplify the signal at ω_S because the filter prevents the laser at ω_L from pumping the amplifier. In (b) the amplifier is pumped by the laser at ω_L and amplifies the signal as ω_S

function of frequency in Fig. 5.10. This curve is characterized by a tuning range of $\sim 300\ cm^{-1}$, deep dips at the frequencies corresponding to on-resonance pumping, and sharp structural features due to atmospheric H_2O absorption. The $6p$ resonances are 77.5 cm^{-1} apart and the overall curve looks like two separate resonances brought close together. The fact that the central dip between resonances does *not* go to zero *is* significant, in that theory predicts a zero, as will be seen below.

Table 5.1. Transmission T and experimental and theoretical gain g_S/I for various laser frequencies ν_L in Rb vapor at a density of $1.45 \times 10^{17}/cm^3$

ν_L (cm^{-1})	$T(\nu_L)$-long pipe	Exp. $g_s/I(\nu_L)$ (cm/Watt)	Theory $g_s/I(\nu_L)$ (cm/Watt)
23659.4	0.96	4.2×10^{-7}	5.6×10^{-7}
23674.9	0.83	4.1×10^{-7}	9.1×10^{-7}
23740.0	0.63	3.6×10^{-7}	2.9×10^{-9}
23826.8	0.51	3.3×10^{-7}	2.5×10^{-6}
23845.4	0.79	3.0×10^{-7}	1.1×10^{-6}
23885.7	1.00	6.5×10^{-7}	4.1×10^{-7}

By focusing into Rb vapor, the tuning range may be widened, but the unfocused geometry facilitated gain measurements. To measure the gain a second Rb vapor oven was employed. This oven had only a 6 cm long Rb vapor region, and while it would "oscillate" with 40 kW input laser power, it would not "oscillate" with input powers of ~ 10 kW. A longer oven (with 1 torr Rb) was used as an oscillator to provide an input Stokes signal for the shorter oven (with 10 torr Rb) which then served as an amplifier. The experimental configuration is shown in Fig. 5.11. The filter is an infrared, long-pass filter which blocks all wavelengths shorter than 2.5 μ (including visible light). Thus in Fig. 5.11a, the Stokes light from the oscillator passes through the amplifier cell and is not amplified because the filter blocks the laser pump at ω_L. When the filter is moved to the configuration of Fig. 5.11b, the amplifier is pumped by the laser and the Stokes light is amplified. Since the infrared linear transmission is the same in both configurations, the signal increase is a direct measure of the gain. The measured gain was a strong function of frequency since the nonlinear susceptibility is strongly frequency dependent and since the transmission of laser light at ω_L is also strongly frequency dependent (due to direct absorption of the laser). From these data we have extracted measured values for gain for various values of frequency and the results are presented in Table 5.1. Discussion is deferred until a theoretical treatment of the gain has been presented.

By using a pyroelectric detector with flat response between the visible and infrared, we were able to measure the conversion efficiency from ω_L to ω_S in the short oven with 4 torr of Rb vapor. The vapor region was 6 cm long. With an input power of 70 kW at 4196 Å (23832 cm^{-1}) in a collimated beam 2 mm in diameter, we measured a power conversion efficiency of 2.8% to Stokes light at 2.704 μ (3698.4 cm^{-1}). Thus we have a multikilowatt source of infrared in a well-defined tuning range near 2.7 μ.

To test the utility of the infrared Stokes output for spectroscopic purposes, we constructed a 3 m long evacuable pipe with sapphire windows and used it to measure the absorption spectrum of CO_2 vapor. The Stokes beam was divided by a Ge beam splitter with one beam going to the pyroelectric detector and the other passing through the CO_2 filled pipe to the InSb de-

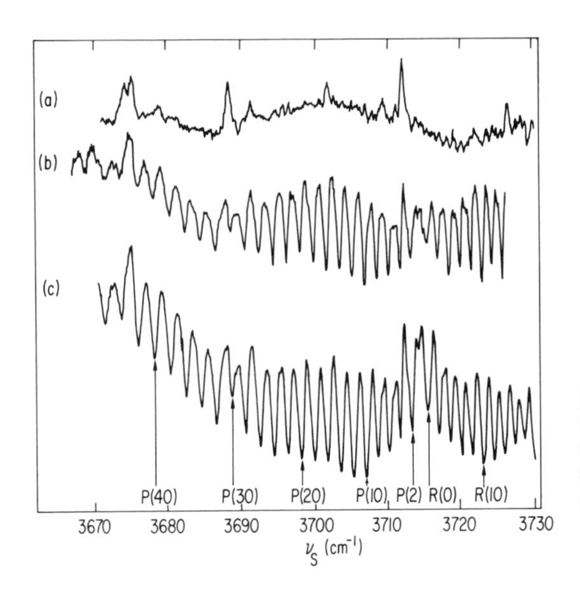

Fig. 5.12. Stokes light transmitted by a 3 m pipe containing (a) no CO_2, (b) 15 torr CO_2, and (c) 50 torr CO_2. (See text for explanation of structure)

tector. Corning 7–69 filters were used to attenuate the signal to the InSb detector, and the signal from the InSb was normalized by dividing by the signal from the pyroelectric detector. The laser was adjusted for a spectral width of $\sim 0.4\ \mathrm{cm}^{-1}$ (as measured with a Fabry-Perot), and the Stokes output was scanned across some *P* and *R* rotational branches of the $10^\circ 1$–$00^\circ 0$ combination band in CO_2. These bands are well studied [5.20, 21] and the various lines are well resolved and known to be much narrower than $\sim 0.4\ \mathrm{cm}^{-1}$ (the laser linewidth). Thus they serve to determine the effective Stokes linewidth.

The data are presented in Fig. 5.12. Curve (a) is the baseline curve taken with no CO_2 in the pipe. Due to the use of two very different types of detectors with different spectral responses and because of different optics in the two paths, there is some structure on the curve in addition to the background noise. The sharp structure does serve as an aid in checking the calibration points on the different curves. For curves (b) and (c), 15 torr and 50 torr, respectively, were admitted to the pipe. These curves clearly show the *P* and *R* branches. The FWHM linewidth for curve (b) is $\sim 0.4\ \mathrm{cm}^{-1}$ while curve (c) shows slight pressure broadening to $\sim 0.5\ \mathrm{cm}^{-1}$.

It should be possible to further narrow the spectral width of the laser to obtain a spectrally narrower Stokes width if higher resolution is desired and power can be sacrificed. However, recent results [5.22] in which the above experiment was duplicated, indicate that the Raman linewidth was $\sim 0.4\ \mathrm{cm}^{-1}$ even though the laser was only $\sim 0.1\ \mathrm{cm}^{-1}$ wide. Nevertheless, the utility of this infrared source will be in its inherent high peak power in combination with tunability. Compared to a tunable lead-salt diode laser, the Stokes source is spectrally broader but also of much higher (pulsed) power. The high power is useful for nonlinear optical applications.

Limiting Effects

The strong dips in Stokes output from both K and Rb as the laser is tuned onto resonance are correlated with increased linear absorption of the pump power. There is also another important effect with which these dips are correlated. Referring to Fig. 5.8, note the position of the $3d$ levels. We have failed to observe K atoms taking part in SERS with the atoms making a transition form $4s-3d$. Only the $4s-5s$ transition participated in SERS. However, as the laser is tuned closer to the $5p$ resonance lines, optically pumped stimulated emission (OPSE) is observed on the $5p-3d$ transitions. This OPSE is maximum when the laser is tuned on resonance, whereas SERS shows a minimum for both the 1/2 and 3/2 lines. It is unclear why the transition to the $5s$ level should result in SERS whereas that to the $3d$ level results in OPSE. *Wynne* and *Sorokin* [5.23] have studied the pumping of higher lying np levels in K, and they observed OPSE to both s and d levels. We do not understand the reasons for OPSE to be favored over SERS to different final states. The same effect is observed in Rb, with the atoms undergoing OPSE on the $6p-4d$ transitions.

Another factor which affects the SERS output is multiphoton ionization. The laser photons have enough energy to ionize the atoms by single photon absorption from the $5s$ excited state of K or the $6s$ excited state of Rb, or by two photon absorption from the ground state. This latter process is resonantly enhanced when the laser is tuned near the $5p$ (in K) or $6p$ (in Rb) resonances. The resulting ions and free electrons will broaden the discrete levels by the Stark effect (Sec. 5.4) causing a decrease in the SERS gain. The magnitude of these effects remains to be calculated.

As the Stokes output power increases, more atoms make transitions from the ns ground state to the $(n+1)s$ excited state. The importance of this population change is manifested by the occurrence of OPSE from the $(n+1)s$ state to the np states. The output in K occurs at 1.253 μ and 1.244 μ and is observed to be quite intense. This signals a saturation phenomenon which has been studied and explained by *Tyler* et al. [5.24]. The OPSE process represents a rapid relaxation of the $(n+1)s$ state population which effectively limits the contribution of this state to the nonlinear polarization. This effect is most important because of the limitations it places on infrared generation by four-wave mixing (Subsect. 5.3.3).

When the $(n+1)s$ state population is not negligible compared to the ground state population, the effect is to decrease the gain and correspondingly decrease the Stokes output from what it would be if the gain did not saturate. A simple way to determine whether this effect is important is to measure the total number of Stokes photons emitted and compare it to the total number of atoms in the strongly pumped region. With a focused input beam having a focal volume of $1.5\times10^{-3}\,\mathrm{cm}^3$, and a density of $6\times10^{16}/\mathrm{cm}^3$ (corresponding to $T=400\,^\circ\mathrm{C}$ and $p=4$ torr), there are 9×10^{13} K atoms in the focal region. This number of Stokes photons at $3700\,\mathrm{cm}^{-1}$ represents an energy of

6.6×10^{-6} Joule, which, emitted in 7.5 ns (the duration of the laser pulse), gives a power of ~ 1 kW. So our observation of several kW of output power implies near saturation and means that atoms in the less intensely pumped region of the focused laser beam are contributing to the output.

Calculation of Stokes Gain and Output Power

It is a difficult problem to calculate the Stokes output power with any accuracy. However, it is relatively straightforward to calculate the Stokes gain. The procedure using coupled waves is fairly standard [Ref. 5.11, pp. 102–110], and we shall briefly present a derivation of the relationship between the gain and $\chi^{(3)}(\omega_S)$. We start with the driven wave equation

$$\nabla^2 E_x(\omega_S) + (\omega_S^2/c^2) E_x(\omega_S) = -(4\pi\omega_S^2/c^2) P_x^{NL}(\omega_S) \tag{5.8}$$

where

$$P_x^{NL}(\omega_S) = 6\chi_{xxxx}^{(3)}(\omega_S)|E(\omega_L)|^2 E(\omega_S)\,. \tag{5.9}$$

The dielectric constant has been set equal to 1. With the laser in the form of a plane wave of constant amplitude (i.e., neglecting absorption) propagating in the z direction, we seek an exponentially growing Stokes wave of the form

$$E_x(\omega_S) = E_{x0}(\omega_S) e^{i(\boldsymbol{k}_S \cdot \boldsymbol{r} - \omega_S t)} \tag{5.10}$$

where $E_{x0}(\omega_S)$ is independent of $\boldsymbol{r}$. Then (5.8) yields

$$-k_S^2 + k_{S0}^2 = -(4\pi\omega_S^2/c^2) 6\chi_{xxxx}^{(3)}(\omega_S)|E(\omega_L)|^2 \tag{5.11}$$

where $k_{S0}^2 = \omega_S^2/c^2$. Neglecting transverse components we write $k_S = k_{S0} + \Delta\kappa$ and we expect $\Delta\kappa \ll k_{S0}$ so that

$$k_S^2 \simeq k_{S0}^2 + 2k_{S0}\Delta\kappa\,. \tag{5.12}$$

Putting (5.12) into (5.11) yields

$$\Delta\kappa = (2\pi\omega_S/c) 6\chi_{xxxx}^{(3)}(\omega_S)|E(\omega_L)|^2\,. \tag{5.13}$$

Since $\chi^{(3)}(\omega_S)$ is negative imaginary (5.7), $i\Delta\kappa$ is positive real and the Stokes wave shows exponential growth. We remind the reader that the nonresonant contributions to $\chi^{(3)}(\omega_S)$ are negligible. The Stokes power gain coefficient is just

$$g_S = 2|\Delta\kappa| = (4\pi\omega_S/c)|6\chi_{xxxx}^{(3)}(\omega_S)|\,|E(\omega_L)|^2\,. \tag{5.14}$$

Equation (5.13) is identical to the result given by (4.68) of [5.11].

Table 5.2. Dipole matrix elements and energies of various states for the alkali metals

Alkali metal	Li	Na	K	Rb	Cs
n	2	3	4	5	6
$\langle ns0\,1/2\|x\|n+1p0\,1/2\rangle$ (10^{-8} cm)	0.0651	−0.124	−0.138	−0.177	−0.195
$\langle n+1p0\,1/2\|x\|n+1s0\,1/2\rangle$ (10^{-8} cm)	−3.16	−3.12	−3.54	−3.61	−3.80
$\nu_{(n+1)p_{1/2}}$ (cm^{-1})	30925.38	30266.88	24701.44	23715.19	21765.65
$\nu_{(n+1)p_{3/2}}$ (cm^{-1})	30925.38	30272.51	24720.20	23792.69	21946.66
$\langle n+1s0\,1/2\|x\|np0\,1/2\rangle$ (10^{-8} cm)	0.905	1.33	1.42	1.53	1.59
$\langle np0\,1/2\|x\|ns0\,1/2\rangle$ (10^{-8} cm)	−1.24	−1.33	−1.57	−1.62	−1.74
$\nu_{np_{1/2}}$ (cm^{-1})	14903.66	16956.18	12985.17	12578.96	11178.24
$\nu_{np_{3/2}}$ (cm^{-1})	14904.00	16973.38	13042.89	12816.56	11732.35

To calculate g_S we will first calculate values for $\chi^{(3)}(\omega_S)$. *Miles* and *Harris* [5.13] have calculated $\chi^{(3)}(3\omega)$ in alkali metal vapors for third-harmonic generation. Using their procedure, and being guided by *Miles* [Ref. 5.25, pp. 124–130] to account for the $np_{1/2}$ and $np_{3/2}$ levels separately, one can calculate $\chi^{(3)}(\omega_S)$ from (5.7). In the alkali metals, taking $ns\ {}^2S_{1/2}$ as the g state, $(n+1)p\ {}^2P_{1/2,3/2}$ as the j states, and $(n+1)s\ {}^2S_{1/2}$ as the j' state, (5.7) reduces to

$$|\chi^{(3)}_{xxxx}(\omega_S)| = 3.145\times 10^7 \frac{N}{\Gamma_{(n+1)s,\,ns}}[2(\nu_{(n+1)p_{3/2}}-\nu_L)^{-1} + (\nu_{(n+1)p_{1/2}}-\nu_L)^{-1}]^2 \times |\langle ns0\,1/2|x|n+1p0\,1/2\rangle \cdot \langle n+1p0\,1/2|x|n+1s0\,1/2\rangle|^2 \text{esu}\,, \tag{5.15}$$

where $\Gamma_{(n+1)s,\,ns}$ and ν_i are now in units of cm^{-1}, N is atomic density in atoms/cm^3, and the matrix elements are taken between states expressed in the $|nlm_lm_s\rangle$ notation, with x the axis of quantization. These matrix elements have been tabulated by *Miles* and *Harris* [5.13], and in Table 5.2 we list the appropriate values for the alkali metals. The $(n+1)p$ state energies are taken from *Moore* [5.26]. For Rb atoms at a pressure of 10 torr at 665 K, $N = 1.45\times 10^{17}/\text{cm}^3$. Assuming a laser linewidth of 0.5 cm^{-1} (for Γ), we calculate $\chi^{(3)}(\omega_S)$ from (5.15). With the relationship

$$|E(\omega)|^2 = 8\pi I(\omega)/c \text{ esu} \tag{5.16}$$

we can calculate $g_S/I(\omega_L)$ from (5.14). The results are presented in Figs. 5.13 and 5.14.

As mentioned above, $\chi^{(3)}(\omega_S)$ exceeds values found in very nonlinear solids. Measurements in Ge with a 10 μ laser give the result $\chi^{(3)} = 10^{-10}$ esu [5.27]. Figure 5.13 shows that this value is exceeded as ν_L is tuned close to resonance. The theoretical values for the gain are compared to experiment at several frequencies in Table 5.1. The values are in good agreement when ν_L is far enough from resonance so that the Rb is highly transparent. Closer to

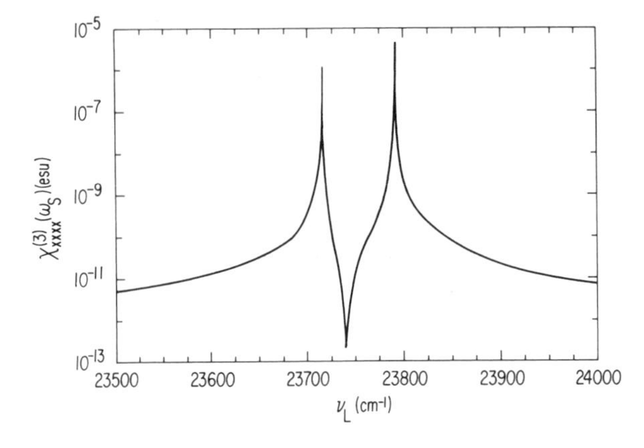

Fig. 5.13. Theoretical values for $\chi^{(3)}_{xxxx}(\omega_S)$ as a function of ν_L for Rb vapor at a density of 1.45×10^{17} cm^{-3}

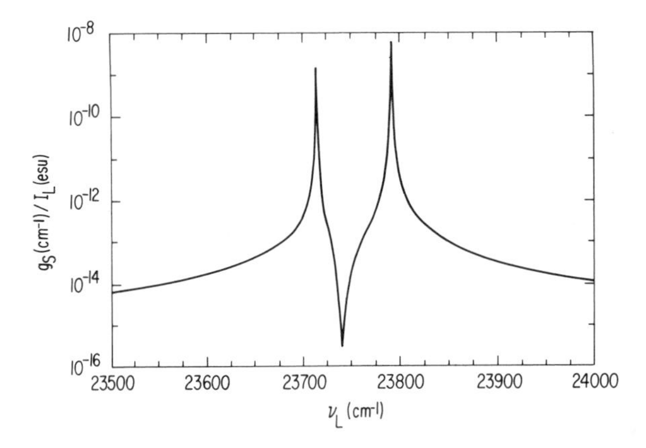

Fig. 5.14. Theoretical values for g_S/I_L as a function of ν_L for Rb vapor at a density of 1.45×10^{17} cm^{-3}

resonance (but not between the two resonance lines), the experimental value for gain is less than that calculated. This is indicative of the loss, saturation, and OPSE processes discussed above. Between resonance lines there is an anomaly. The theory presented above predicts a zero value for $\chi^{(3)}(\omega_S)$ and g_S at $\nu_L = 23741.02$ cm^{-1} where the resonance contributions of the $6p_{3/2}$ and $6p_{1/2}$ levels cancel. This cancellation is a predicted feature for all the alkali metal atoms. It is consequence of the sign of the matrix element product $x_{ns,(n+1)p_j} x_{(n+1)p_j,(n+1)s}$ being independent of j(1/2 or 3/2). However the experimental results show no such cancellation in Rb or K. We have no sensible explanation for this anomaly. We recognize that laser absorption is not negligible close to resonance. The theoretical treatment just presented needs to be modified to account for exponential decay of $|E(\omega_L)|^2$ when absorption is important. The net effect is to replace the factor displaying exponential growth of the Stokes wave, $\exp(\Delta\kappa z)$, with a modified growth factor of the form $\exp\{-A\chi^{(3)}(-\omega_S)\omega_S[\exp(-\alpha(\omega_L)z)-1]/\alpha(\omega_L)\}$ where A is a constant which is proportional to the initial laser power, and $\alpha(\omega_L)$ is the laser power absorption coefficient. The effect of laser absorption is to round off the sharp peaks in gain shown in Fig. 5.14. But the minimum, corresponding to the cancellation discussed above, is not modified significantly.

The theory predicts that, for wider tuning ranges, higher laser power is required, and this is experimentally confirmed. For lower power, SERS may still be achieved by tuning closer to resonance, and we have seen SERS in K with pump powers as low as ~ 100 W when tuning close to the $5p$ resonance lines.

From Table 5.2 there should not be very large differences between the SERS gains for a given input power near resonance in K, Rb, or Cs. Yet over an order of magnitude higher laser power is required for observable Stokes output in Cs. One possible explanation is that dimers, which are present in relative concentrations of a few percent at these pressures, absorb the Stokes light via an electronic transition between the singlet ${}^1\Sigma_g^+$ ground state and the dissociating ${}^3\Sigma_u^+$ triplet state of the dimer molecules [5.28, 29]. Cs, being much heavier than Rb or K, will have a weaker selection rule prohibiting such absorption. This explanation can be tested by doing infrared transmission measurements on alkali metals, in particular Cs, in heat-pipe ovens at pressures in the range 1–10 torr.

The calculation of total power radiated by the Stokes process is difficult to carry out. It is instructive to use a simplified model based on the notion that the output is due to amplified spontaneous Raman scattering. In this model the Stokes beam grows from noise in a single pass through the pumped nonlinear medium. Thus we must determine an appropriate value for $E_{x0}(\omega_S)$, or of the initial Stokes photon density resulting from spontaneous Raman scattering. The differential cross section $d\sigma/d\Omega$ for Stokes scattering (where $N(d\sigma/d\Omega)Adzd\Omega$ is the fraction of laser photons scattered by an area A of thickness dz into the solid angle $d\Omega$) is given by [Ref. 5.11, Eq. (2.93)]

$$\frac{d\sigma}{d\Omega}=\frac{e^4\omega_L\omega_S^3}{\hbar^2c^4}\left[\sum_j\frac{|x_{gj}x_{jj'}|}{(\omega_{jg}-\omega_L)}\right]^2. \tag{5.17}$$

Damping has been ignored. With $n_L(=I(\omega_L)/\hbar\omega_L)$ and n_S defined as the photon flux (photons/s/unit area) of the laser and Stokes waves, respectively, one has

$$d^2n_S/d\Omega dz=n_LN(d\sigma/d\Omega)+g_S(dn_S/d\Omega). \tag{5.18}$$

A simple first-order approximation is based on the geometry of Fig. 5.15. It is assumed that the laser uniformly pumps a cylindrical volume of length L and radius R. Following the approach described by *Dreyfus* and *Hodgson* [5.30], only Stokes light emitted into a solid angle such that it passes out the end of the pumped region experiences maximum gain. Thus, slice dz contributes

$$dN_S=(d^2n_S/d\Omega dz)Adzd\Omega=n_LN\frac{d\sigma}{d\Omega}\frac{\pi^2R^4}{(L-z)^2}dz+g_S(dn_S/d\Omega)\frac{\pi^2R^4}{(L-z)^2}dz \tag{5.19}$$

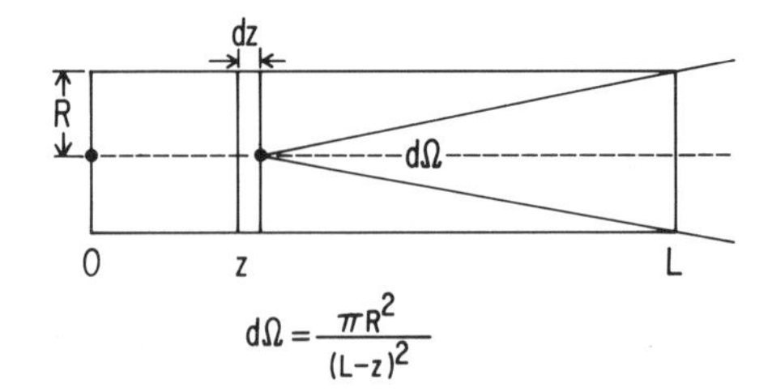

Fig. 5.15. Cylindrical geometry for calculation of total Stokes power

Stokes photons propagating towards the right-hand end of the cylinder. The total Stokes photon number is found by integrating (5.19) over dz. A simplification follows from the observation that g_S will be large enough so that only photons emitted spontaneously near $z=0$ are important. Then the factor $(L-z)^2$ in the denominator of (5.19) may be approximated by L^2 and the total Stokes output becomes

$$N_s \simeq n_L N \frac{d\sigma}{d\Omega} \frac{\pi^2 R^4 e^{g_S L}}{L^2 g_S} \text{ photons/s}. \tag{5.20}$$

This expression may be simplified by relating $d\sigma/d\Omega$ to g_S through (5.7), (5.14), (5.16), and (5.17). We find

$$d\sigma/d\Omega = (\Gamma_{j'g}\omega_S^2/8\pi^2 N n_L c^2) g_S \tag{5.21}$$

(Γ_{jg} has been neglected) which upon substitution into (5.20) yields

$$N_S \simeq \frac{R^4}{L^2} \frac{\Gamma_{j'g}\omega_S^2}{8c^2} e^{g_S L}. \tag{5.22}$$

The dispersion is given by the inverse square dependence of g_S on $\omega_{jg} - \omega_L$.

Putting in values for the various parameters in (5.22) one finds very poor agreement with experiment. The most flagrant disagreement is the magnitude of the predicted output power. For $I_L \sim 10^6$ W/cm^2 the predicted output power is ~ 5 orders of magnitude smaller than measured experimentally for laser frequencies >50 cm^{-1} away from the resonances. The problem is that this calculated process starts from much too small a noise level to build up to significant powers in a single pass. The situation is essentially identical if one uses quantum noise in the form of zero-point fluctuations (1/2 photon per mode) as the effective input[2]. Black-body radiation is a much smaller noise source, since $kT < \hbar\omega_S$. One probable explanation is that feedback must be important. Fresnel reflections of the windows terminating the heat-pipe oven

[2] In using zero-point fluctuations as the noise source, one must count the number of modes of one polarization in the bandwidth and solid angle corresponding to the laser beam and then attribute 1/2 photon for each such mode. The number of modes for this case is $(\pi R^2) \cdot (\omega_S^2/8\pi^3 c^2)\, d\Omega \Gamma_{j'g} = (R^4/L^2)(\Gamma_{j'g}\omega_S^2/8\pi c^2)$.

are not negligible (4%/face for quartz, 8%/face for sapphire). Even though no care was taken to align these windows, their reflective feedback is much more important than noise as an input for amplification. For the longest pipes used, light could make about 2 round trips during the pump pulse. So we probably have a low-Q cavity with high enough amplification to produce a Raman oscillator in the conventional sense-amplification plus feedback. Careful experiments to study this Raman oscillator need to be carried out. In particular, the use of aligned windows with high infrared reflectivity should lower the threshold for Raman oscillation and allow a much wider tuning range since much lower single pass gain would still result in oscillation.

The SERS oscillator described in this section is characterized by relatively wide tuning ($>500\ \mathrm{cm}^{-1}$), narrow linewidth ($<0.5\ \mathrm{cm}^{-1}$), high power (>1 kW) and excellent collimation compared to conventional infrared sources.

5.3.2 Optical Mixing Between One Laser and Two Stokes Photons

The observations which provided the key to the methods of infrared generation discussed in this chapter were made by *Sorokin* and *Lankard* [5.31] in Cs vapor. They irradiated Cs with the 3470 Å frequency doubled light from a mode-locked ruby laser and saw efficient parametric conversion to longer wavelength light near 3600 Å. Their results were interpreted as being due to the four-wave mixing process of the type $\omega_P = \omega_L - 2\omega_S$ where ω_L corresponds to the 3470 Å input light and ω_S is the 20 μ Stokes light generated by SERS with Cs atoms making a transition from the 6s ground state to the 10s excited state. The 10p states act as the near-resonant intermediate states to enhance the nonlinear susceptibility.

In our studies in K, Rb, and Cs, upon irradiation with laser light tuned near the $ns-(n+1)p$ transitions, we observed parametrically generated light at ω_P which was colored yellow, orange, and red, respectively. This light at ω_P was most intense when ω_L was tuned on the high-frequency side of the $(n+1)p$ resonance lines and close enough to resonance so that the Stokes generation was strong and the phase-matching condition for the four-wave mixing process was satisfied. With an input beam power of ~ 10 kW, the output at ω_P from Rb was ~ 5 W, easily observable by eye on a target in a well-lit room. On the low-frequency side, weaker emission at ω_P could be detected, but it was impossible to phase match. In between the $(n+1)p$ lines, weak ω_P generation was observed, and, although it is possible in principle to phase match by tuning between the lines, other processes such as multiphoton ionization seem to limit severely the conversion efficiency.

We observed that, in general, the light generated at ω_P emerged from the alkali metal vapor in a cone. The angle of the cone was a function of ω_L and increased as ω_L was tuned toward the resonances from the high-frequency side. Within several cm^{-1} of the resonances, the cone became diffused. Still closer, the ω_P generation disappeared. This behavior can be understood in terms of phase matching. Depending on ω_L, the ω_P emission is phase matched for

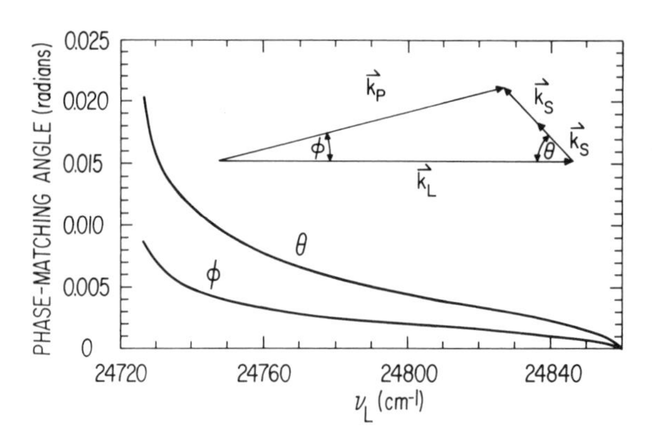

Fig. 5.16. Geometry for optical mixing of the form $\omega_P = \omega_L - 2\omega_S$, with phase-matching angles ϕ and θ plotted as a function of ν_L

different angles. In the earlier discussion of the calculation of g_S, transverse components of k_S were neglected. However, there is always gain for a Stokes wave traveling at an angle to the z axis. The angle is limited only by the aperture effects of the region pumped by the focused laser beam. For Stokes light at too large an angle, the path length over which the Stokes wave experience gain is diminished and the Stokes output falls off rapidly with increasing angle.

For a given value of ω_L and ω_S, emission at ω_P will occur most strongly in a direction which satisfies the phase-matching condition

$$\boldsymbol{k}_P = \boldsymbol{k}_L - 2\boldsymbol{k}_S \tag{5.23}$$

provided that Stokes emission at the necessary angle is also present. The frequency dependence of the angles depends on the linear dispersion properties of the medium. Here it is no longer appropriate to set $\varepsilon = 1$, since it is the small difference $\varepsilon - 1$ which completely determines the phase-matching conditions. The dispersion comes in through

$$k = \varepsilon(\omega)^{1/2}\, \omega/c = n(\omega)\, \omega/c\,. \tag{5.24}$$

The index of refraction $n(\omega)$ has a frequency dependence which may be described by a Sellmeier formula of the type

$$\Delta n(\omega) = n(\omega) - 1 = \frac{2\pi N\, e^2}{m} \sum_i \frac{f_i}{\omega_i^2 - \omega^2} \tag{5.25}$$

where f_i is the oscillator strength of a transition from the ground state to state i, and ω_i is the resonance frequency of that transition.

To analyze this in greater detail refer to Fig. 5.16. Equation (5.23) may be expressed as

$$\cos\phi = (k_P^2 + k_L^2 - 4k_S^2)/2k_P k_L \tag{5.26a}$$

or

$$\cos\theta = (k_L^2 + 4k_S^2 - k_P^2)/4k_S k_L \, . \tag{5.26b}$$

The magnitude of the power emitted at ω_P depends on the amount of Stokes light generated and the size of the nonlinearity $\chi^{(3)}(-\omega_P, \omega_L, -\omega_S, -\omega_S) = \chi^{(3)}(\omega_P)$ describing the polarization

$$P_x^{NL}(\omega_P) = 3\chi_{xxxx}^{(3)}(\omega_P)\, E_x(\omega_L)\, E_x^2(-\omega_S) \, . \tag{5.27}$$

From a consideration of the important processes contributing to $\chi^{(3)}(\omega_P)$, the proper expression for the alkali metals is

$$\begin{aligned}\chi_{xxxx}^{(3)}(\omega_P) = {} & i\, 6.289 \times 10^{17} \frac{N}{\Gamma_{(n+1)s,\, ns}} \\ & \cdot \langle ns0\,1/2|x|n+1p0\,1/2\rangle \langle n+1p0\,1/2|x|n+1s0\,1/2\rangle \\ & \cdot \langle n+1s0\,1/2|x|np0\,1/2\rangle \langle np0\,1/2|x|ns0\,1/2\rangle \\ & \cdot [2(\nu_{(n+1)p3/2} - \nu_L)^{-1} + (\nu_{(n+1)p1/2} - \nu_L)^{-1}] \\ & \cdot \{[2(\nu_{np_{3/2}} - \nu_P)^{-1} + (\nu_{np_{1/2}} - \nu_P)^{-1}] \\ & \qquad + [2(\nu_{np_{3/2}} - \nu_S)^{-1} + (\nu_{np_{1/2}} - \nu_S)^{-1}]\} \text{ esu} \, . \end{aligned} \tag{5.28}$$

($\Gamma_{(n+1)s,\, ns}$ is in cm^{-1}). This expression gives smaller values than (5.15) because ν_P and ν_S are not near resonance with the *np* states. However, the product of matrix elements (Table 5.2) in (5.28) is larger than that of (5.15). The nonlinear polarization (5.27) acts as the driving term for the field $E(\omega_P)$. The theoretical treatment in Subsection 5.3.1 offers an approach for calculating the angular dependence of the emitted power at ω_P. We write

$$E_x(\omega_P) = E_{x0}(\omega_P)\, e^{i(\boldsymbol{k}_L - 2\boldsymbol{k}_{S0})\cdot \boldsymbol{r} - i\omega_P t}\, e^{2|\Delta\kappa| z} \tag{5.29}$$

and define the phase mismatch Δk by

$$k_L - 2k_{S_{z0}} = k_{P_z} - \Delta k \, . \tag{5.30}$$

We have assumed that the angles the various wave vectors make with the z axis are small so that only a longitudinal phase match is considered. The solution for $E_{x0}(\omega_P)$, using the analog of (5.8) along with (5.10), (5.27), (5.29), and (5.30) is

$$E_{x0}(\omega_P) \simeq \frac{-2\pi\omega_P}{c(\Delta k + 2i|\Delta\kappa|)} [3\chi_{xxxx}^{(3)}(\omega_P)\, E_x(\omega_L)\, E_{x0}^2(-\omega_S)] \tag{5.31}$$

where $|\Delta\kappa|$ is given by (5.13). This result shows that the ω_P emission is conical with a maximum of intensity at the angle for which $\Delta k = 0$. The intensity falls off about this angle as Δk grows. For an angle such that $|\Delta k| = |2\Delta\kappa|$, the intensity is half its maximum.

We observed that the angular spread of the conical emission increased as we tuned ω_L closer to resonance. This is consistent with (5.31) since $\Delta\kappa$ is expected to increase so that the half intensity angles would correspond to a larger value for $|\Delta k|$. Detailed studies were not carried out. Note that measurements of the angle for phase matching would give a method for determining oscillator strengths of the second-resonance lines relative to the first-resonance line. The oscillator strength of the *ns-np* transition dominates the dispersion of the index of refraction except for frequencies tuned very close to another resonance, such as the second-resonance line. Thus the indices for ω_P and ω_S can be accurately calculated from (5.25) using only the *ns-np* transitions. Measurements of the phase-matching angle for various values of ω_L would with (5.25) and (5.26) determine values for the oscillator strengths of the $ns-(n+1)p$ transitions.
Values of f_i for K atoms have been tabulated [5.32] and the phase-matching angles ϕ and θ are plotted as a function of laser frequence in Fig. 5.16. This plot only covers a range with $\omega_L > \omega_{(n+1)p\,^2P_{3/2}}$, since this is the region where the conical emission pattern was observed.

5.3.3 Infrared Generation by Four-Wave Optical Mixing Using Stokes Photons

We have seen in Subsection 5.3.2 how the Stokes wave can participate in a four-wave mixing process to generate a wave at a new frequency, in that case $\omega_P = \omega_L - 2\omega_S$. In this subsection we shall consider the generation of an infrared (IR) wave of frequency $\omega_{IR} = \omega_L - \omega_S - \omega_R$, where ω_L and ω_S are the same as before, and ω_R corresponds to a second laser beam. In the alkali metal vapor systems, $\omega_L - \omega_S = \omega_{(n+1)s}$ so that $\omega_{IR} = \omega_{(n+1)s} - \omega_R$, and it is clear that ω_{IR} tunes in the opposite sense to ω_R. Also as ω_R approaches $\omega_{(n+1)s}$, ω_{IR} tunes far into the infrared. The frequency relationships are clarified in Fig. 5.17 where the two important processes which contribute to $\chi^{(3)}(\omega_{IR})$ are depicted. We can immediately write an expression for the nonlinear susceptibility, which is very similar to (5.28):

$$\begin{aligned}\chi^{(3)}_{xxxx}(\omega_{IR}) = \chi^{(3)}_{xxxx}(-\omega_{IR}, \omega_L, -\omega_S, -\omega_R) = \mathrm{i}\, 3.145 \times 10^7 \frac{N}{\Gamma_{(n+1)s,ns}} \\ \cdot \langle ns0\,1/2|x|n+1p0\,1/2\rangle \langle n+1p0\,1/2|x|n+1s0\,1/2\rangle \\ \cdot \langle n+1s0\,1/2|x|np0\,1/2\rangle \langle np0\,1/2|x|ns0\,1/2\rangle \\ \cdot [2(\nu_{(n+1)p3/2} - \nu_L)^{-1} + (\nu_{(n+1)p1/2} - \nu_L)^{-1}] \\ \cdot \{[2(\nu_{np_{3/2}} - \nu_R)^{-1} + (\nu_{np_{1/2}} - \nu_R)^{-1}] \\ + [2(\nu_{np_{3/2}} - \nu_{IR})^{-1} + (\nu_{np_{1/2}} - \nu_{IR})^{-1}]\}\ \mathrm{esu}\,.\end{aligned} \quad (5.32)$$

($\Gamma_{(n+1)s,ns}$ is in cm^{-1}).

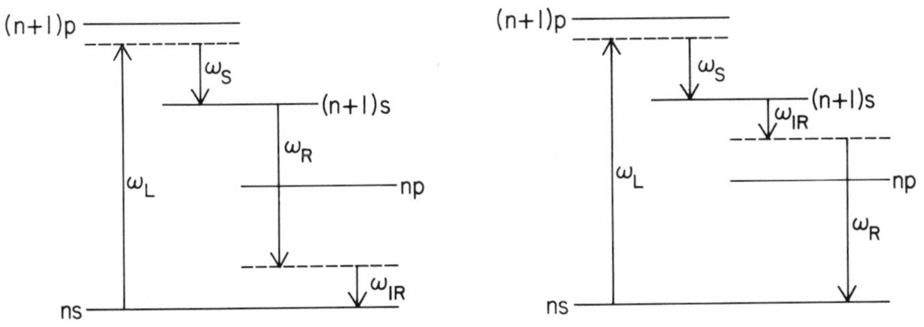

Fig. 5.17. Important processes contributing to infrared generation by four-wave mixing of the form $\omega_{IR}=\omega_L-\omega_S-\omega_R$

The nonlinear polarization is given by

$$P_x^{NL}(\omega_{IR})=6\chi_{xxxx}^{(3)}(\omega_{IR})\,E_x(\omega_L)\,E_x(-\omega_S)\,E_x(-\omega_R)\,. \tag{5.33}$$

For the alkali metals $\omega_{(n+1)s}$ is a frequency that can be reached by tunable dye lasers; so, in principle, ω_{IR} can be tuned from the near IR out to very long wavelengths ($>100\,\mu$). In practice the amount of IR generation may be seriously affected by various other factors. The most significant factors which severely reduce the efficiency of ω_{IR} generation as one tunes towards longer wavelengths are 1) the decreasing radiative efficiency of a macroscopic phased nonlinear polarization (with radiated power going as ω_{IR}^2) and 2) the diffractive losses suffered by the infrared from the region of the nonlinear medium pumped by the focused laser beam. As the infrared wavelength increases, diffraction effects increase. The effective interaction length is proportional to the infrared frequency. This interaction length is also proportional to the area of the focused laser beams, so that looser focusing also decreases the detrimental effects of diffraction. But looser focusing also decreases the intensity of the lasers at the focus, so that the nonlinear generation per unit length is not as great. The solution seems to be to use longer pipes containing the alkali metal vapors and not to focus too tightly. This makes the phase-matching requirement more critical. The laser intensity cannot be too low or else the threshold for SERS will not be reached. The detailed behavior of the infrared generation, including the effects of these conflicting requirements, has not been fully worked out. However, a related problem, that of parametric generation of the form $\omega_4=\omega_1-\omega_2-\omega_3$, where ω_1, ω_2, and ω_3 are present as input beams, has been studied theoretically by *Bjorklund* [5.33] for the case of single transverse mode lasers. His results are not strictly applicable because of the dependence of our process on a SERS process, but he does provide some guidance as to how to treat the diffraction loss problem.

Although alkali metal atoms are completely transparent in the IR, a significant number of dimers exist at the pressures with which we work. These dimers can absorb in the IR as pointed out near the end of Subsection 5.3.1. This effect is probably not very important in K or Rb.

When ω_R is tuned close to the *np* resonance lines, there are stronger linear absorptive and dispersive effects as well as resonance enhancement of $\chi^{(3)}(\omega_{IR})$. This effect is most important for short wavelength ($<2\,\mu$) IR generation. This wavelength range can be spanned more simply by the methods discussed in Chapter 3. Of greater concern are the dispersive effects of the *np* and $(n+1)p$ resonance lines on phase matching. This shall be discussed in greater detail below where two methods of phase matching are presented.

Experimental Results

Working with K and Rb vapors, *Sorokin* et al. [5.19] observed infrared light generated by four-wave mixing. The IR was continuously tunable from 2–4 μ in K and from 2.9–5.4 μ in Rb. In this experiment the lasers (ω_L and ω_R) were collinear and focused into a heat-pipe oven containing the atomic vapors. The output light from the oven passed through a prism monochromator and was detected using either a PbS or an InSb photoconductive detector. Phase matching was maintained as ω_R (and ω_{IR}) were tuned by fine tuning ω_L, thus varying the strong dispersive influence of the $(n+1)p$ resonance lines. The longest IR wavelength obtainable from pure K was limited by the fact that it was not possible simultaneously to generate SERS and have phase matching for $\lambda \gtrsim 4\,\mu$ in K. The longest wavelength detected from Rb was limited by the InSb detector. The shortest wavelength generated was determined by the dyes used to produce ω_R.

Recently, *Tyler* et al. [5.24] discovered a saturation effect which limits the IR output power. But by taking account of this saturation effect, they determined an optimum method of achieving higher IR output powers. The effect, to be discussed in more detail below, is that the IR power does not increase for laser power at ω_L above a certain maximum value. This value (2.6 kW in K for the case studied by *Tyler* et al. [5.24]) corresponds to the onset of OPSE from the $(n+1)s$ level to the *np* levels. Exciting two dye lasers (ω_L and ω_R) with a common nitrogen laser, they pumped the laser at ω_L only strongly enough to produce 2.6 kW, pumping the laser at ω_R with the rest of the nitrogen laser output. With powers of ~ 40 kW at ω_R they obtained an IR power output of 9.5 W at 2.2 μ. This represents a hundredfold increase in IR power over that reported by *Sorokin* et al. [5.19].

Phase Matching

The method of phase matching in the collinear geometry by tuning ω_L will now be discussed. As was the case for ω_P generation discussed in Subsection 5.3.2, it is possible to phase match the ω_{IR} generation process by introducing appropriate angles between the various waves. But the generation is more efficient if the waves are all collinear. Then phase matching takes the simple form

$$k_{IR} = k_L - k_S - k_R \tag{5.34}$$

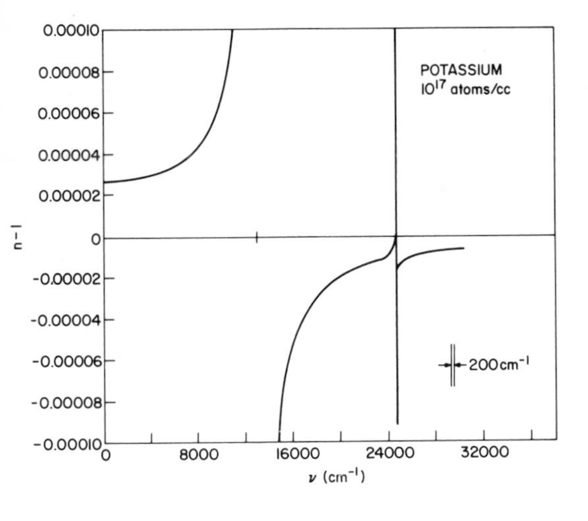

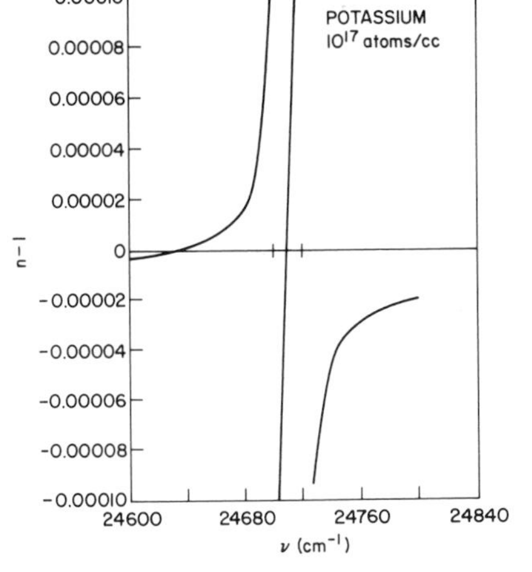

Fig. 5.18. Index of refraction of K vapor as a function of frequency. The curves are drawn from the case of no damping

or using (5.24)

$$n(\omega_{\mathrm{IR}})\,\omega_{\mathrm{IR}} = n(\omega_{\mathrm{L}})\,\omega_{\mathrm{L}} - n(\omega_{\mathrm{S}})\,\omega_{\mathrm{S}} - n(\omega_{\mathrm{R}})\,\omega_{\mathrm{R}}\,. \tag{5.35}$$

Using the relationship between the frequencies, (5.35) may be written as

$$\Delta n(\omega_{\mathrm{IR}})\,\omega_{\mathrm{IR}} = \Delta n(\omega_{\mathrm{L}})\,\omega_{\mathrm{L}} - \Delta n(\omega_{\mathrm{S}})\,\omega_{\mathrm{S}} - \Delta n(\omega_{\mathrm{R}})\,\omega_{\mathrm{R}} \tag{5.36}$$

with $\Delta n = n-1$. We may plot the dependence of $\Delta n(\omega)$ on ω using (5.25); the results for K are present in Fig. 5.18 for $N = 10^{17}/\mathrm{cm}^3$. With ν_{S} and ν_{IR} both less than 5000 cm^{-1}, the contribution to (5.36) of the terms involving these

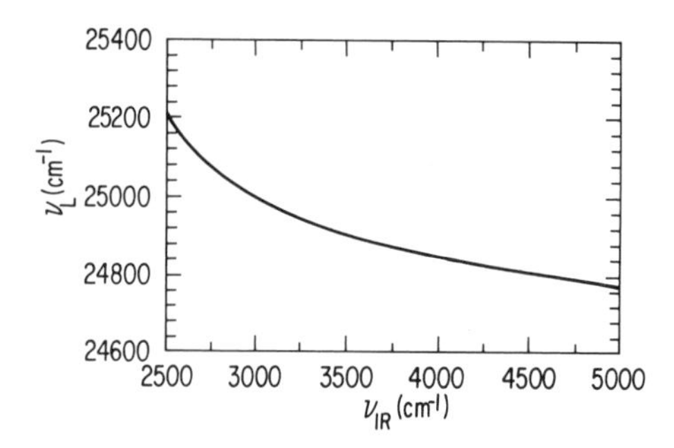

Fig. 5.19. Laser frequency (ν_L) required for phase matching the infrared generation process in K vapor plotted as a function of generated infrared frequency (ν_{IR})

frequencies is, in the first approximation, negligible. Thus phase matching roughly corresponds to

$$\Delta n(\omega_R)\,\omega_R \simeq \Delta n(\omega_L)\,\omega_L\,. \tag{5.37}$$

For infrared generation at wavelengths $>2\,\mu$, $\nu_R > \nu_{4p} \cong 13000\,\text{cm}^{-1}$ and $\nu_R < \nu_{5s} = 21026.8\,\text{cm}^{-1}$. Then $\Delta n(\omega_R) < 0$ and, for phase matching, ω_L must be chosen so that $\Delta n(\omega_L) < 0$ in order to satisfy (5.37). Since it is not suitable to tune ω_L between the $5p$ levels, ω_L must be set on the high-frequency side of the $5p$ levels. From Fig. 5.18 it is clear that from some value of $\omega_L > \omega_{5p_{3/2}}$ phase matching is achievable for values of ω_R appropriate for IR generation. But to tune to longer infrared wavelengths ω_R must increase, and ω_L must correspondingly increase to maintain phase matching. The long wavelength IR limit is reached when ω_L is tuned too far from the $5p$ levels to pump SERS. This occurred in K for IR generation near 4μ with the pumping powers used by *Sorokin* et al. [5.19]. In Fig. 5.19 the phase matching condition is plotted as a relationship between ν_L and ν_{IR}. This relationship is determined by (5.25) and (5.36).

While in pure atomic vapors collinear phase matching can be achieved over limited ranges by "fine tuning" ω_L, an obvious alternative is to change the linear dispersion characteristics of the metal vapor. One simple way to do this is to add a second component which changes the linear dispersion while having negligible effect upon the nonlinear optical properties. *Wynne* et al. [5.34] added Na to K and found that they could phase match for any value of ω_{IR} with ω_L set at a value on the high-frequency side of the $5p$ lines where SERS was strong. The addition of Na changes the linear dispersion of the vapor primarily through the effect of the main resonance lines, the Na "D" lines, on the index of refraction. Figure 5.20 shows the dispersion for equal densities of Na and K. By comparing to Fig. 5.18, it is seen that in the region between 14000 and $21000\,\text{cm}^{-1}$ Δn is more negative in the mixed system than in the pure system. Then, in general, (5.36) can be satisfied by adjusting the Na:K ratio. The presence of Na has the greatest effect on ω_R, with the smaller effects on ω_L, ω_S and ω_{IR}. Figure 5.21 plots the ratio of Na:K for exact phase matching as a function of ν_{IR}, with ν_L set at $24795\,\text{cm}^{-1}$. This range of concentration can be achieved through the use of a concentric heat-pipe oven

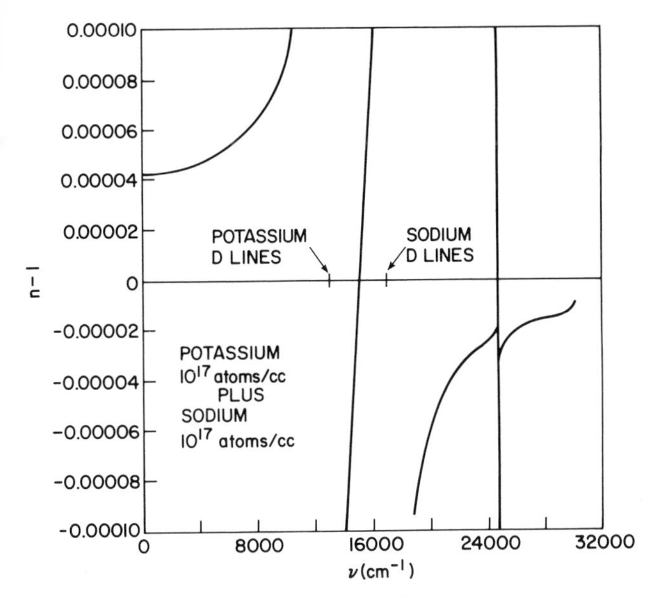

Fig. 5.20. Index of refraction of mixture of K and Na

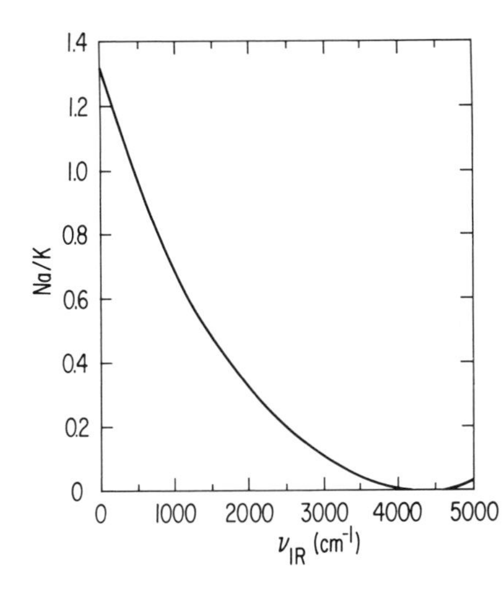

Fig. 5.21. Ratio Na:K for phase matching as a function of generated infrared frequency

[5.9]. Note that for ν_{IR} between 4000 cm^{-1} and 4700 cm^{-1} this process cannot be phase matched with Na for $\nu_L = 24795$ cm^{-1}. It is important that *exact* phase matching is *not* necessary for the most efficient generation of infrared. The dependence of IR generation on the phase mismatch shall be derived below. In practice, with the laser beams focused so that the region of infrared generation was ~5 cm long, a given mixture of Na and K suffices for a tuning

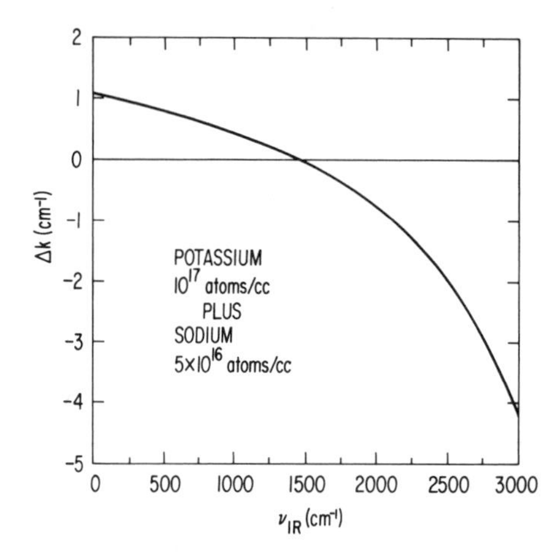

Fig. 5.22. Phase mismatch, $\Delta k = k_{IR} + k_S + k_R - k_L$, as a function of infrared frequency, for a given Na:K ratio

range of $\sim 1000\,\mathrm{cm}^{-1}$. This is also a typical tuning range for dyes used to produce radiation at ω_R. The phase mismatch, defined as

$$\Delta k = k_{IR} + k_S + k_R - k_L \tag{5.38}$$

is plotted as a function of ν_{IR} for a given Na:K ratio in Fig. 5.22.

The criterion to be satisfied by adding another component is that its dispersion be such as to reduce $|\Delta k|$ as defined by (5.38). When the second component is another alkali metal vapor, its dispersion is dominated by the first resonance lines (the *ns-np* transitions), also known as the "*D*" lines. While it may not be obvious without exact calculations whether a second component will achieve phase matching, the general effect is to alter the dispersion curves in the manner shown in Figs. 5.18 and 5.20. Since ω_R is likely to be in the region between the *D* lines and the second resonance line of the component responsible for the nonlinear effects, the second component of the mixture should have its *D* lines at higher frequency than those of the nonlinear component. (Both components are, of course, nonlinear, but the second component is far from resonance so its nonlinear properties are not resonantly enhanced and it behaves in a "linear" fashion.) If one is using K, Rb or Cs as the nonlinear component, Na will serve to phase match infrared generation for wavelengths longer than 3 μ.

Experimental Results

Using a mixture of Na and K in a concentric heat-pipe oven, *Wynne* et al. [5.34] observed phase-matched IR generation from 2 to 25 μ. In subsequent work using the same system and stronger pumping power, IR generation to 30.5 μ has been detected. This limit is set by the Cu:Ge photoconductive detector used to detect the infrared. This detector has peak detectivity near 20 μ and the detectivity falls rapidly at longer wavelengths, being down by

2 orders of magnitude at 30 μ. Using another detector, such as In:Ge or Ga:Ge, we would expect to be able to generate and detect radiation with wavelengths >30 μ. However, due to the problems discussed earlier in this subsection, the power falls rapidly at longer wavelengths. We estimate that the power generated at 25 μ was only ~0.1 mW. Phase matching was not exact at each IR frequency. For each dye used to produce ω_R, a Na:K ratio was chosen to maximize IR generation when the dye laser was set in the middle of its tuning range. When the dye was changed to tune to a different region, the Na:K ratio was suitably adjusted.

Through better understanding of the saturation effect discovered by *Tyler* et al. [5.24], it is believed that IR output powers ~0.1 W may be obtained at wavelengths as long as 20 μ. Such power levels will make this source a useful spectroscopic tool for such studies as surface wave generation and propagation in the infrared, and long path length, moderate resolution, absorption spectroscopy.

Theory

The growth of the infrared at ω_{IR} resembles the growth of the anti-Stokes wave which often accompanies stimulated Raman scattering [Ref. 5.11, pp. 110–119]. Since the ω_{IR} wave is weak relative to the Stokes wave, the influence of the ω_{IR} wave back on the Stokes wave is negligible. Using the theoretical approach of Subsection 5.3.1, we write

$$E_x(\omega_{IR}) = E_{x0}(\omega_{IR})\, e^{i(\boldsymbol{k}_L - \boldsymbol{k}_S - \boldsymbol{k}_R)\cdot \boldsymbol{r} - i\omega_{IR}t} e^{|\Delta\kappa|z} . \tag{5.39}$$

The longitudinal phase mismatch is given by

$$k_L - k_R - k_{S0} = k_{IR} - \Delta k \tag{5.40}$$

where, as in Subsection 5.3.2, we have assumed that the angles of k_{IR} and k_S relative to the z axis are small. The solution for $E_{x0}(\omega_{IR})$, using the analog of (5.8) along with (5.10), (5.33), and (5.39) is

$$E_{x0}(\omega_{IR}) = \frac{-2\pi\omega_{IR}}{c(\Delta k + i|\Delta\kappa|)}\, 6\chi^{(3)}_{xxxx}(\omega_{IR})\, E_x(\omega_L)\, E_x(-\omega_R)\, E_{x0}(-\omega_S) \tag{5.41}$$

with $|\Delta\kappa|$ given by (5.13). Saturation effects limit the usefulness of (5.39) and (5.41), but (5.41) is expected to give some insight into the IR generation process. In particular, the dependence of $E(\omega_{IR})$ on Δk given by (5.39), and (5.41) predicts a halving of the IR output power for $|\Delta k| = |\Delta\kappa|$ as compared to its exact phase-matched value. As an example, for $N_K = 10^{17}/\mathrm{cm}^3$ and a focused laser ($\nu_L = 24795\ \mathrm{cm}^{-1}$, $\Gamma = 0.2\ \mathrm{cm}^{-1}$) intensity of $5\times10^6\ \mathrm{W/cm}^2$, calculation gives $\Delta\kappa \simeq 1\ \mathrm{cm}^{-1}$ in K. Then for ν_{IR} between 0 and 2200 cm^{-1}, Fig. 5.22 shows that $|\Delta k| \leq 1\ \mathrm{cm}^{-1}$, with Na:K = 0.5. Readjusting the phase-matching mixture would at most increase the ω_{IR} power by a factor 2 over the entire tuning range. We may calculate the ratio of the ω_{IR} to Stokes power if we assume that the ratio of the corresponding electric fields is given by (5.41). This assumption ignores the very real diffraction effects and possible ω_{IR} and ω_S absorption processes. Converting (5.41) to a relationship between intensities

$$I(\omega_{IR})/I(\omega_S) \simeq \frac{1024\pi^6 \nu_{IR}^2}{c^2[(\Delta k)^2+(\Delta\kappa)^2]} |6\chi^{(3)}(\omega_{IR})|^2 \, I(\omega_L)\, I(\omega_R)\,. \tag{5.42}$$

With $I(\nu_L = 24795\,\mathrm{cm}^{-1}) = 5\times 10^6\,\mathrm{W/cm}^2$, and $\nu_{IR} = 1490\,\mathrm{cm}^{-1}$, $|\Delta\kappa| \simeq 1\,\mathrm{cm}^{-1}$ and $\Delta k \simeq 0$ for $N_K = 10^{17}\,/\mathrm{cm}^3$ and $N_{Na} = 5\times 10^{16}\,/\mathrm{cm}^3$. From (5.32) with $\Gamma = 0.2\,\mathrm{cm}^{-1}$, $|6\chi^{(3)}(\omega_{IR})| = 2.725\times 10^{-11}$ esu, and with $I(\omega_R) = 10^7\,\mathrm{W/cm}^2$, (5.42) gives $I(\omega_{IR})/I(\omega_S) \simeq 0.01$. We estimate the experimental ratio at ~ 0.001 to within a factor 10 when experimental conditions matched these values for the various parameters. Allowing for diffraction and absorption losses, the agreement is reasonable.

The theoretical treatment just described strongly resembles that of *Giordmaine* and *Kaiser* [5.35] who treated the analagous problems of mixing of the form $\omega_b = \omega_a - (\omega_L - \omega_S)$ and $\omega_b = \omega_a + (\omega_L - \omega_S)$. They described the problem as one in which a nonlinear material is coherently driven at the frequency $(\omega_L - \omega_S)$ by stimulated Raman scattering. A second probe laser beam at frequency ω_a is coherently scattered from the material excitation to produce light of ω_b. For the case of phase matching, they present results [Ref. 5.35, Eqs. (30) and (43)] which are analogous to (5.42) with $\Delta k = 0$. To make this analogy more explicit, for $\Delta k = 0$, using (5.13), (5.42) may be written as

$$\frac{I(\omega_{IR})}{I(\omega_R)} = \frac{\omega_{IR}^2}{\omega_S^2} \left|\frac{\chi^{(3)}(\omega_{IR})}{\chi^{(3)}(\omega_S)}\right|^2 \frac{I(\omega_S)}{I(\omega_L)}\,. \tag{5.43}$$

This shows that the IR power is expected to increase with $I(\omega_R)$ as well as with the efficiency of the SERS process.

As mentioned earlier *Tyler* et al. [5.24] have observed in K that for $I(\omega_L)$ above an optimum level, $I(\omega_{IR})$ does not increase with $I(\omega_L)$. Above the optimum power level for ω_L, $I(\omega_{IR})$ depends only on $I(\omega_R)$ and this dependence is observed to be linear. Furthermore for $I(\omega_L)$ above the optimal power, OPSE at 1.25 μ commences. This corresponds to a transition from 5*s* to 4*p* (Fig. 5.8). These observations indicate that rapid relaxation of the 5*s* population caused by OPSE effectively limits $P^{NL}(\omega_{IR})$. Equation (5.33) expresses $P^{NL}(\omega_{IR})$ in terms of a susceptibility and electric fields. Due to the OPSE process $\chi^{(3)}(\omega_{IR})$ is not power independent. If one calculates $P^{NL}(\omega_{IR})$ by time-dependent perturbation theory [cf. Ref. 5.11, pp. 26–30], the coherent superposition of states resulting from the applied fields is limited in magnitude by the rapid relaxation resulting from OPSE. Further increases in fields ($E(\omega_L)$ and $E(\omega_S)$) above the OPSE threshold do not increase the coherent superposition of states which is responsible for the induced phased nonlinear polarization, $P^{NL}(\omega_{IR})$.

The contribution to $P^{NL}(\omega_{IR})$ of the coherent induced superposition of the ns and $(n+1)s$ states (represented by $\varrho_{ns,(n+1)s}(\omega_L-\omega_S)$ in the density matrix formalism) is maximum at the OPSE threshold and stays constant for further increases in $E(\omega_L)$ and $E(\omega_S)$. There is no restriction placed by this effect on the dependence of $P^{NL}(\omega_{IR})$ on $E(\omega_R)$.

To complete this discussion, we must mention one more loss mechanism. In the concentric heat-pipe oven K_2, Na_2, and Na–K dimers form, and these linearly absorb the ω_R light due to rotational-vibrational bands of electronic transitions. Since this absorption is not intensity dependent, it is active throughout the vapor region in the oven. To minimize its effects it is best to focus into a region close to the cell entrance so that the light at ω_R is not significantly absorbed before the region of strong nonlinear interaction. But too tight focusing is detrimental because of diffraction effects; so, in practice, an empirical compromise is reached.

Detailed experimental and theoretical dependences of the generation on the various factors await a careful, lengthy investigation.

5.3.4 Infrared Generation by Other Four-Wave Mixing Processes

Our efforts have been concentrated on processes which utilize SERS and the accompanying strong resonant enhancement of the nonlinearities. Alternatives, using three input waves (two of which may correspond to the same frequency), exist. Still utilizing the third-order nonlinearities of atomic vapors, consider the process depicted in Fig. 5.23. If the sum $\omega_1+\omega_2$ is tuned close to $\omega_{j'g}$, the process is resonantly enhanced. Then ω_3 can be tuned to approach $\omega_{j'g}$, pushing ω_{IR} farther into the IR. There are many atomic systems in which this process may be studied. For highest efficiency the matrix elements should be large, and levels j and j'' (which may be the same level) should not be too distant from ω_1 (or ω_2) or ω_3. One likely candidate for study is Sr, which we have used in a different way for ultraviolet generation (Sect. 5.4). There are many suitable levels connected to the ground state by two-photon transitions in Sr.

Returning to the alkali metals, the $3d$, $5s$, or $4d$ levels in Na may be used for level j'. In a related process (Fig. 5.24), *Bloom* et al. [5.36] utilized four-wave mixing to detect IR by up-conversion. They pumped Na with 10^7 W/cm^2 at 6856 Å (ω_1), and observed as much as 58% quantum efficiency of conversion from infrared (ω_{IR}) near 10 μ to the ultraviolet (ω_4) in a $\omega_4=2\omega_1+\omega_{IR}$ process. This high quantum efficiency occurred when the IR was at 9.26 μ, putting ω_4 very close to $\omega_{4p,3s}$. As a means of IR generation this process could be used with tunable ω_4 as the input to generate ω_{IR}.

Bloom et al. [5.36] also reported the observation of optical parametric oscillation (OPO) pumped by the two input photons. This process, depicted in Fig. 5.25, of the type $\omega_S+\omega_I=2\omega_P$ where the signal (ω_S) and idler (ω_I) both build up from noise, has been theoretically treated by *Tomov* [5.37] and *Popov* [5.38]. Such a process may result in a useful, tunable source in the infrared or ultraviolet but much experimental work remains to be done. *Wynne*

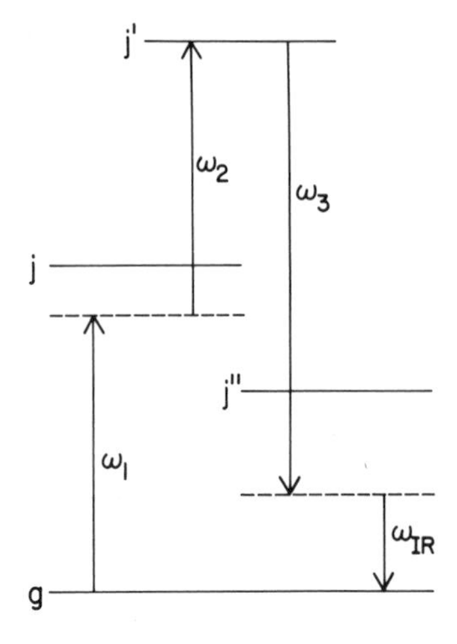

Fig. 5.23. Elementary process for generation of $\omega_{IR}=\omega_1+\omega_2-\omega_3$

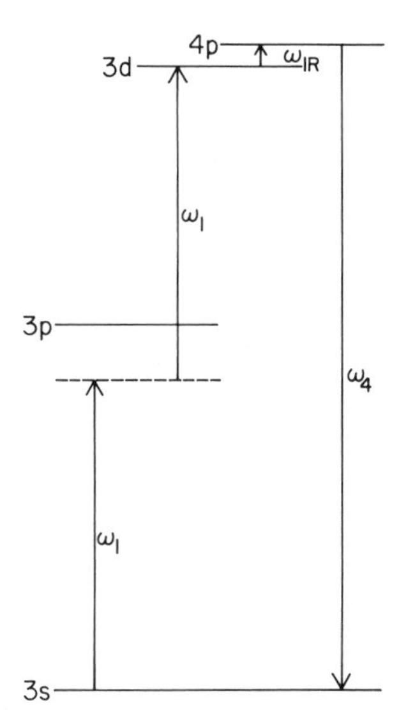

Fig. 5.24. Elementary process for detecting ω_{IR} by detecting generation of $\omega_4=2\omega_1+\omega_{IR}$

et al. [5.39] observed OPO of this type from Na when ω_P was provided by a rhodamine 6 G dye laser at 5787 Å, so that $2\omega_P=\omega_{4d,\,3s}$. The signal and idler outputs occurred at 2.34 μ and 3303 Å, respectively.

Finally, the type of process depicted in Fig. 5.17, where there is no reliance on SERS but where ω_L, ω_S, and ω_R are all present as input waves, may also serve as an infrared generation process. Some theoretical work has been done on this problem by *Bjorklund* [5.33].

As seen by this brief survey, and by the work reported on at some length in this section, infrared generation in atomic vapors may be pursued from several different but related approaches. There is nothing to preclude the use of

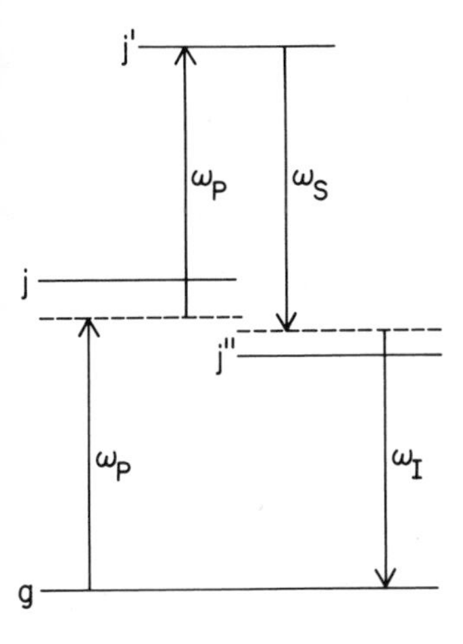

Fig. 5.25. Optical parametric oscillation process where signal at ω_S and idler at $\omega_I(\omega_I+\omega_S=2\omega_P)$ are generated

molecular or ionic vapors, which will certainly enlarge the number of material systems to be studied. Care must be taken to avoid molecules with strong absorption in the infrared range of interest.

5.4 Ultraviolet Generation

The simplest four-wave parametric process is third-harmonic generation (THG), which was discussed in Section 5.2 and depicted in Fig. 5.4. In this section we shall consider THG and other four-wave parametric sum mixing processes which lead to generation of coherent, tunable, ultraviolet light.

The earliest experiments in THG in vapors were carried out with noble gas atomic vapors. These experiments consisted in measuring the so-called "hyper-polarizabilities" [5.3, 4]. The researchers at Stanford University who pioneered the use of THG in gases as a practical source of coherent ultraviolet light studied atomic systems other than the noble gases. Specifically, they studied the alkali metals, using fixed-frequency input beams [5.13, 40, 41]. They obtained useful third-harmonic conversion efficiencies and calculated that the large nonlinearities were due to resonant enhancement stemming from a general reduction of the energy denominators ΔE_1, ΔE_2, and ΔE_3 (see Fig. 5.4). The advantage of using alkali metal vapors stems from their generally low first ionization potentials. Consequently, various bound excited states lie in an energy range where resonance enhancement is possible with the visible input or ultraviolet third-harmonic output beams. This work stimulated the interest of various other groups [5.42–44] who independently focused attention on the further advantages to be gained in attaining the exact resonance condition $\mathrm{Re}(\Delta E_2)=0$. These advantages, which have already been discussed in Section 5.2, are based upon the lack of linear absorption or dispersion associated with a two-photon resonance. The focus in this section will be on our own approach, which is characterized experimentally by the use of nitrogen-laser-pumped dye lasers to provide the input beams.

5.4.1 Multiphoton Ionization

It is worthwhile to begin by describing an effect which graphically illustrates the importance of two-photon resonance in nonlinear phenomena in gases. With the aim of investigating its usefulness as a tunable IR source, we had been attempting to produce SERS in Sr vapor, with expected Stokes emission in the vicinity of the $5s5p\,^1P_1^\circ - 5s4d\,^1D_2$ transition at 6.45 μ (Fig. 5.26). A nitrogen-laser-pumped dye laser, continuously tunable over a wavelength range that encompassed the $5s^2\,^1S_0 - 5s5p\,^1P_1^\circ$ resonance line (4607 Å), was focused into a vapor cell containing a few torr of Sr and a few hundred torr of He buffer gas. No SERS was observed, even at small detunings of the laser frequency from the resonance line. Later we verified that the Stokes component was absent even when the He buffer gas pressure was reduced to the point where the vapor cell operated in the heat-pipe mode. We still have no convincing explanation as to why the Raman threshold was not exceeded in this case. At the higher He buffer gas pressures strong OPSE did occur on the 6.45 μ line, as well as on transitions ($5s4d\,^3D_i - 5s5p\,^3P_j$) lying within the triplet manifold, when the dye laser was tuned anywhere within a few hundred cm^{-1} of the Sr resonance line. This effect had been observed before, but not studied as a function of the primary laser wavelength [5.45]. When we now monitored the intensity of any of the above IR lasing transitions, and scanned the primary laser wavelength, we recorded a trace similar to that shown in Fig. 5.27. Strikingly sharp dips are seen to occur at certain wavelengths which correspond either to Sr two-photon transitions from the ground state, or to single-photon transitions from the $5s5p\,^1P_1^\circ$ or $5s4d\,^1D_2$ states to various excited states. We immediately speculated that the occurrence of the dips was a manifestation of three-photon photoionization, resonantly enhanced at the two-photon energies. However, it was *Carlsten* [5.46] who actually suggested to us the plausible model that the Sr levels involved in the IR lasing would be Stark broadened by the multiphoton ionization created at the focus of the dye laser, thus causing the gain to drop.

Bebb [5.47] has calculated the three-photon photoionization rate for the alkali metals. We shall make use of his work to estimate the amount of ionization in the Sr experiment discussed above. With the use of time-dependent perturbation theory, one may show that

$$w_{f,g}^{(3)} \propto I_\mathrm{L}^3 |M^{(3)}|^2 \,, \tag{5.44}$$

where $w_{f,g}^{(3)}$ is the rate of three-photon photoionization. Here I_L is the laser intensity, and

$$M^{(3)} = \sum_{j,j'} \frac{\mu_{fj'}\mu_{j'j}\mu_{jg}}{(\omega_{j'g} - 2\omega_\mathrm{L} - \mathrm{i}\Gamma_{j'g})(\omega_{jg} - \omega_\mathrm{L} - \mathrm{i}\Gamma_{jg})} \,. \tag{5.45}$$

Equation (5.45) shows that resonant enhancement occurs when $\omega_{j'g} = 2\omega_\mathrm{L}$, that is, when the laser is tuned to two-photon resonance with one of the inter-

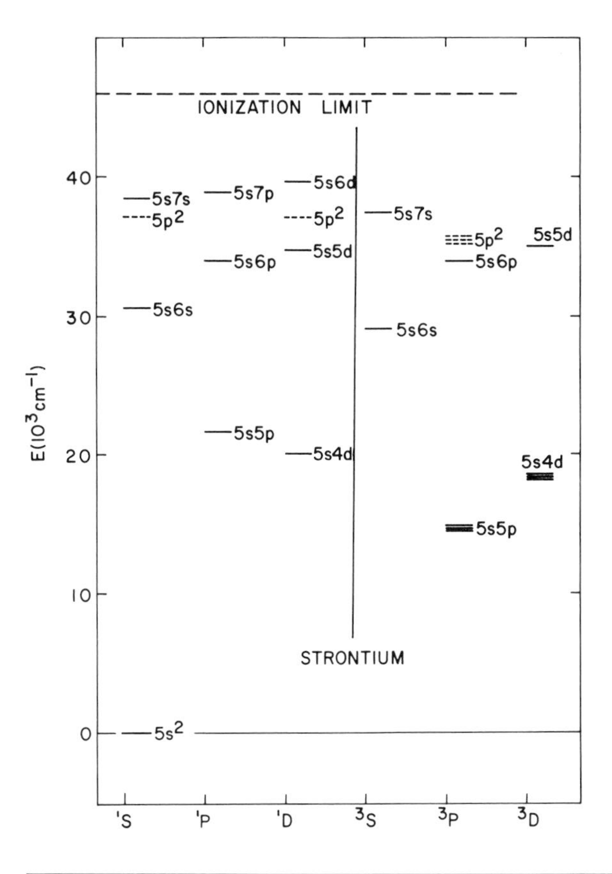

Fig. 5.26. Energy-level diagram for Sr atoms

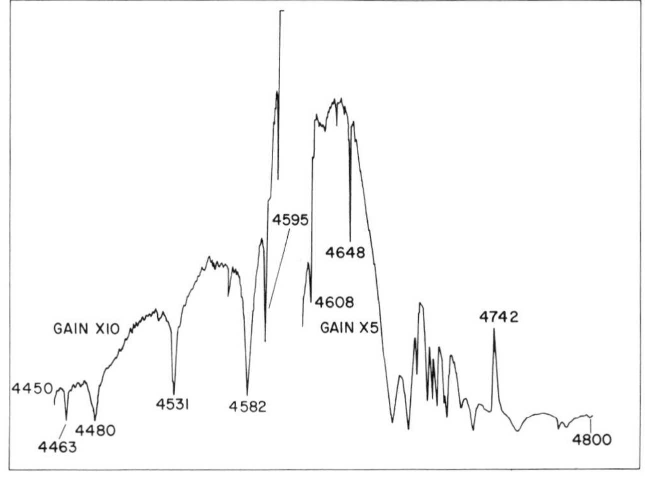

Fig. 5.27. Intensity of infrared laser ($5s4d\,^3D_i - 5s5p\,^3P_j^\circ$) in Sr vapor as a function of pumping dye laser wavelength. Wavelengths are given in units of Å. The dye laser is tuned in the vicinity of the Sr resonance line at 4607 Å. Helium buffer gas pressure is several hundred Torr

mediate states j'. The other intermediate states may then be neglected in the sum over j'. If the laser frequency is, in addition, nearly resonant with one of the states included in the sum over j, this state can likewise be singled out. Both these conditions apply in the Sr experiment, since the dye laser was tuned in the vicinity of the resonance line. *Bebb* [5.47] noted that the occurrence of simultaneous near resonances in first and second order leads to unusually large transition rates ($w^{(3)}_{f,g} \sim 10^{-72}$ F^3, where F is photon flux) in Rb and Cs for photon energies near 1.58 eV and 1.38 eV, respectively. Making the assumption that the same large rate coefficient applies in the Sr case, one finds that at the focus (100 μ spot size) of a 10 kW beam, at ~4607 Å, the ionization rate $w^{(3)}_{f,g} \sim 10^7\,\mathrm{s}^{-1}$. Hence, at the end of a 10 ns laser pulse the probability that an atom will be ionized is ~0.1. Neglecting the recombination that occurs in a 10 ns time interval, a plasma density at the focus as high as $10^{16}/\mathrm{cm}^3$ would result, since typical Sr atom densities in our experiments are $10^{17}/\mathrm{cm}^3$.

What is the magnitude of the Stark broadening expected at a plasma density of $10^{16}/\mathrm{cm}^3$? Let us consider, for simplicity, only the expression for linear Stark broadening of outer electron transitions by ion fields in the quasi-static limit [5.48]

$$\delta\omega \sim 8.8(n_u^2 - n_1^2) Z_i N_i^{2/3}/S_e\,\mathrm{s}^{-1}\,. \tag{5.46}$$

Equation (5.46) applies to transitions between shells of principal quantum numbers n_u and n_1. S_e is the effective nuclear charge experienced by electrons in the n_u shell, and N_i and Z_i are the number density and charge, respectively, of the ions. If we consider the 6.45 μ lasing transition of Sr, we may set $n_u=5$, $n_1=4$, $Z_i=1$, and $S_e=1$. Thus we obtain $\delta\omega \sim 80\,N_i^{2/3} = 3.7 \times 10^{12}\,\mathrm{s}^{-1}$, (i.e., $\delta\nu > 10\,\mathrm{cm}^{-1}$) and it is seen that even without adding contributions from other sources of plasma broadening (such as electron impact, ion impact, turbulence, etc.), the Stark broadening can be significantly greater than even the $\sim 1\,\mathrm{cm}^{-1}$ collisional broadening that would exist with ~1 atm of He buffer gas present [5.49]. Carlsten's conjecture about the cause of the dips thus seems reasonable.

5.4.2 Ultraviolet Generation by Third-Harmonic Generation

Our colleague, *R. T. Hodgson*, had planned to detect two-photon resonantly enhanced THG in atomic systems. As a result of our observation of strong two-photon resonant effects in Sr we combined efforts. Our first observations immediately revealed the effect of two-photon resonance enhancement of THG in Sr. With a single input dye laser, and using a solar-blind photomultiplier to detect the output light, we observed that the intensity of the THG in the vacuum ultraviolet (VUV) increased by many orders of magnitude when the fundamental frequency was tuned to the half frequency of a two-photon allowed transition. For example, when a single, linearly polarized input beam from a sodium fluorescein dye laser was tuned from 5337 to 5710 Å, the THG light signal displayed enormous resonant enhancements—increasing $\sim 10^5$ times from its weak, off-resonance value—at four different wavelengths: 5380, 5409, 5605, and 5681 Å. These wavelengths correspond to exact half frequencies of the following doubly excited, even parity states of Sr: $(5p^2)\ ^1S_0$, $^1D_2$, $^3P_2$, and $^3P_0$, respectively (see Fig. 5.26). Strong THG was also observed when a rhodamine 6 G laser was tuned to 5757 Å, the wavelength for two-photon resonance of the singly excited $5s5d\,^1D_2$ state. Smaller resonantly enhanced signals were noted when a 7-diethylamino-4-methyl coumarin (7D4MC) laser was tuned to the half frequencies of the $5s8s\,^1S_0$, $5s7d\,^1D_2$, $5s9s\,^1S_0$, and $5s8d\,^1D_2$ states, i.e., to frequencies corresponding to some of the dips in Fig. 5.27. We observed that, at maximum input powers, the two-photon absorption rate to each of the $5p^2$ singlet states was sufficiently great to produce visible OPSE laser beams easily detected by the naked eye. The transitions on which OPSE was observed were $5p^2\,^1D_2 - 5s5p\,^1P_1^\circ$ (6550 Å) and $5p^2\,^1S_0 - 5s5p\,^1P_1^\circ$ (6566 Å).

Using light at 5757 Å for THG, we noted that the VUV signal increased about 5 times when a few hundred Torr of xenon gas was added to the Sr vapor maintained at ~ 10 torr. The addition of more Xe resulted in a sharp decrease of the VUV signal. This behaviour is in accord with the expected dependence of phase matching on Xe pressure [5.13]. The relatively small increase observed shows that the THG process with the fundamental at 5757 Å was not too badly phase mismatched in the absence of Xe. However, for the case in which the input beam is supplied by the 7D4MC laser and wavelengths are in the vicinity of the 4607 Å resonance line, a bad phase mismatch occurs which cannot conveniently be compensated by the addition of Xe gas. Consequently the frequency tripling is much less efficient.

Thus far we have described resonant enhancement of THG that occurs for a few select frequencies within the tuning range of a single dye laser. These resonances always appear to be quite sharp, but only in one instance (5757 Å) have we actually performed a careful scan over the two-photon resonance while monitoring the VUV output in order to determine the exact width of the THG resonance. The half width measured in this case ($\delta v \sim 0.5\ \mathrm{cm}^{-1}$) reflects both the spectral width of the dye laser used as well as collisional broadening due to the He buffer gas. The measurement was done with the input beam attenuated considerably, since several mechanisms will broaden the resonance at the higher powers. *Chang* [5.44] has calculated that for the two-photon enhanced THG, saturation due to the ac Stark effect will occur for high input intensities. He predicts that the width of the THG resonance should increase, eventually becoming proportional to I_L^2. Also the third-harmonic intensity should eventually vary as I_L, rather than as I_L^3. In addition the plasma Stark broadening mechanism advanced in Subsection 5.4.1 will broaden THG resonances observed with dye lasers operating at high powers in the vicinity of the 4607 Å resonance line.

Table 5.3. Tuning ranges in the VUV obtainable by sum frequency mixing in Sr with the use of four dye solutions

Dye for laser at λ_1	7-Diethylamino-4-methylcoumarin	7-Diethylamino-4-trifluoro-methylcoumarin	Sodium fluorescein	Rhodamine 6 G
Resonant state	$5s7d\,^1D_2$	$5s6d\,^1D_2$	$5p^2\,^1D_2$	$5s5d\,^1D_2$
λ_1	4779 Å	5032 Å	5409 Å	5757 Å
Dye for laser at λ_2 and its tuning range				
7-Diethylamino-4-methylcoumarin 4648–4925 Å	1578–1609 Å	1632–1666 Å	1710–1746 Å	1778–1817 Å
7-Diethylamino-4-trifluoro-methylcoumarin 4900–5450 Å	1607–1662 Å	1662–1722 Å	1743–1808 Å	1814–1884 Å
Sodium fluorescein 5337–5710 Å	1657–1685 Å	1710–1747 Å	1795–1836 Å	1870–1914 Å
Rhodamine 6 G 5680–6111 Å	1682–1718 Å	1744–1783 Å	1833–1875 Å	1907–1957 Å

5.4.3 Ultraviolet Generation by Four-Wave Parametric Sum Mixing

To the same extent that a THG signal becomes resonantly enhanced by tuning the input laser to the half frequency of a double-quantum transition, a *sum mixing* signal, obtained with the introduction of a second input laser, will be also enhanced, provided that the first laser remains tuned to the double-quantum transition. There is no constraint upon the frequency of the second laser. This becomes the basis for a tunable VUV source. The third-order nonlinear susceptibility which applies to the $\omega_{\mathrm{VUV}}=2\omega_1+\omega_2$ sum mixing process, when a single state j' is near resonance with $2\omega_1$, can be written as

$$\chi^{(3)}(\omega_{\mathrm{VUV}})=\frac{N}{12\hbar^3}\sum_{j,j',j''} \cdot \frac{\mu_{gj''}\mu_{j''j'}\mu_{j'j}\mu_{jg}}{(\omega_{j''g}-\omega_{\mathrm{VUV}}-\mathrm{i}\Gamma_{j''g})(\omega_{j'g}-2\omega_1-\mathrm{i}\Gamma_{j'g})(\omega_{jg}-\omega_1-\mathrm{i}\Gamma_{jg})}\,. \tag{5.47}$$

From (5.47) one sees that resonance enhancement of the sum mixing process will always occur when the first laser is tuned to the double-quantum transition. Additional resonance enhancements should occur when $\omega_{j''g}-\omega_{\mathrm{VUV}}=0$. We shall shortly see that these, indeed, are observed.

A single-frequency, tunable VUV light source can be made based on the fact that circularly polarized light cannot be frequency tripled in isotropic media since angular momentum would not be conserved [5.50]. In the configuration of Fig. 5.1, a linearly polarized dye laser (ω_1) is tuned to be two-photon resonant with an even parity $J=2$ state such as $5s5d\,^1D_2$ (5757 Å) or $5p^2\,^1D_2$ (5409 Å), and the THG is nulled using a $\lambda/4$ plate to convert the light to circular polarization. A second dye laser (ω_2) is then added collinearly by means of the Glan prism. The $\lambda/4$ plate converts some part of this light to circular polarization in the opposite sense, allowing strong VUV generation at $\omega_{\mathrm{VUV}}=2\omega_1+\omega_2$. Tuning ω_2 sweeps ω_{VUV} over a tuning range as wide as that of ω_2. This tuning has been verified from various combinations of dyes with the use of a VUV monochromator and also, in the case of direct photomultiplier detection, by sweeping the generated output wavelength over various known absorption lines.

In Table 5.3 the tuning ranges that can be covered using combinations of four commonly used dyes are shown. In each case, one laser is fixed at a double-quantum resonance while the other is tuned over its complete tuning range. These four dyes allow one to cover the range from 1578 to 1957 Å. All of the tuning ranges shown in Table 5.3 correspond to energies that exceed the first ionization limit in Sr (Fig. 5.26). They are thus co-energetic not only with the continuum states of Sr, but also with various so-called autoionizing levels which exist in this range. These levels can play the roles of states j'' in (5.47) and can thus lead to additional resonance enhancement of the generated VUV signal when ω_{VUV} becomes resonant with any of them. Such resonant

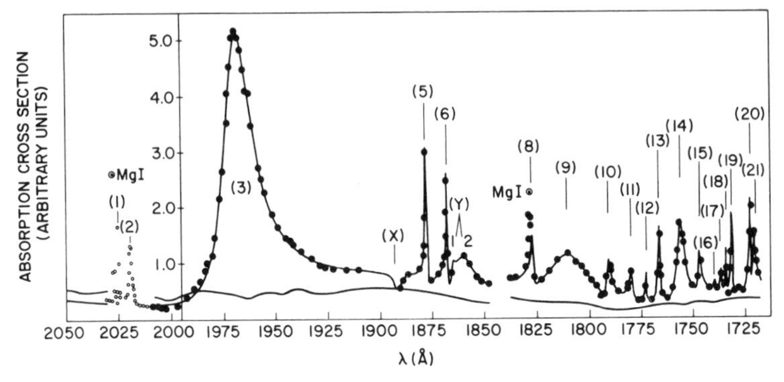

Fig. 5.28. Autoionization resonances seen in absorption in Sr. Points are experimental measurements. Solid curve is summation of Fano profiles determined by parameters from Table 5.4 (after [5.52])

enhancements are, in fact, a dramatic feature of the VUV generation observed by us in Sr. Before presenting these experimental results, a brief discussion of *Fano*'s [5.51] model of these states will be presented.

Theory of Autoionizing States

Autoionizing levels occur in atoms like Sr because the simultaneous excitation of the two outer (here 5*s*) electrons can result in a set of discrete states, some of which have energies in excess of that required for normal ionization. Thus, these states can give rise to discrete absorption lines seen in the midst of the photoionization continuum. If there is no interaction between the electron configurations involved in the discrete state and the various continuum states, the absorption lines have a normal appearance, notwithstanding their location in energy. More generally, there *is* an interaction between the discrete state and the continuum. The continuum states correspond to orbits in which the electron moves to infinity. As a result of its interaction with the continuum, the discrete state acquires to some degree the property of spontaneous ionization with one of its electrons moving out to infinity. This property is called *autoionization*. The width of an autoionizing level, as seen in absorption from the ground state, will be proportional to the rate of spontaneous ionization, in accordance with the Uncertainty Principle. Observed widths of autoionizing levels typically range from a few cm^{-1} to a few thousand cm^{-1}. Besides their broadened appearance, absorption spectra involving these states are observed to have characteristically asymmetric, dispersion-like shapes. This may be seen for the case of Sr in Fig. 5.28. *Fano* [5.51] pointed out that the shape of autoionization lines result from the sharp variation in the coefficients of admixture of the various configurations as the energy value passes through the discrete state. The result of his calculation is that the ratio of the actual absorption cross section in the vicinity of an auto-

Table 5.4. Results of analysis of Sr absorption spectrum (Fig. 5.28). After Table 1 of [5.52]

Resonance no.	Transition	λ_r (Å)	Γ (cm^{-1})	$-q$
	$5\,^1S_0$ to			
1	$4d6p\,^3D_1^\circ$	2024	too narrow for analysis	
2	$^3P_1^\circ$	2018	53.1	
3	$^1P_1^\circ$	1970	415	− 5.2
X	$5p6s\,^3P_1^\circ$	1891	83.8	0.10
5	$4d(^2D_{3/2})4f[3/2]_1^\circ$	1878.0	17.4	4.6
6	$4d(^2D_{5/2})4f[3/2]_1^\circ$	1867.9	20.3	3.5
Y_1	$5p6s\,^1P_1^\circ +$	1865.5	49.5	− 1.9
Y_2	$4d(^2D_{5/2})4f[3/2]_1^\circ?$	1859.6	291	−13
8	$4d7p\,^3P_1^\circ$	1827.2	51.1	2.5
9	$^1P_1^\circ$	1810.1	700	
10	$4d(^2D_{3/2})5f[3/2]_1^\circ$	1791.4	68.4	− 2.6
11	$4d(^2D_{5/2})5f[3/2]_1^\circ$	1780.1	75.3	4.8
12	$4d8p\,^3D_1^\circ$	1772.8	19.0	7.1
13	$^3P_1^\circ$	1766.7	36.7	27
14	$^1P_1^\circ$	1756.8	132	− 6.2
15	$4d(^2D_{3/2})6f[3/2]_1^\circ$	1747.1	47.6	− 3.1
16	$4d(^2D_{5/2})6f[1/2]_1^\circ$	1740.3	53.1	
17	$4d(^2D_{5/2})6f[3/2]_1^\circ$	1737.4	26.5	
18	$4d9p\,^3D_1^\circ$	1735.1	14.3	
19	$^3P_1^\circ$	1732.2	24.4	
20	$4d(^2D_{3/2})7f[3/2]_1^\circ$	1723.5	31.0	3.7
21	$4d9p\,^1P_1^\circ$	1722.2	59.9	− 2.5

ionizing line to that of the unperturbed photoionization continuum can be written as the square of the ratio of dipole matrix elements

$$\left|\frac{\mu_{\Psi_\omega g}}{\mu_{\psi_\omega^0 g}}\right|^2 = \frac{(q+\varepsilon)^2}{1+\varepsilon^2}\,. \tag{5.48}$$

In (5.48), Ψ_ω is the perturbed continuum state at energy $\hbar\omega$, ψ_ω^0 is the unperturbed continuum state, q is a dimensionless parameter describing the lineshape, and ε is the frequency offset of the probing radiation from the center of the resonance, normalized to the halfwidth $\Gamma/2$ of the resonance.

It has become customary to analyze the shapes of autoionizing absorption lines by fitting them with "Fano profiles", that is, finding appropriate q values for the transitions involved. Table 5.4 presents results of just such a procedure done for the case of the Sr absorption spectrum in Fig. 5.28 [5.52]. A wide range of both q values and linewidths is seen to be represented. Note the interesting $q \simeq 0$ case represented by the resonance "X". This represents the case for which the transition moment from the ground state to the discrete state is nearly zero. Nonetheless, there is a strong interaction between the discrete and continuum states which causes a spectral repulsion of the latter away from the former and the results in the appearance of a "hole" in the spectrum. Note also the relatively equal numbers of lines with positive and negative q's. Lines a, b for which $q_a = -q_b$ are mirror images of each other.

One feature of Fig. 5.28 which requires some explanation is the following. It will be noted that the "zeros" of the Fano profiles do not drop completely to the baseline, in seeming contradiction to (5.48). This is because additional absorption continua, which do not interact with the discrete states through the process of autoionization, overlap the region of zero absorption. In Fig. 5.28 the noninteracting part of the continuum is indicated by the quasi-smooth line drawn above the baseline. The curve delineating the resonances is actually the result of a computer fit using the choice of parameters shown in Table 5.4.

Sum Mixing Enhanced by Autoionizing Levels

Let us now focus on the wavelength dependence of generated VUV signal (Figs. 5.29–33), comparing the results to Fig. 5.28. The relatively uniform sum mixing spectrum depicted in Fig. 5.29 spans a range of wavelengths (1911 Å–1959 Å) for which there is seen to be no sharp structure in the corresponding absorption spectrum. (The falloff of intensity away from the middle of the trace in Fig. 5.29 is due primarily to the gradual falloff of the dye laser power (at ω_2) away from its maximum value. In none of the plots shown in Figs. 5.29–33 is the VUV signal normalized, that is, divided by the intensity of the laser at ω_2.) For the case of Fig. 5.29, the VUV sum mixing process may be regarded as being resonantly enhanced by the big autoionizing resonance at 1970 Å (labeled as #3 in Fig. 5.28) whose high-frequency wing extends out to at least the short wavelength limit of Fig. 5.29. For cases in which the VUV intensity varies smoothly, as in Fig. 5.29, the sum mixing process directly provides a usable, tunable VUV source whose spectral resolution is equal to that of the input lasers.

Figure 5.30 shows the intensity variation of the VUV signal (solid line) in a wavelength range that includes both a relatively sharp absorption peak (#5) as well as the Fano resonance "X". Part of the absorption spectrum from Fig. 5.28 has been superimposed to show clearly the exact correspondence between the two Fano resonances and the two large observed VUV resonance enhancement peaks. The importance of resonance enhancement due to autoionizing states is dramatically shown in this figure. The ratio of the heights of the two peaks is $\sim 10:1$, and the smaller peak itself is roughly ten times greater than the VUV signal in the range 1896 Å–1917 Å. The strength of signal in the latter range corresponds to the VUV intensity for the case of Fig. 5.29, since the wavelength spans of Figs. 5.29 and 5.30 partially overlap. The same two-photon resonant state ($5s5d\,^1D_2$) was employed in both instances. The distortion which appears on the low-frequency side of the smaller peak in Fig. 5.30 is attributable to the presence of a small amount of Ba impurity in the Sr vapor cell (both being alkaline earths). As ω_2 scans

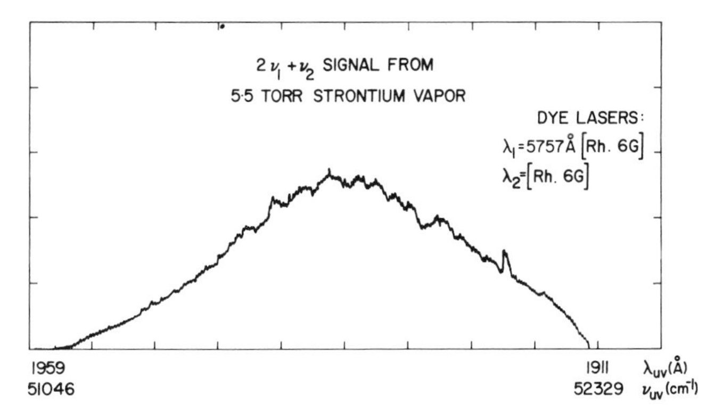

Fig. 5.29. Intensity of VUV generated in Sr

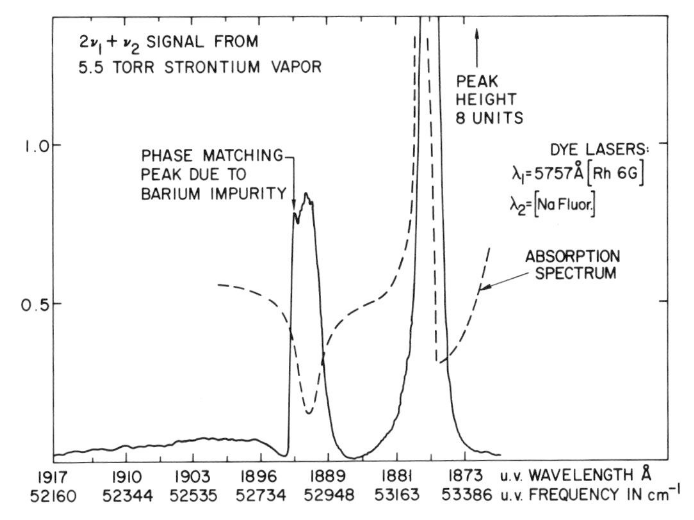

Fig. 5.30. Intensity of VUV generated in Sr (solid curve) superposed on part of the absorption curve (dotted curve) taken from Fig. 5.28

through the spectral region which corresponds to a VUV wavelength of 1893 Å, it encounters the main Ba resonance line ($6s^2\,{}^1S_0-6s6p\,{}^1P_1^0$) at 5535 Å. This causes the sum-mixing process to become less well phase matched on the low-frequency side of the Ba line, to drop almost to zero as ω_2 is largely absorbed at exact resonance, and, finally, to become better phase matched on the high-frequency side of the Ba line.

Figures 5.31 and 5.32 show VUV generation traces for a range of somewhat shorter output wavelengths. The two traces differ only in that for Fig. 5.32 the transitional section between the LiF output window and solar-blind photomultiplier was not quite properly evacuated. The resulting absorption, seen

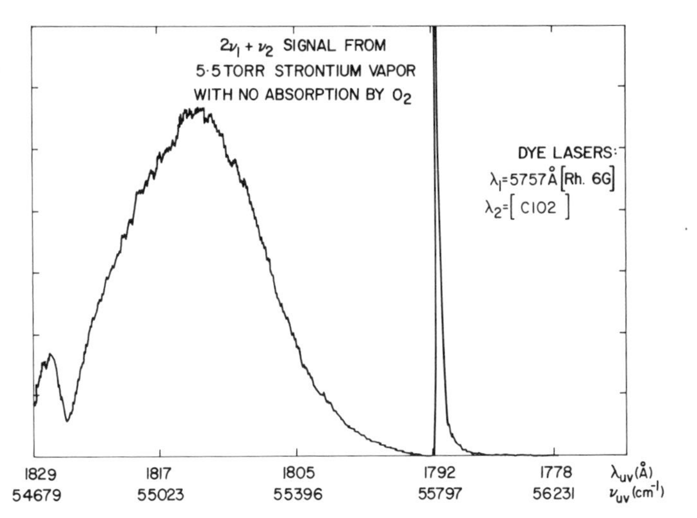

Fig. 5.31. Intensity of VUV generated in Sr. The sharp peak near 1792 Å arises from an accidental two-photon resonance (see text)

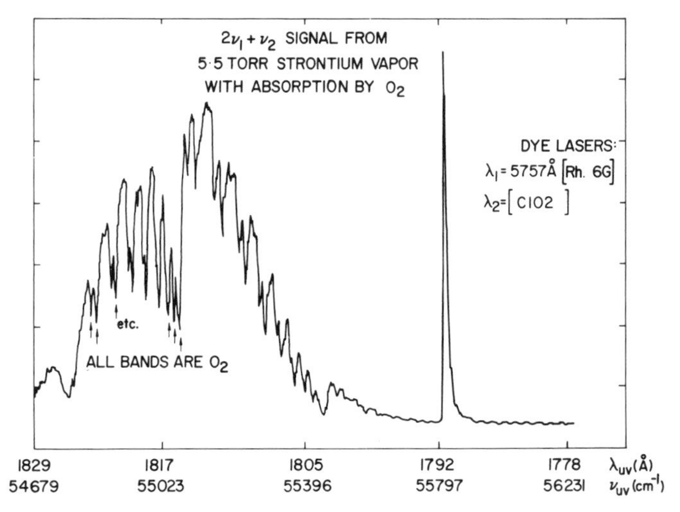

Fig. 5.32. Signal detected by photomultiplier after VUV generated in Sr passes through a small amount of air. The tuning range is identical to that of Fig. 5.31. The additional structure is due to absorption by the Schumann-Runge band system of O_2

by comparing the two figures, is due to the Schumann-Runge band system of O_2 and indicates how the present experimental configuration can be utilized as a VUV spectrometer. The intrinsic features of the sum mixing process shown in Fig. 5.31 include a broad band that obviously corresponds to the Fano resonance #9 and a very sharp peak that appears just to the long wavelength side of 1792 Å. This sharp peak results not from an autoionizing state,

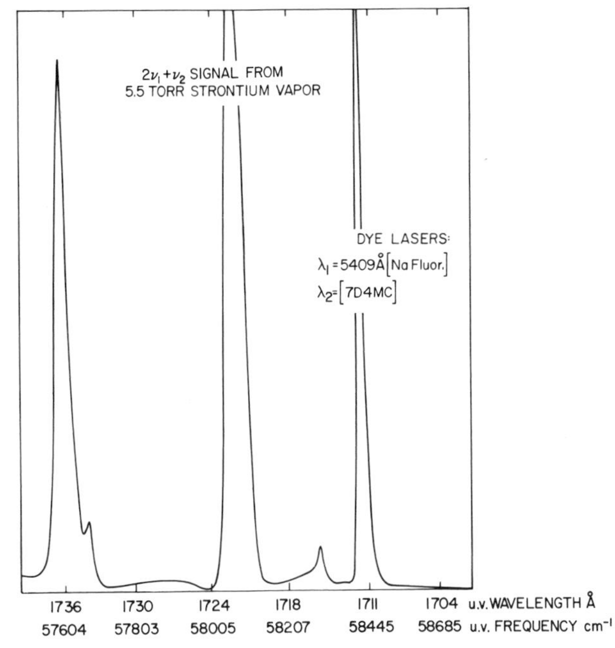

Fig. 5.33. Intensity of VUV generated in Sr in a range that includes the extreme right-hand portion of Fig. 5.28

but from additional two-photon resonance enhancement accidentally achieved as ω_2 is swept through its range of frequencies. Specifically, with v_1 fixed at the half frequency of the $5s5d\,^1D_2$ state ($17\,363.75\ \mathrm{cm}^{-1}$), when v_2 happens to attain the value $21\,080.25\ \mathrm{cm}^{-1}$, the sum v_1+v_2 becomes resonant with the $5s7s\,^1S_0$ state. A new, resonantly enhanced contribution to the signal at $2v_1+v_2$ ($55\,807.75\ \mathrm{cm}^{-1}$) occurs. At this point the $5s7s$ two-photon resonance enhancement appears more important than that of the $5s5d$ state. This is attributed not to intrinsically larger matrix elements involving the former state, but rather to the fact that ΔE_1 in the former case is $618.25\ \mathrm{cm}^{-1}$, whereas in the latter case it is $4334.75\ \mathrm{cm}^{-1}$. Since the VUV signal varies as $[\chi^{(3)}]^2$, i.e., as $(\Delta E_1)^{-2}$, one expects the $5s7s$ enhancement in the case of Fig. 5.31 to be ~ 50 times greater than that due to the $5s5d$ state, with other factors assumed to be the same. Figure 5.31 thus illustrates the importance of minimizing all *three* quantities ΔE_1, ΔE_2, and ΔE_3 in order to generate a maximum amount of VUV.

For the trace shown in Fig. 5.33 the $5p^2\,^1D_2$ state was chosen as the two-photon resonant state. Strong, well-defined resonances in the VUV signal are observed. In particular, the two peaks at the shortest wavelengths (1712.3 Å and 1715.5 Å) are seen to be well resolved. In the work reported by *Garton* et al. [5.52], from which Fig. 5.28 has been taken, a 1-meter, normal incidence, vacuum spectrometer was utilized in order to obtain photometric scans.

Garton et al. [5.52] remark that below 1722 Å the spectrum was too crowded for reliable data to be obtained with the dispersion available. Thus, at the shorter wavelengths, autoionizing state studies performed with the present technique of sum frequency mixing can be made with greater accuracy than with currently available, commercial scanning spectrometers. In the original high-resolution photographic work of *Garton* and *Codling* [5.53] a 3-meter, normal incidence grating was used, and the plates presented in that work do show fully resolved autoionizing lines of Sr out to ~1650 Å.

Comparing Fig. 5.33 with Fig. 5.28 and with various measurements listed in [5.53] one arrives at the following conclusions: The large peak at the longest wavelength shown in Fig. 5.33 corresponds to peak #17 in Fig. 5.28. The much smaller peak on its high-frequency wing corresponds to peak #18. Note that in Fig. 5.28 peak #18 is larger than peak #17. Even more remarkable is the fact that there is absolutely no sign of a peak in the VUV generation spectrum corresponding to peak #19, which in Fig. 5.28 is seen to be larger than either #17 or #18! In Fig. 5.33 the off-scale central peak is actually a double peak (measured wavelengths 1722.9 Å, 1722.3 Å) which would appear to correspond to the pair of peaks #20 and #21, although there is some discrepancy in wavelengths. (The separation between the peaks is greater in the case of Fig. 5.28 than in Fig. 5.33.) Finally, in [5.53] the next two shorter wavelength autoionizing lines listed are the $5s^2\,{}^1S_0 - 4d({}^2D_{5/2})\,7f[1/2]_1^0$ transition at 1715.65 Å and the $5s^2\,{}^1S_0 - 4d({}^2D_{3/2})\,10p[1/2]_1^0 : {}^3D_1^0$ transition at 1712.44 Å. These are seen to correlate well with our measured values of 1715.5 Å and 1712.3 Å for the two shortest wavelength lines in Fig. 5.33. The relative sharpness of these two lines as seen in VUV generation, particularly the one at 1712.3 Å, is also apparent in absorption in the plates of *Garton* and *Codling* [5.53]. As a general rule, both in absorption and in VUV generation, the resonances get sharper for the higher members of a sequence approaching a series limit. The two transitions described above fall in this category.

Armstrong and *Wynne* [5.54] have shown that it is possible to express the frequency dependence of the generated VUV signal in terms of Fano q parameters. In the calculation it is assumed that because of the two-photon resonance ($2\omega_1 \cong \omega_{j'g}$) and the near resonance of the VUV generated frequency with a Fano state, only one type of term is important in the four-wave susceptibility, viz.,

$$\chi^{(3)}(\omega_{\mathrm{VUV}}) \cong \frac{N}{12\hbar^3} \frac{\mu_{j'j}\mu_{jg}}{(\omega_{j'g} - 2\omega_1 - \mathrm{i}\Gamma_{j'g})(\omega_{jg} - \omega_1 - \mathrm{i}\Gamma_{jg})} \cdot \int_{-\infty}^{\infty} \frac{\mu_{g\Psi_\omega}\mu_{\Psi_\omega j'}\,d\omega}{\omega - \omega_{\mathrm{VUV}} - \mathrm{i}\Gamma(\omega)}\,. \tag{5.49}$$

The limits of integration reflect the fact that all continuum states contribute for which the matrix elements do not vanish. In (5.49) the integration automatically takes into account both the contribution to $\chi^{(3)}$ of the autoionizing state and that of the continuum with which it interacts if one uses for $\mu_{g\Psi_\omega}$ and $\mu_{\Psi_\omega j'}$ the matrix elements [5.51]

$$\mu_{g\Psi_\omega} = \mu_{g\psi^0_\omega}(q_g \sin\Delta - \cos\Delta) \tag{5.50}$$

and

$$\mu_{\Psi_\omega j'} = \mu_{\psi^0_\omega j'}(q_{j'} \sin\Delta - \cos\Delta)\,. \tag{5.51}$$

In (5.50) $\mu_{g\psi^0_\omega}$ is, as in (5.48), the matrix element of the dipole moment operator from the ground state to the unperturbed continuum state ψ^0_ω; q_g is the same Fano q parameter appearing in (5.48). Analagously defined quantities for the two-photon resonant state j' appear in (5.51). The quantity Δ is a function only of the autoionizing state, not states g, j':

$$\Delta = -\arctan(1/\varepsilon) \tag{5.52}$$

where $\varepsilon = (\omega - \bar{\omega})/(\Gamma/2)$ is the frequency offset from the Fano resonance at $\bar{\omega}$, normalized to its halfwidth. From (5.50) and (5.52) one can, of course, derive (5.48). As can be seen from (5.50), the rapid variation of Δ near a Fano resonance is responsible for the interference that leads to a cancellation of the transition probability on one side of the resonance. The inclusion of the term $-\mathrm{i}\Gamma(\omega)$ in the denominator of the integrand of (5.49) accounts for relaxation processes.

Since most of the contribution to the integral in (5.49) comes from the region of the singularity $\omega = \omega_{\mathrm{VUV}}$, upon substitution of (5.50–52) one can remove the slowly varying functions $\mu_{g\psi^0_\omega}$ and $\mu_{\psi^0_\omega j'}$ from under the integral sign yielding

$$\chi^{(3)}(\omega_{\mathrm{VUV}}) = \frac{N}{12\hbar^3} \frac{\mu_{g\psi^0_\omega}\mu_{\psi^0_\omega j'}\mu_{j'j}\mu_{jg}F}{(\omega_{j'g} - 2\omega_1 - \mathrm{i}\Gamma_{j'g})(\omega_{jg} - \omega_1 - \mathrm{i}\Gamma_{jg})} \tag{5.53}$$

with

$$F = \int_{-\infty}^{\infty} \frac{[q_g q_{j'} + \varepsilon^2 + \varepsilon(q_g + q_{j'})]\, d\varepsilon}{[\varepsilon - x - 2\mathrm{i}\Gamma(\varepsilon)/\Gamma](1+\varepsilon^2)} \tag{5.54}$$

where $x = (\omega_{\mathrm{VUV}} - \bar{\omega})/(\Gamma/2)$. Equation (5.54) can be evaluated by the methods of contour integration. With $\Gamma(\varepsilon)$ set equal to zero *after* the integration, $\chi^{(3)}$ is shown to have both real and imaginary parts, as first pointed out by *Armstrong* and *Beers* [5.55]. The spectral shape of the VUV-generated light is computed from the square magnitude of $\chi^{(3)}$, provided that linear absorption at ω_{VUV} is negligible, and is given by

$$|\chi^{(3)}|^2 \sim |\mu_{g\psi^0_\omega}|^2 |\mu_{\psi^0_\omega j'}|^2 \left\{\frac{q_g^2 q_{j'}^2 + [x + (q_g + q_{j'})]^2}{1+x^2}\right\}. \tag{5.55}$$

Note that the expression for $|\chi^{(3)}|^2$ in (5.55) reduces to $|\mu_{g\psi\mathcal{Q}}\mu_{\psi\mathcal{Q}j'}|^2$ as x becomes large. This is a measure of the VUV signal strength away from the autoionizing resonances. For large values of q_g, $q_{j'}$ the VUV signal becomes enhanced at resonance ($x=0$) by a factor $\sim q_g^2 q_{j'}^2$ which may easily reach a value $\sim 10^4$. *Armstrong* and *Wynne* [5.54] investigated the $4d4f$ level (#6) of Sr at 1867 Å by VUV generation, using both the $5s5d\ ^1D_2$ and the $5p^2\ ^1D_2$ states for j'. They observed two very different lineshapes. The pressure of Sr was kept sufficiently low so that linear absorption at ω_{VUV} had no observable effect on the line shapes. Since the q_g are available from ground state VUV absorption data (Table 5.4), a fit of (5.55) to the experimental line shapes of the VUV-generated light serves to determine values for $q_{j'}$. The results for this case are $q_{5p^2}=-0.6$ and $q_{5s5d}=2.1$, with $q_g=-3.5$ [5.55]. The linewidths are found to be in good agreement with [5.52]. Recently we have succeeded in obtaining preliminary results on resonantly enhanced, three-photon photoionization spectra from which the $q_{j'}$ can be directly measured. In these experiments the solar-blind photomultiplier is replaced by an ionization probe. As in the case of VUV generation, one laser is tuned to a double-quantum resonance, and the other laser is scanned, scanning ω_{VUV} over the autoionizing states. According to (5.44), (5.45), (5.51), and (5.52), the rate of three-photon ionization is proportional to $|\mu_{\psi\mathcal{Q}j'}|^2(q_{j'}+\varepsilon)^2/(1+\varepsilon^2)$. Thus, there is provided an independent measure of $q_{j'}$ from the variations in ionization observed as ω_2 is scanned.

VUV Generation in Other Atomic Systems

We conclude with some remarks on studies in systems other than Sr, intensities of VUV beams generated, and future prospects for study. As the reader can probably judge, the work done thus far in Sr by no means exhausts what can be profitably studied with this system.

With the same techniques used in Sr, we were *unable* to generate a detectable VUV signal in Na vapor. Two particular schemes were tried. In the first scheme a single linearly polarized rhodamine 6G laser was tuned to two-photon resonance with the $4d$ states. In the second, two dye lasers were employed. One laser (ω_1) was tuned between the $p_{1/2}$, $p_{3/2}$ components of the $3p$ state, roughly twice as close to the $p_{1/2}$ component as to the $p_{3/2}$ component in order to achieve phase matching. The other laser (ω_2) was then adjusted so that the sum frequency $\omega_1+\omega_2$ again was equal to the frequency of the $4d$ states. We attribute the lack of signal in both cases to two facts: 1) Na does not have any autoionizing levels in the region at interest, and 2) at the wavelengths of the expected VUV generation (~ 1900 Å) the smooth photoionization curve for Na happens to have almost zero value [5.56]. It should be noted, however, that *Bloom* et al. [5.57] have successfully used Na vapor for THG of Nd (1.06 μ) laser radiation.

In the case of two rare earth vapors, Eu and Yb, we were successful in generating strong VUV signals comparable to those in Sr. Eu and Yb both have ($6s^2$) outer electron configurations and thus satisfy the requirement of having autoionizing states in the vicinity of the generated VUV frequency.

In the case of Eu, however, there is a half-filled $4f$ shell ($4f^7\ ^8S_{7/2}$) that interacts with the $6s^2$ electron shell, splitting its various terms and thus "diluting" the matrix elements which determine the nonlinear susceptibility $\chi^{(3)}$. By comparison to Sr, there are many more two-photon resonances encountered during the scan of a single dye laser. Fortunantly, most of the states lying below the ionization limit are well characterized in the remarkable early work of *Russell* and *King* [5.58], who were the ones to "crack" the Eu spectrum. With Eu we were able to demonstrate a further principle of resonance enhancement, namely, that the intensity of the VUV output generated by sum mixing can be enhanced by ~ 100 in the case of relatively heavy (two electron) atoms by tuning one of the input beam frequencies close to a relatively forbidden transition from the ground state to a triplet state. This may be understood by considering the general expression for $\chi^{(3)}$ given in (5.47). If it is assumed that the oscillator strength f_{gj} characterizing the transition from state g to state j is $\sim 1/100$, instead of ~ 1, then $\omega_{jg} - \omega_1$ can be reduced by ~ 100 without increasing the difficulty of phase matching. Since $\mu_{jg} \sim f_{gj}^{1/2}$, $\chi^{(3)}$ will be increased by ~ 10 and the VUV power output, which varies as $|\chi^{(3)}|^2$, will be increased by ~ 100. In this argument it is assumed that the other matrix elements retain their same order of magnitude when the relevant states lie in the triplet manifold. This assumption can be justified from the known data. Using the above principle, in Eu we observed VUV power increases of roughly two orders of magnitude by tuning one laser ν_1 very close to a triplet state (e.g., $6s6p\ ^6P_{7/2}$ at 17340.65 cm^{-1}) while the other laser (ν_2) was tuned so that $\nu_1 + \nu_2$ equalled the frequency of a two-photon resonant state (e.g., $6s6d\ ^6D^0_{9/2}$ at 36566.6 cm^{-1}). Of course, to take advantage of this technique and still have a tunable output would require the introduction of a third input beam. This technique should work even better in Sr, whose levels are not diluted by the interaction with another shell, as in Eu. In Sr one would tune the first laser close to the $5s5p\ ^3P^0_1$ state (Fig. 5.26). The required wavelength (~ 6892.6 Å) can be provided by a nitrogen-laser-pumped dye laser using a mixture of rhodamine B and nile blue perchlorate A.

Yb has a completely filled $4f$ shell and so does not have as complex a spectrum as Eu. The difficulty in its use is one of a technical nature. Its melting temperature is close to the temperature at which suitable vapor pressure for VUV generation is developed. This makes it impossible to operate in a heat-pipe mode where capillary action would return the Yb to the hot zone. Hence, the operation of the vapor cell for any but the shortest period of time is impossible without cooling and recharging it. For Yb, in contrast to the case of most atoms including Sr, very few of the states lying below the first ionization limit are known [5.59]. We believe we have located the first excited singlet state of Yb. We observe a strong two-photon resonantly enhanced THG signal at $2\nu_1 = 34360$ cm^{-1}. By comparing its position with the few other states known for Yb, and deducing that it is a $J=0$ state, we arrived at the conclusion that the two-photon resonant state is the $6s7s\ ^1S_0$ state.

5.5 Concluding Remarks

Nonlinear optical mixing is an important technique for generating coherent light outside of the range of existing lasers. The theme of this book is the generation of light in the infrared. In this chapter we have shown how to use atomic vapors as the nonlinear medium to generate both infrared and ultraviolet light, tunable over wide regions of the spectrum, through the use of four-wave interactions. Although this book is primarily about the infrared, the physics of the nonlinear interactions we have studied in atomic vapors is common to both infrared and ultraviolet generation.

The reader may find it interesting to know the course we have followed in carrying out the research discussed in this chapter. We started studying infrared generation via stimulated Raman scattering in Cs. Infrared generation by four-wave difference mixing was an outgrowth of these initial studies. In the course of trying to observe stimulated Raman scattering tunable over wider regions in the infrared, we observed the strong photoionization effects in Sr discussed in Subsection 5.4.1. The failure to observe SERS from Sr, coupled with the obvious strength of the resonant enhancement of the three-photon ionization, led us to look for THG as *Harris* and co-workers [5.13, 40] had done in the alkali metals. The result was that we shifted our attention and interest to ultraviolet generation. In the course of tuning the ultraviolet output, we observed interesting resonance effects due to autoionizing levels and realized how this method could be useful to study the spectroscopic properties of the nonlinear medium. By undertaking the essentially applied physics problem of extending the tuning range of coherent sources of light, we have been led back to the basic physics problem of spectroscopy of atomic energy levels.

Although the work reported in this chapter has dealt exclusively with atoms as the nonlinear medium, ions and molecules should show the same effects. Multiphoton photoionization studies in molecules have been carried out [5.60]. Attempts in our laboratory to observe THG from benzene and some benzene derivatives were unsuccessful because of the large molecular extinction coefficients at the ultraviolet wavelengths we tried to detect. Recently, we observed THG from Ca ions which were produced in the focal region of a dye laser beam by two-photon photoionization, the same laser both producing the ions and generating the third-harmonic output. Experiments of this nature may be carried out using one laser to produce the ions and another to interact nonlinearly with these ions. Alternatively the ions may be produced by running an electrical discharge in the pipe containing the vapor of interest.

Clearly there is a wealth of different phenomena which can now be studied using tunable dye lasers to probe the nonlinear optical properties of atoms, ions and molecules. We have discussed in detail some of those phenomena which lead to the generation of tunable, coherent light.

Note Added in Proof

Recently, *Cotter* and *Hanna* [5.61] have cleared up the anomaly concerning the failure of the SERS gain to go to zero between resonance lines (Section 5.3.1, p. 172). They point out that the theoretical gain for Stokes emission polarized perpendicular to the laser polarization does not vanish. They experimentally verify that between the resonance lines the polarization of the Stokes emission does indeed switch from parallel to perpendicular.

Acknowledgements. Our colleagues, Dr. *John Armstrong*, Dr. *Rodney Hodgson*, and Mr. *John Lankard* have contributed to various aspects of the research discussed in this chapter. In addition we have benefitted from discussions with Dr. *Daniel Grischkowsky*, Dr. *Michael Loy*, Dr. *Russell Dreyfus*, and Professor *Y.-R. Shen*. We thank *Harold Lynt*, *Louis Manganaro*, and *Clinton Wood* for technical assistance. We also thank Mrs. *Pauline Webb* for her patience and perserverance in typing this manuscript.

References

5.1 P.A. Franken, A.E. Hill, C.W. Peters, G. Weinreich: Phys. Rev. Lett. **7**, 118 (1961)
5.2 R.W. Terhune, P. Maker, C.M. Savage: Phys. Rev. Lett. **8**, 21 (1962)
5.3 G.H.C. New, J.F. Ward: Phys. Rev. Lett. **19**, 556 (1967)
5.4 J.F. Ward, G.H.C. New: Phys. Rev. **185**, 57 (1969)
5.5 F.P. Schafer: *Dye Lasers*, Topics in Applied Physics, Vol. 1 (Springer-Verlag, Berlin, Heidelberg, New York 1973)
5.6 T.W. Hänsch: Appl. Opt. **11**, 895 (1972)
5.7 E.D. Stokes, F.B. Dunning, R.F. Stebbings, G.K. Walters, R.D. Rundel: Opt. Commun. **5**, 267 (1972);
F.B. Dunning, F.K. Tittel, R.F. Stebbings: Opt. Commun. **7**, 181 (1973)
5.8 C.R. Vidal, J. Cooper: J. Appl. Phys. **40**, 3370 (1969)
5.9 C.R. Vidal, M.M. Hessel: J. Appl. Phys. **43**, 2776 (1972)
5.10 C.R. Vidal, F.B. Haller: Rev. Sci. Instr. **42**, 1779 (1971)
5.11 N. Bloembergen: *Nonlinear Optics* (W.A. Benjamin, New York 1965)
5.12 P.D. Maker, R.W. Terhune: Phys. Rev. **137**, A 801 (1965)
5.13 R.B. Miles, S.E. Harris: IEEE J. QE-**9**, 470 (1973)
5.14 N. Bloembergen: Am. J. Phys. **35**, 989 (1967)
5.15 P.P. Sorokin, N.S. Shiren, J.R. Lankard, E.C. Hammond, T.G. Kazyaka: Appl. Phys. Lett. **10**, 44 (1967)
5.16 M. Rokni, S. Yatsiv: Phys. Lett. **24** A, 277 (1967)
5.17 Y.R. Shen, N. Bloembergen: Phys. Rev. **137**, A 1787 (1965);
Y.R. Shen: Phys. Rev. **138**, A 1741 (1965)
5.18 R. Srinivasan: private communication (1973)
5.19 P.P. Sorokin, J.J. Wynne, J.R. Lankard: Appl. Phys. Lett. **22**, 342 (1973)
5.20 W.L. France, F.P. Dickey: J. Chem. Phys. **23**, 471 (1955)
5.21 J.R. Gordon, T.K. McCubbin, Jr.: J. Mol. Spectr. **19**, 137 (1966)
5.22 D. Cotter, D.C. Hanna, P.A. Kärkkäinen, R. Wyatt: Opt. Commun. **15**, 143 (1975)
5.23 J.J. Wynne, P.P. Sorokin: J. Phys. B. L **37**, (1975)
5.24 I.L. Tyler, R.W. Alexander, R.J. Bell: Appl. Phys. Lett. **27**, 346 (1975)
5.25 R.B. Miles: Stanford Univ. M. L. Rept. No. 2069 (1972)
5.26 C.B. Moore: National Bureau of Standards Circular No. 467, Vol. I (1949), II (1952), and III (1958)

5.27 J.J.Wynne: Phys. Rev. **178**, 1295 (1969)
5.28 P.Kusch, M.M.Hessell: J. Mol. Spectr. **32**, 181 (1969)
5.29 H.O.Dickinson, M.R.H.Rudge: J. Phys. B **3**, 1448 (1970)
5.30 R.W.Dreyfus, R.T.Hodgson: Phys. Rev. A **9**, 2635 (1974)
5.31 P.P.Sorokin, J.R.Lankard: IEEE J. QE-**9**, 277 (1973)
5.32 W.I.Wiese, W.M.Smith, B.M.Miles: *Atomic Transition Probabilities*, NSRDS-National Bureau of Standards 22 (1969)
5.33 G.C.Bjorklund: IEEE J. QE-**11**, 287 (1975)
5.34 J.J.Wynne, P.P.Sorokin, J.R.Lankard: In: *Laser Spectroscopy*, ed. by R.G.Brewer, A. Mooradian (Plenum Publishing Corp., New York 1974), pp. 103–111
5.35 J.A.Giordmaine, W.Kaiser: Phys. Rev. **144**, 676 (1966)
5.36 D.M.Bloom, J.T.Yardley, J.F.Young, S.E.Harris: Appl. Phys. Lett. **24**, 427 (1974); S.E.Harris, D.M.Bloom: Appl. Phys. Lett. **24**, 229 (1974)
5.37 I.V.Tomov: Phys. Lett. **48** A, 153 (1974)
5.38 V.I.Barantsov, A.K.Popov, G.K.H.Tartakovsky: *Quasi-Resonant Four-Photon Parametric Processes in Gases and the Possibility of the Steady-State Oscillation in the* UV *and* IR *Regions* (to be published)
5.39 J.J.Wynne, P.P.Sorokin, M.M.T.Loy, Y.R.Shen: private communication
5.40 J.F.Young, G.C.Bjorklund, A.H.Kung, R.B.Miles, S.E.Harris: Phys. Rev. Lett. **27**, 1551 (1971)
5.41 A.H.Kung, J.F.Young, G.C.Bjorklund, S.E.Harris: Phys. Rev. Lett. **29**, 985 (1972)
5.42 R.T.Hodgson, P.P.Sorokin, J.J.Wynne: Phys. Rev. Lett. **32**, 343 (1974)
5.43 K.M.Leung, J.F.Ward, B.J.Orr: Phys. Rev. A **9**, 2440 (1974)
5.44 C.S.Chang: Phys. Rev. A **9**, 1769 (1974)
5.45 P.P.Sorokin, J.R.Lankard: Phys. Rev. **186**, 342 (1969)
5.46 J.L.Carlsten: private communication
5.47 H.B.Bebb: Phys. Rev. **153**, 23 (1967)
5.48 R.A.McCorkle, J.M.Joyce: Phys. Rev. A **10**, 903 (1974)
5.49 S.Y.Ch'en, M.Takeo: Rev. Mod. Phys. **29**, 20 (1957)
5.50 P.P.Bey, H.Rabin: Phys. Rev. **162**, 794 (1967)
5.51 U.Fano: Phys. Rev. **124**, 1866 (1961)
5.52 W.R.S.Garton, G.L.Grasdalen, W.H.Parkinson, E.M.Reeves: J. Phys. B **1**, 114 (1968)
5.53 W.R.S.Garton, K.Codling: J. Phys. B **1**, 106 (1968)
5.54 J.A.Armstrong, J.J.Wynne: Phys. Rev. Lett. **33**, 1183 (1974)
5.55 L.Armstrong,Jr., B.L.Beers: Phys. Rev. Lett. **34**, 1290 (1975)
5.56 R.D.Hudson: Phys. Rev. **135**, A 1212 (1964)
5.57 D.M.Bloom, G.W.Bekkers, J.F.Young, S.E.Harris: Appl. Phys. Lett. **26**, 687 (1975)
5.58 H.H.Russell, A.S.King: Astrophys. J. **90**, 155 (1939)
5.59 W.F.Meggers, B.F.Scribner: J. Res. Nat. Bur. Std. **19**, 651 (1937)
5.60 P.M.Johnson: J. Chem. Phys. **62**, 4562 (1975)
5.61 D.Cotter, D.C.Hanna: J. Phys. B **9**, 2165 (1976)

6. Optical Pumping in Gases

T. Y. Chang

With 24 Figures

6.1 Background

Optical pumping of gases with known laser sources is gaining increasing importance as an efficient and versatile technique to obtain stimulated radiation at new infrared wavelengths. For example, hundreds of optically pumped far-infrared (FIR) laser lines have been discovered since 1970 [6.1] yielding new laser wavelengths throughout a region where previously relatively few laser lines were known to exist. These optically pumped FIR lasers can now provide cw output at the mW level or pulsed output at the kW-to-MW level. Optical pumping has also made possible the operation of very short (1 mm), very high pressure (> 30 atm.) CO_2 [6.2] and N_2O [6.3] lasers. This eliminates both the rotational fine structure of the gain profile and the close spacing of competing longitudinal resonator modes in these molecular lasers and leads to the possibility of continuous wavelength tuning and ultrashort-pulse generation. In these and other similar optically pumped gas lasers, two criteria must be met: the availability of powerful pump lasers and the close coincidence of the pump wavelengths with absorption lines of the gas being pumped. A good spectral coincidence is a rather stringent requirement for low-pressure gases. Even so, the total number of useful coincidences that can be expected between all the known medium-to-high-powered lasers and all the molecules available in gas or vapor form is already in the thousands if not in the tens of thousands, and the wavelength tunability of pump sources is steadily improving. Therefore, optical pumping with its advantages of being highly selective, relatively predictable, and non-destructive holds great promise of contributing in a major way to the generation of new wavelengths throughout the infrared region.

In this chapter, we shall review the development of optically pumped infrared gas lasers up to mid-1976. The scope of this review will be largely limited to a consideration of optical and collisional processes only. Lasers involving photodissociation or other photochemical processes will be touched on only when they are closely related to an optically pumped laser. In the remainder of this introductory section, we shall discuss the general principles and certain common features of optically pumped gas lasers. This will be followed in subsequent sections by discussions of specific systems based on electronic transitions (primarily in the near IR), vibrational-rotational transitions (primarily in the middle IR), and rotational or inversion transitions

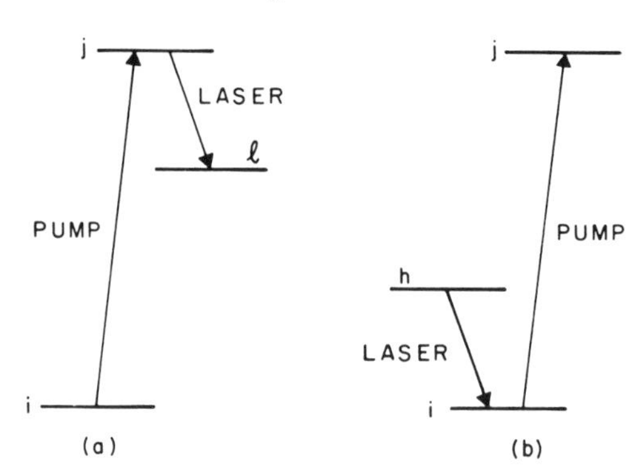

Fig. 6.1a and b. Three-level schemes in lasers: (a) normal, (b) reversed

(primarily in the FIR). Since the development of optically pumped FIR lasers has been the most extensive, more than half of the chapter will be devoted to a discussion of these new long-wavelength lasers.

The term optical pumping refers to the general technique of creating large nonequilibrium population differences in atomic or molecular systems by the combined means of resonant or near-resonant optical excitation and various decay processes. It is a technique that was used and studied well before the invention of the laser [6.4]. It is also the technique that was suggested for the first proposed laser systems (optically pumped infrared gas lasers) [6.5]. Optical pumping soon became the dominant excitation method for solid-state and liquid-state lasers. However, since incoherent pumping radiation used in the early period of laser development could not be absorbed efficiently by the narrow absorption lines in gases, optical pumping failed to become an important method of excitation for gas lasers until recent years. Now the use of various high-power lasers as pump sources has produced many important optically pumped gas lasers. Interestingly, the monochromaticity of the pump source makes these new optically pumped lasers very similar in principle to three-level maser devices that predated lasers. In the discussions to follow, we shall take advantage of this similarity and borrow useful concepts from maser theory from time to time.

There are many different energy-level schemes for which the achievement of population inversion by optical pumping is possible. We shall discuss some of the more basic ones here. Shown in Fig. 6.1a is an energy-level diagram for a normal three-level laser. In this level scheme, the following energy relation is usually satisfied: $\varepsilon_j > \varepsilon_l \gg \varepsilon_i + kT$, which means that the levels l and j are essentially unpopulated before pumping. Population inversion between levels j and l is readily achieved by optical pumping of molecules from level i to j. If pumping occurs in a time short compared to the decay time of level j, then the total inverted population is essentially equal to the total number of pump photons absorbed during the pulse. For cw operation, population inversion is possible only if the following condition is met [6.6]:

$$(n_{i0} - n_{l0})\,\tau_{li}^{-1} > (n_{l0} - n_{j0})\,\tau_{jl}^{-1}\,, \tag{6.1}$$

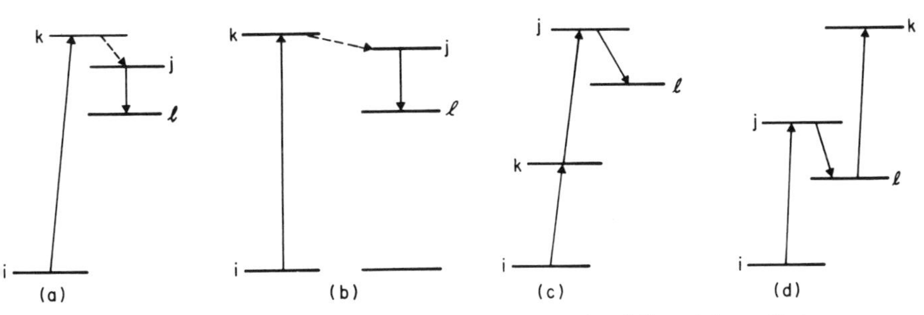

Fig. 6.2a—d. Four-level schemes in lasers: (a) optical pumping followed by radiative or non-radiative decay, (b) optical pumping followed by energy-transfer (optical-transfer), (c) two-step (push-push) optical pumping, (d) differential (push-pull) optical pumping

where n_{i0}, n_{l0}, and n_{j0} are equilibrium populations before pumping, and τ_{li} and τ_{jl} are characteristic decay times. In most real systems, (6.1) does not strictly apply because of the involvement of many other levels in the decay process. This condition nevertheless remains an instructive guideline.

Population inversion can also be achieved in a "reversed" three-level system as shown in Fig. 6.1b. It is as though the diagram in Fig. 6.1a has been turned upside-down with all the arrows reversed. Optical pumping is used in this case to pump out the lower laser level rather than feed the upper laser level. Usually we have $\varepsilon_j \gg \varepsilon_i + kT > \varepsilon_h$, and the necessary condition for cw operation now becomes

$$(n_{h0} - n_{j0})\,\tau_{jh}^{-1} > (n_{i0} - n_{h0})\,\tau_{hi}^{-1}\,. \tag{6.2}$$

For the achievement of population inversion, this scheme is more difficult and less efficient than the "normal" three-level scheme and is therefore less commonly used.

Excitation of atoms or molecules can also take place through multiple-step and multiphoton processes. Shown in Fig. 6.2a is a four-level scheme in which the atoms or molecules are optically pumped into level k before decaying, radiatively and/or nonradiatively, into the upper laser level j. The transitions from k to j can also be due to laser action, in which case the laser transition from j to l is a cascade laser transition. The scheme shown in Fig. 6.2b is an optically pumped energy-transfer (sometimes referred to as optical-transfer) scheme. Optical pumping is used to excite "seed" molecules into a level k. The excitation energy is then transferred by collisions to the upper laser level j of the active molecules. The energy-transfer process is much less specific with regard to energy-level matching than direct optical pumping, with the advantage that the requirement of precise spectral coincidence between the pump source and the active molecule is waived. Figure 6.2c illustrates a two-step pumping process known as the "push-push" pumping in maser terminology [6.7]. A special case of this is two-photon pumping for which the intermediate level k is not necessarily in resonance with the pump radiation.

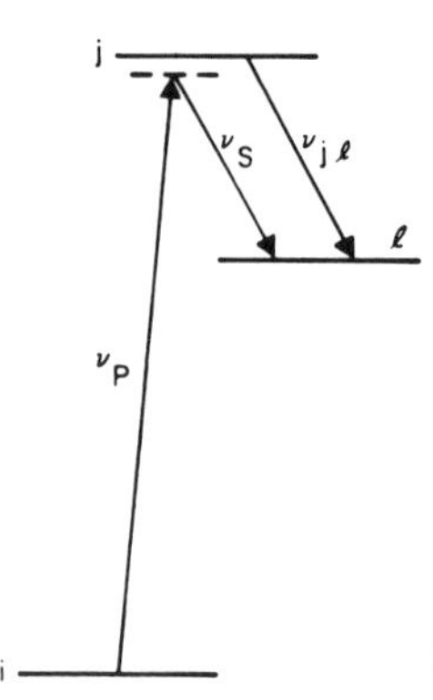

Fig. 6.3. Off-resonance pumping and Raman or Raman-like transition

The scheme shown in Fig. 6.2d combines a normal three-level scheme with a reversed three-level scheme and is known as the push-pull pumping in maser terminology. Many further variations on the level schemes described above are possible and will be described as we encounter them later in the chapter.

Nonlinearities can arise not only in the pumping process, such as in two-photon pumping, but also in the emission process. When the pump radiation is off-resonance from the absorbing transition as in the case illustrated in Fig. 6.3, the laser medium may exhibit gain both near the transition frequency v_{jl} and near the Stokes frequency $v_s = v_p - v_{il}$. When the level j is an excited electronic state the Stokes gain is of course due to the stimulated Raman effect. When the level j is not an excited electronic state, the Stokes gain can still arise from an analogous Raman-like process [6.8–9]. As the pump frequency becomes near or on resonance, the Raman or Raman-like process becomes coalesced with another nonlinear effect-resonant modulation splitting [6.8, 10]. The emission line becomes split, and the shape of the line which now depends on the pump intensity no longer resembles either a gaussian or a Lorentzian line. These nonlinear effects must be considered in addition to the usual population inversion effects when the quantity $\mu_{ij} E_p \tau_2/\hbar$ (where μ_{ij} is the dipole moment for the pump transition $i \to j$, E_p is the electric field intensity of the pump radiation, and τ_2 is the collisional dephasing time) becomes comparable to or greater than unity.

We note in passing that when inversion symmetry in an active gas is destroyed by the application of an external dc field, an optically pumped laser can in principle exhibit a second-order nonlinear susceptibility and behave as a parametric oscillator with gain at both the signal and the idler frequencies [6.11, 12].

The basic expression for the gain γ (in cm^{-1}) in a laser medium is given by [6.13]

$$\gamma(v) = F_{jl}(n_j - n_l g_j/g_l)\, S(v, v_{jl}), \tag{6.3}$$

where n_j and n_l are population densities (in cm^{-3}), g_j and g_l are statistical weights of levels j and l, and $S(v, v_{jl})$ is the line-shape function (in Hz^{-1}). The leading factor F_{jl} is related to the oscillator strength f_{jl} (dimensionless), the Einstein coefficients A_{jl} (in s^{-1}) and B_{jl}, the transition dipole moment μ_{jl} (in Debye), and the transition wavelength λ_{jl} (in μm) by

$$\begin{aligned} F_{jl} &= 2.654 \times 10^{-2} f_{jl} \\ &= 3.979 \times 10^{-10} \lambda_{jl}^2 A_{jl} \\ &= hB_{jl}/\lambda_{jl} \\ &= 1.2478 \times 10^{-4} \mu_{jl}^2/\lambda_{jl} \,. \end{aligned} \tag{6.4}$$

The value of f_{jl} is 0.77 for the 7.18 μm electronic transition of Cs (see Subsect. 6.2.1), 5.7×10^{-7} for the 10.4 μm vibrational-rotational band of CO_2 (see Subsect. 6.3.1), and 1.6×10^{-5} for the 496 μm pure-rotational line of CH_3F (see Subsect. 6.4.3). From these representative values, it is clear that electronic transitions are by far the strongest transitions. Vibrational-rotational and pure-rotational transitions are much weaker, with the latter being the stronger of the two.

In an optically pumped laser system, the pump radiation can be introduced into the laser cavity either from the end or from the side. When the absorption length for the pump radiation is much longer than the gas cell, it is favorable to introduce the pump beam into the system in such a way that the pump radiation will be trapped between resonator mirrors for a large number of round trips (see Subsect. 6.4.4). On the other hand, when the absorption length is extremely short, it may be advantageous to pump the gas cell transversely. The amount of pump power (or energy) that can be absorbed by the gaseous medium depends on the population n_i available at the initial level i which in turn depends on the operating gas pressure and on the number of levels sharing the total population. Simpler molecules have the advantage that n_i is higher for a given pressure.

The laser gain exhibited by any given optically pumped gaseous medium depends on the gas pressure, the pump intensity, and the frequency offset of the pump radiation from the center of the absorption line. The qualitative relationships among these parameters are illustrated in Fig. 6.4 where the laser gain is plotted as a function of gas pressure for different values of pump intensity I_p. We denote by p_1 the pressure at which pressure broadening becomes equal to the Doppler width of the emission line. Since the pump frequency is higher than the emission frequency, pressure broadening becomes equal to the Doppler width of the pump transition at a higher pressure which we denote by p_2. For the purpose of discussing off-resonance pumping, we use p_3 to denote the pressure at which the half width of the pressure-broadened pump line becomes equal to the frequency offset $\delta\nu$ of the pump radiation from the center of the absorption line. For the purpose of present discussion, we define the low-pressure range as $p<p_1$, the medium-pressure range as $p_1<p<p_2$, and the high-pressure range as $p>p_2$. In the low-pressure range, the maximum attainable gain is proportional to the gas pressure (i.e., the

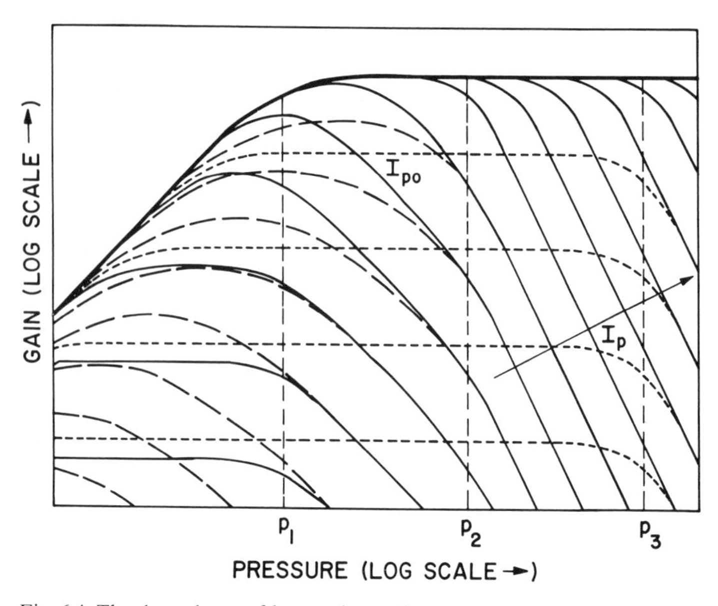

Fig. 6.4. The dependence of laser gain on the gas pressure and pump intensity (I_p), showing the effects of hole-burning (dashed curves) and off-resonance pumping (dotted curves). p_1: the pressure at which pressure broadening is equal to the Doppler width of the emission line. p_2: the pressure at which pressure broadening is equal to the Doppler width of the pump transition. p_3: the pressure at which the pressure broadened halfwidth is equal to the frequency deviation in off-resonance pumping

number of molecules available) as is depicted by the heavy solid curve. In the medium- and high-pressure ranges, any further increase in gain due to increases in the molecular density is offset by the broadening of the emission line, and the maximum attainable gain (the heavy solid curve) becomes independent of the gas pressure. For any finite value of I_p the pump radiation will become too weak to saturate the absorption transition beyond a certain value of gas pressure, and the gain will drop below the maximum attainable value represented by the heavy curve. For unsaturated pumping, the pumping rate is given by $w_p = I_p\alpha \propto I_p n \Delta\nu_p^{-1}$ where n is the molecular density, and $\Delta\nu_p$ is the linewidth of the pump transition. The laser gain γ is proportional to $w_p\tau_j\Delta\nu_L^{-1} \propto I_p n\tau_j\Delta\nu_p^{-1}\Delta\nu_L^{-1}$, where $\tau_j \propto n^{-1}$ is the upper-level lifetime, and $\Delta\nu_L$ is the linewidth of the laser transition. In the high-pressure range, $\Delta\nu_p \cdot \Delta\nu_L \propto n^2$, and the gain under unsaturated pumping is $\propto I_p n^{-2}$. In the medium- and low-pressure ranges, the gain available depends on whether the pump source is a single-frequency or a broadband (or multifrequency) source. For a broadband, on-resonance pump source, we have $\Delta\nu_p\Delta\nu_L \propto n$ and $\gamma \propto I_p n^{-1}$ in the medium-pressure range, while in the low-pressure range $\Delta\nu_p\Delta\nu_L \propto n^0$ and $\gamma \propto I_p n^0$, as indicated by light solid curves in Fig. 6.4. The maximum laser gain can be achieved with a minimum pump intensity of

$I_p \sim I_{p0}$ at a pressure of $p \sim p_1$. A single-frequency pump radiation leads to "hole"-burning in the Doppler broadened absorption line. The resulting gain is indicated by dashed curves in Fig. 6.4. For $p \ll p_1$ and $I_p \ll I_{p0}$, hole-burning results in a narrowing of the emission line due to a reduced velocity spread of excited molecules. Consequently, a higher gain is obtained than with a broadband pump of the same intensity. Note, however, that the velocity spread increases with power broadening of the hole. For $p \sim p_1$ and $I_p \sim I_{p0}$, hole-burning leads to a restriction of the number of molecules that can be excited, but does not result in a significant narrowing of the emission line. Consequently, a lower gain is obtained than with a broadband pump of the same intensity. For $I_p \gg I_{p0}$, a single-frequency pump source behaves much like a broadband pump source due to a greatly power-broadened hole width. In off-resonance pumping, the maximim gain (as shown by dotted curves) is usually lower than can be obtained by resonance pumping with the same pump intensity. However, if I_p is sufficiently large that power broadening is equal to or greater than the frequency offset δv, then on- and off-resonance pumping yield the same result. A more quantitative discussion of pump saturation, power broadening, and off-resonance pumping will be given in Subsection 6.4.5.

It can be seen from Fig. 6.4 that if the value of I_p relative to I_{p0} and the value of δv are unknown, then one is most likely to obtain the highest gain for a given I_p at $p \sim p_1$. This is a good rule-of-thumb for lasers based on vibrational-rotational or pure-rotational transitions. However, if $I_p \ll I_{p0}$ and the pump source is a single-frequency one, then the optimum operating pressure would occur at some value of $p < p_1$. This latter situation is most likely to occur when pumping an electronic transition with a single-frequency source. By way of reference, the value of p_1 is 8 Torr for the 10 μm pure CO_2 laser and 30 mtorr for the 496 μm CH_3F laser.

Additional refinements in the above qualitative description of laser gain are necessary when the pump radiation is linearly polarized. Initially, the molecules in level i are randomly oriented in space. However, when they are irradiated with the pump beam, some will have either very weak or no interaction at all with the pump radiation due to an unfavorable molecular orientation. Molecules excited into the upper laser level j will therefore exhibit a macroscopic orientation and produce a polarized gain pattern. This phenomenon of molecular polarization was well recognized in optical pumping experiments before the invention of the laser.

It is understood in the gain expression (6.3) that the excited molecules are randomly oriented and the gain produced in two orthogonal polarizations is equal. When the molecules are oriented, we obtained for the same inversion density somewhat higher gain for the favored polarization and lower gain for the other polarization. Conversely, for the same peak gain, the pump power requirement is somewhat lower when a polarized pump beam is used. The automatic polarization of optically pumped lasers is of practical importance and will be discussed further in Subsection 6.4.1.

With optical pumping it is usually possible to achieve the necessary population inversion with less pump power than by any other method. Thus a good overall efficiency can be achieved if the pump source is an efficient one. The condition for maximum efficiency, however, does not necessarily coincide with the condition for the highest output power. Partly due to this reason, typical photon conversion efficiencies in practical systems are on the order of only a few percent. Most of the energy is lost to various competing decay processes.

6.2 Lasers Based on Electronic Transitions

Due to their large oscillator strengths, many electronic transitions lend themselves well to optical pumping. Unfortunately, the fact that most electronic absorption bands occur in the UV region seems to have limited the amount of work done on this type of lasers. Further development of powerful UV sources may bring about more vigorous research in this area.

6.2.1 Atomic Cs, Sr, and K Vapor Lasers

The Cs laser is of historic significance since it is one of the first laser systems proposed by *Schawlow* and *Townes* in 1958 [6.5]. Observation of optical gain in this system was first reported in 1961 [6.14], and the operation of the first cw optically pumped Cs laser was reported in 1962 by *Rabinowitz* et al. [6.15].

An energy-level diagram for the Cs laser is shown in Fig. 6.5. Cesium atoms are excited from the ground state $6S_{1/2}$ to the upper laser level $8P_{1/2}$ by 388.8 nm line radiation from a helium discharge lamp. This He line is known to be in close coincidence with the Cs absorption line. The laser can be made to oscillate either at 7.18 μm ($8P_{1/2} \rightarrow 8S_{1/2}$) or at 3.20 μm ($8P_{1/2} \rightarrow 6D_{3/2}$). However, the gain for the 3.20 μm line was found to be an order of magnitude lower than that for the 7.18 μm line. Experimentally, the Cs vapor at a pressure of 20 mTorr was contained inside a 1 cm ID glass tube. The pump radiation from a He lamp was collected and imaged onto the laser cell by cylindrical aluminum reflectors. With BaF_2 windows, an output of 50 μW was obtained at 7.18 μm from a 92 cm active length. With sapphire windows, an output of 40 μW was obtained at 3.20 μm from an active length of 50 cm [6.16].

The Sr vapor laser reported by *Sorokin* and *Lankard* in 1969 [6.17] was made possible by the earlier development of organic dye lasers. As the energy-level diagram of Fig. 6.6 shows, Sr atoms are excited from the ground state $5s^1S_0$ to the $5p^1P_1$ state in the singlet manifold by a tunable dye laser at 460.73 nm. The expected laser transition, $5p^1P_1 \rightarrow 4d^1D_2$, is at 6.45 μm. An output of 0.4 kW at this wavelength was obtained by longitudinally pumping a He buffered Sr vapor cell with a 300 kW pulsed dye laser. In addition, a total of 4 kW output was observed on three transitions in the triplet manifold (see Fig. 6.6). The authors attributed laser action in the triplet manifold to

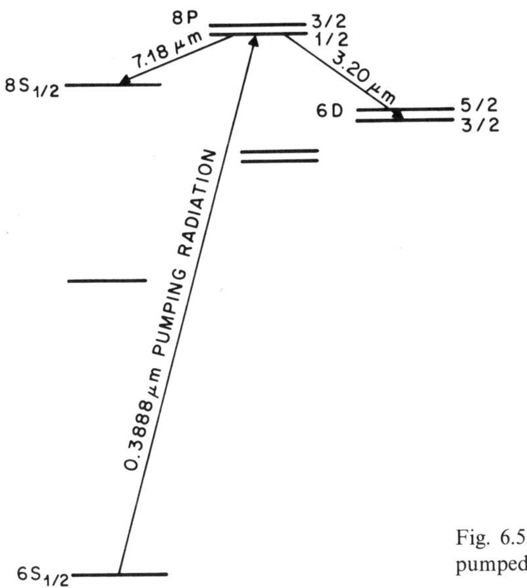

Fig. 6.5. Energy-level diagram for the optically pumped Cs laser

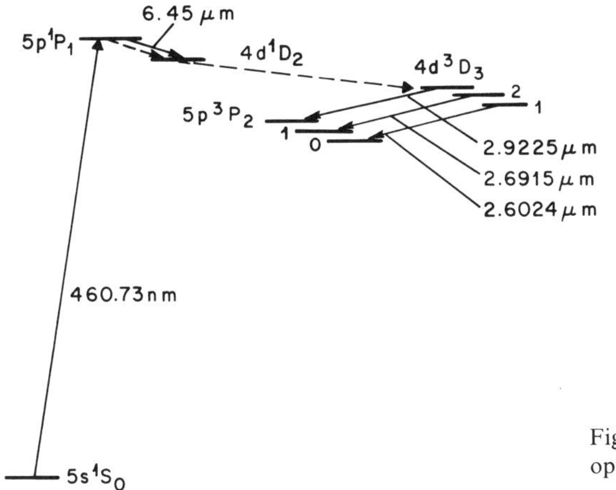

Fig. 6.6. Energy-level diagram for the optically pumped Sr laser

collisional de-excitation of some optically pumped Sr atoms from $5p^1P_1$ to $4d^1D_2$ and then further down to the three $4d^3D$ states by the presence of He buffer gas at atmospheric pressure. The same authors also obtained infrared laser action in He-buffered Rb_2, Cs_2, and K_2 vapors by optical excitation [6.18, 19]. Here, the excitation process involved two-photon absorption by the molecular species followed by dissociation into various excited atomic states. These lasers are therefore photodissociation lasers rather than optically pumped lasers.

In their original proposal for infrared and optical masers, *Schawlow* and *Townes* [6.5] analyzed the possibility of obtaining laser action in potassium vapor by optical pumping. Specifically, they recommended pumping the K atoms from $4s$ (ground state) to $5p_{3/2}$ and looking for laser action on the $5p_{3/2}-3d$ transition at 3.14 µm. However, they concluded that the resonance radiation available from a potassium lamp is not quite sufficient for the purpose. Recently, essentially the same scheme, but extended to higher $np_{3/2}$ and $np_{1/2}$ states, has been demonstrated experimentally by *Wynne* and *Sorokin* [6.20]. They used the frequency-doubled output from a dye laser to pump K atoms from $4s$ to $np_{3/2}$ and $np_{1/2}$ states (with $n=7$ to 16), and obtained stimulated emission from a mirrorless K-vapor cell at 19 infrared wavelengths from 11.26 to 223 µm. These wavelengths correspond to $np-ns$ and $np-(n-2)d$ transitions in K. Wavelength measurements of the $np-(n-2)d$ emissions provided improved determinations of d-level energies. They also observed stimulated electronic Raman emission for $n=7$ and $n=8$ at small detunings of the pump from the np resonance lines. The pump source provided an output power of 800 W and could be tuned from 322.6 to 284.1 nm with a linewidth of $\sim 0.2\,\mathrm{cm}^{-1}$. The vapor pressure of K in the "heat-pipe" vapor cell was about 10 Torr.

Infrared generation by stimulated electronic Raman effect was first reported by *Rokni* and *Yatsiv* [6.135] in 1967. In their experiment, potassium atoms are pumped from the $4S$ ground state to the $4P_{3/2}$ state by a Raman shifted (in a liquid) ruby laser beam. The unshifted ruby laser beam then excites the potassium atoms from $4P_{3/2}$ to $5P_{3/2}$ while simultaneously generating stimulated Stokes photons at $2720.6\,\mathrm{cm}^{-1}$ corresponding to an electronic Raman transition. They also observed laser outputs at $2730.4\,\mathrm{cm}^{-1}$ ($6S\rightarrow 5P_{3/2}$) and $2749.2\,\mathrm{cm}^{-1}$ ($6S\rightarrow 5P_{1/2}$) due to two-photon pumping of potassium atoms from the $4S$ ground state to the $6S$ excited state by the combination of Raman shifted and unshifted ruby laser beams.

Recently, *Grischkowsky* et al. [6.136] obtained laser emission in potassium vapor at 15.95 µm ($6D_{3/2}\rightarrow 7P_{1/2}$) and at associated cascade transitions at 12.5 µm ($7P\rightarrow 7S$), 7.9 µm ($7S\rightarrow 6P$) and 6.4 µm ($6P\rightarrow 6S$). In this case, the potassium atoms are first excited to the $4P_{1/2}$ level by electrical discharge and then optically pumped from $4P_{1/2}$ to $6D_{3/2}$ by a pulsed dye laser at 534.31 nm.

In 1973, *Sorokin* and *Lankard* [6.137] reported the observation of stimulated electronic Raman scattering with output at 20 µm. They also inferred from the spectrum of parametrically generated frequencies that eight additional infrared frequencies corresponding to cascade transitions in Cs are also stimulated in the process. The pump wavelength of 347 nm (frequency doubled ruby laser output) is only $46\,\mathrm{cm}^{-1}$ from the $6S\rightarrow 10P_{3/2}$ transition and leads to strong stimulated electronic Raman transition to the $10S$ level with output at 20 µm. The cascade transitions from $10S$ to $9P$ levels occur at 16.18 µm and 15.08 µm. The next cascade transitions from $9P$ levels to $9S$ occur at 12.97 µm and 13.77 µm. These are followed by $9S\rightarrow 8P$ transitions at 8.94 µm and 8.32 µm and $8P\rightarrow 8S$ transitions at 6.78 µm and 7.18 µm.

Recently, tunable infrared signals have been obtained through stimulated electronic Raman scattering from Ba [6.138] and Cs [6.139] vapors by using pulsed, tunable dye lasers as the pump source. The tuning ranges are 2.87–2.97 μm and near 2.78 μm in Ba and 2.5–4.75 μm, 5.67–8.65 μm, and 11.7–15 μm in Cs. The linewidth is typically 0.3–1.0 cm^{-1}.

6.2.2 Molecular I_2, Na_2, and NO Lasers

Byer et al. [6.21] obtained more than 150 laser lines from 0.544 to 1.334 μm in optically pumped I_2 vapor. In principle, approximately 10^6 discrete laser lines can be generated in I_2 by this method. The I_2 molecules are excited by the second harmonic of an etalon-tuned Nd:YAG laser into a specific vibrational-rotational level in the excited electronic state $B^3\Pi^+_{0u}$. Some of the excited molecules then emit stimulated radiation and return to one of many possible vibrational states in the ground electronic state $X^1\Sigma^+_g$. The rotational selection rule $\Delta J = \pm 1$ is strictly observed in the process. The vibrational selection rule, on the other hand, is only loosely governed by the Frank-Condon principle. The relevant energy levels of I_2 and a representative set of transitions involved in this laser are shown in Fig. 6.7.

Although both absorption and emission take place in the same electronic band as in a dye laser, the I_2 laser differs from a dye laser in two important respects. In the first place, the excited I_2 molecules emit without going through vibrational-rotational relaxation. In the second place, the I_2 laser is based on a normally forbidden singlet-to-triplet transition. The usual spin selection rule [6.22] is weakened in I_2 due to the large nuclear charge of iodine atoms.

The laser gain calculated from the observed cavity-length-dependent build-up time is found to be in excellent agreement with the theoretical value of 0.024 cm^{-1}. With 190 ns pump pulses, only a few microjoules were required for the laser to reach threshold. The I_2 pressure used was 0.32 Torr. The maximum laser output was limited to 0.054 μJ by the total number of I_2 molecules available in the cell.

Recently, *Henesian* et al. [6.23] reported laser action at various wavelengths between 0.75 μm and 0.85 μm in Na_2 vapor when pumped by the output from a doubled-Nd:YAG laser near 0.659 μm. Laser action is attributed to the transition between the $X^1\Sigma^+_g$ ground state and the $A^1\Sigma^+_u$ excited electronic state. They also observed superfluorescent emission in the green-yellow region by pumping at 0.473 μm. This visible emission is associated with the $B^1\Pi_u$ state of Na_2. As is the case in the I_2 laser, the Na_2 laser is potentially capable of providing numerous laser lines in the visible and near-IR regions.

Laser action at 1.2237 μm, 1.1069 μm, 2.6072 μm and 2.6380 μm in flash-pumped NO was reported by *Lin* [6.24]. In this system ground state NO molecules are excited by 160–190 nm UV radiation into $C^2\Pi$ and $D^2\Sigma^+$ electronic states as shown in Fig. 6.8. Laser transitions are from $C^2\Pi$ to $A^2\Sigma^+$ (1.2 μm) and from $D^2\Sigma^+$ to $A^2\Sigma^+$ (1.1 μm). The 2.6 μm output which is delayed by about 10 μs has not been positively identified. The experimental device

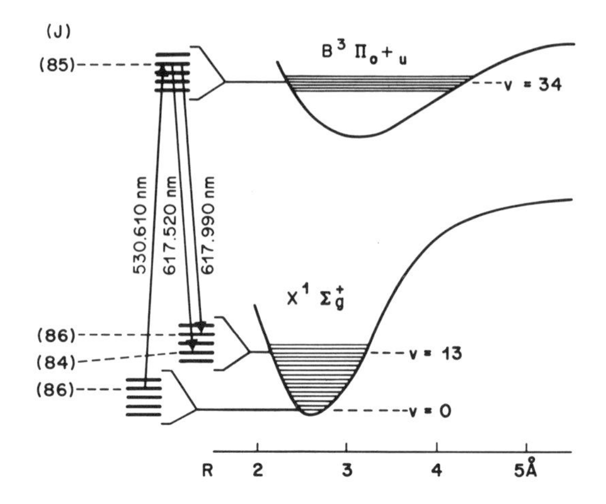

Fig. 6.7. Energy-level diagram for the optically pumped I_2 laser

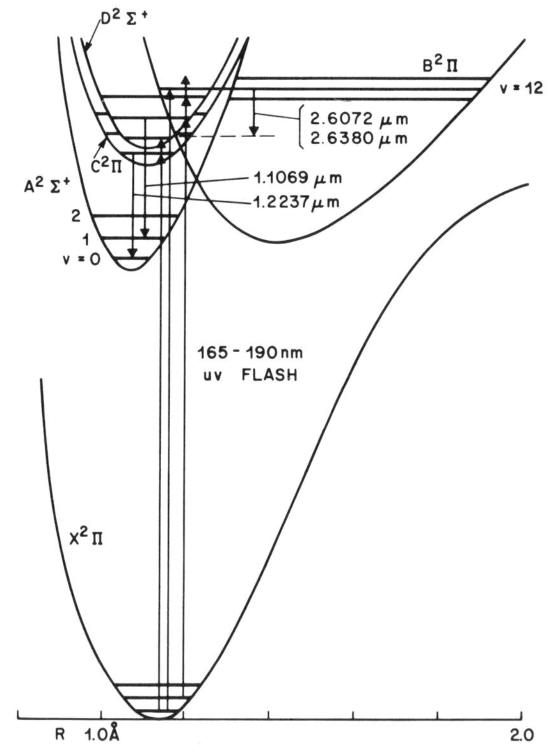

Fig. 6.8. Energy-level diagram for the optically pumped NO laser

was a 1 m long suprasil tube surrounded by a coaxial flash lamp. The flash lamp was typically filled with 10 Torr of a 1% Xe/Ar mixture and produced 45 μs flashes with 1.6 kJ energy input. The NO pressure was from 1 to 4 Torr. The addition of N_2O, CO_2, or SF_6 was found to be beneficial for the 1 μm emission lines. Most other additive gases were found to be detrimental. The addition of CO_2 also led to CO_2 laser emission at 10.6 μm. This effect will be further discussed in Subsection 6.3.1. The same NO laser lines have also been observed in flash photolysis of NO_2.

6.3 Lasers Based on Vibrational-Rotational Transitions

Lasers based on vibrational-rotational transitions can be classified into three groups. For the first group, discussed in Subsection 6.3.1, the laser transitions correspond to the difference between two normal vibrational modes of the active molecule. Very high pressure operation is possible in this case because of the absence of significant self-absorption. For the second group, discussed in Subsection 6.3.2, the laser transition corresponds to a normal vibrational mode of the active molecule which is slightly shifted in frequency from its corresponding fundamental absorption band due to the simultaneous presence of additional vibrational excitations. The frequency shift is often enhanced by a Fermi resonance involving one of the laser states. The transition dipole moment is often higher for this group but operation at high pressure is very unlikely due to self-absorption. In the third group, optical pumping and laser action take place in different parts of the same vibrational band, e.g., with absorption in the *R*-branch and emission in the *P*-branch. Such a scheme is illustrated in Fig. 6.9. The possibility of achieving such a partial inversion between vibrational states was first pointed out by *Patel* [6.25]. The realization of partial inversion by optical pumping was more generally discussed and analyzed by *Golger* and *Letokhov* [6.26]. Recent examples of this type of lasers are discussed in Subsection 6.3.3.

6.3.1 Lasers Based on Difference Bands in CO_2, N_2O, OCS, CS_2, HCN, and C_2H_2

Infrared transitions between different vibrational modes of a molecule arise from the anharmonicity of chemical bonds in the molecule. The corresponding dipole moments are generally very small. A notable exception is the difference frequency between stretching vibrational modes in linear molecules. The well-known 10.6 μm CO_2 laser [6.27] and the 10.9 μm N_2O laser [6.28] are both based on this type of transition. Although the CO_2 and N_2O lasers have been extensively developed as electrical-discharge lasers, their investigation by optical pumping is still of interest for the following reasons. First, an optically pumped system, owing to its simplicity, can be useful for improving the understanding of the basic physical processes involved in such a laser. Second, it is relatively straightforward to scale an optically pumped laser to very high pressures (or even condensed states) where electrical excitation would become extremely difficult. Third, the optical pumping technique is more adaptable to a variety of different molecules than electrical excitation techniques and hence is more useful when searching for new laser wavelengths. Indeed, optical

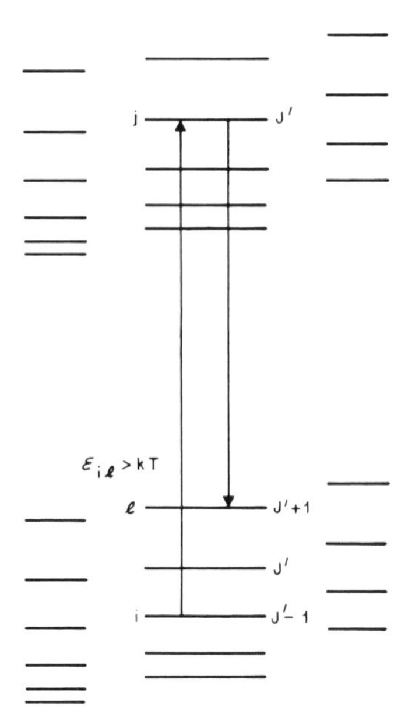

Fig. 6.9. The level scheme for pumping and lasing in the same vibrational band

pumping has already produced continuously tunable high-pressure CO_2 and N_2O lasers and a closely related method has generated new laser wavelengths in HCN.

The CO_2, N_2O, OCS, and CS_2 lasers to be discussed in this subsection are all based on the normal three-level scheme shown in Fig. 6.1a. The upper laser level j corresponds to the ν_3 vibrational state and the lower laser level corresponds to the ν_1 vibrational state. The ν_3 vibrational mode is strongly infrared active and hence can be excited by optical pumping if a pump source of correct wavelength is available.

The first optically pumped CO_2 laser was reported by *Wieder* in 1967 [6.29]. He placed a 4 m long quartz CO_2 cell between two sheets of CO_2 flame which appeared as black-body sources of >3000 K throughout the ν_3 band near 4.3 μm. By using 0.3 to 1.3 Torr of CO_2 in the cell, he was able to obtain 1 mW of cw laser output at 10.6 μm.

Another optically pumped CO_2 laser using an incoherent source was reported by *Bokhan* in 1972 [6.30]. The radiation source was a pulse-heated molybdenum foil placed in a vacuum jacket surrounding a quartz laser tube. The excitation length was 2.6 m. With an atmospheric-pressure flowing mixture of CO_2/N_2 containing 30 Torr of CO_2, 40 mJ laser pulses of 20 ms duration were obtained.

In 1972, *Chang* and *Wood* [6.31] noted a close spectral coincidence between the 4.23 μm $P_{2-1}(6)$ line of HBr laser and the $R(20)$ line of the ν_3 absorption band of CO_2 and used it to optically pump a CO_2 laser. By using a transverse-discharge HBr laser as the pump source, they obtained a pulsed laser output of 80 W from pure CO_2 in their initial experiment. Under

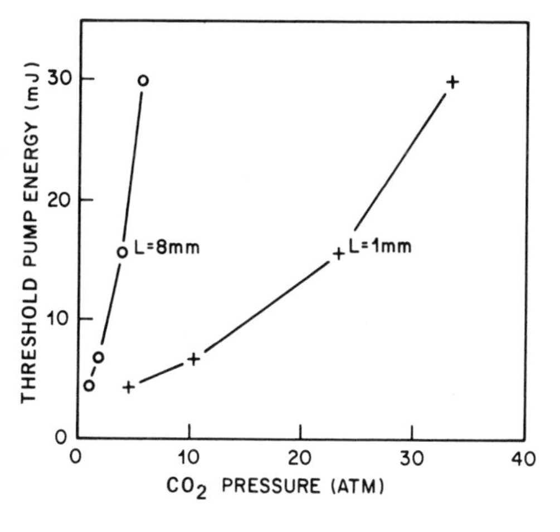

Fig. 6.10. Dependence of maximum operating pressure on the pump energy in 1 mm long and 8 mm long CO_2 lasers (from [6.2])

optimized conditions, substantially complete conversion of 4.23 μm photons to 10.6 μm photons was observed, resulting in a power conversion efficiency of 41%. The maximum operating pressure for a 12 cm cell was found to be one atmosphere.

Subsequently, the same authors were able to increase the operating pressure in pure CO_2 to 33 atmospheres [6.2]. This was achieved by improving the HBr laser and by shortening the CO_2 laser resonator to 1 mm. It should be noted that there are important differences between subatmospheric and very high pressure CO_2 lasers. At very high pressures (above ≈ 10 atm) the absorption bands of CO_2 become broad continuous bands without gaps. Hence many more HBr laser lines with much higher output power are absorbed by the CO_2 gas at very high pressures than at low pressures. It also becomes virtually impossible to saturate the absorption in the 4.3 μm region because of the presence of continuous and heavily overlapping hot bands ($2\nu_3 \leftarrow \nu_3$, $3\nu_3 \leftarrow 2\nu_3$, etc.), coupled with the fact that these nearly harmonic vibrational levels thermalize at an extremely fast rate. Consequently, the penetration depth of 4.3 μm radiation is expected to remain approximately 0.1 mm at the peak of the ν_3 band up to very high pump energy densities. Increasing the pump power only results in heating the ν_3 vibrational manifold to a higher temperature within the short pump-penetration depth.

The 10 μm emission band also becomes continuous at high pressures. Because of the greater overlap of broadened vibrational-rotational lines in the *R*-branch, the gain becomes higher in the *R*-branch (10.3 μm) than in the *P*-branch (10.6 μm). Consequently, it was experimentally observed that at pressures above 17 atm, the CO_2 laser oscillates only near 10.3 μm.

Shown in Fig. 6.10 is the dependence of the maximum operating pressure on the absorbed pump energy for two different resonator lengths. It is seen that for any given pump energy, a much higher operating pressures can be attained in the 1 mm resonator than in the 8 mm resonator. One implication

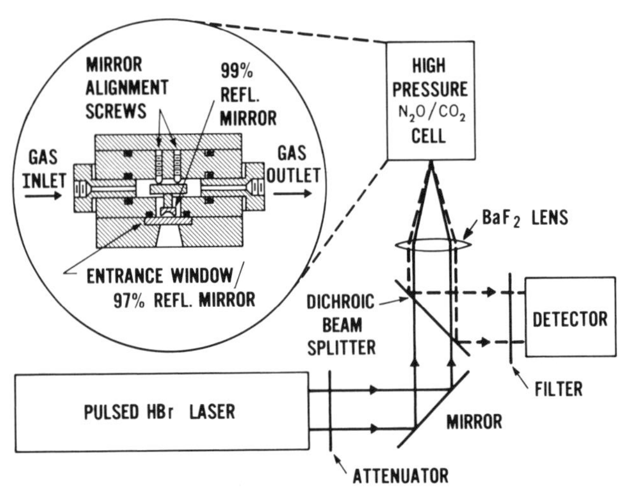

Fig. 6.11. Schematic diagram of an optically pumped high-pressure laser system (from [6.3])

of this is that the pump penetration distance is much less than 8 mm (as has already been pointed out). Optical distortion in the pumped region is probably responsible for the poorer performance of the longer resonator.

Shown in Fig. 6.11 is a schematic diagram of the experimental setup for optically pumped high-pressure CO_2 and N_2O lasers (the latter to be discussed shortly). The output from a transverse-discharge HBr laser is focused by a 20 cm focal length BaF_2 lens into the high-pressure laser device. A totally reflecting concave mirror with a radius of curvature of 5 to 10 cm mounted inside the high-pressure device served as one end-mirror for the high-pressure laser. The other mirror is a dielectric-coated Ge flat which is 97% reflecting at 10.5 μm and approximately 50% transmitting at 4.3 μm. This Ge mirror serves simultaneously as a high-pressure window and as an input-output mirror for the laser. The 10 μm output signal retraces part of the optical path of the input pump beam and is then split off by a dichroic beam-splitter for detection.

The multiline multimode output from the HBr laser consists of two 0.4 μs pulses of approximately equal amplitude spaced 0.6 μs apart. The output from the optically pumped CO_2 laser is typically a single gain-switched pulse which is approximately 2.2 ns at 20 atm and somewhat longer at lower pressures. Multiple output pulses are observed when the pump energy far exceeds the threshold value. The peak output power at 20 atm is approximately one order of magnitude lower than the peak pump power.

Another approach to the optical pumping of a CO_2 laser was demonstrated by *Balykin* et al. [6.32]. As the diagram in Fig. 6.12 indicates, output pulses from a TEA CO_2 laser operating at 9.6 μm are used to pump CO_2 molecules, contained in a separate cell, from the 02^00 state to the 00^01 state. Within the duration of the pump pulse thermal equilibrium between 10^00 and 02^00

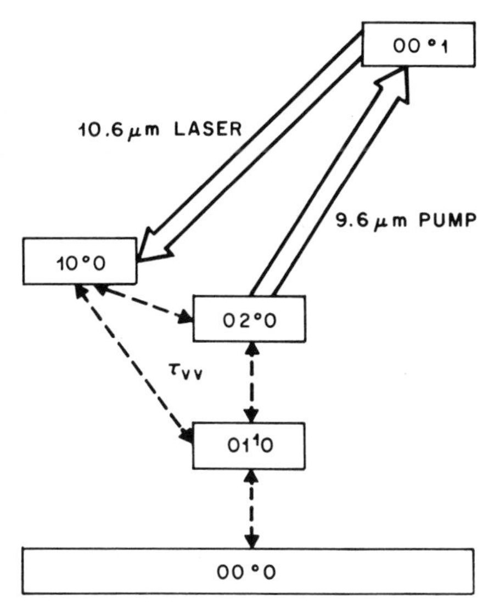

Fig. 6.12. A CO_2 laser optically pumped from the (02°0) state

states is restored by collisions. Since thermal equilibrium dictates that $n_{02^\circ 0} = n_{10^\circ 0} \exp(\Delta\varepsilon/kT)$, where $\Delta\varepsilon = \varepsilon_{10^\circ 0} - \varepsilon_{02^\circ 0}$, and optical saturation leads to $n_{00^\circ 1} \approx n_{02^\circ 0}$, it is possible to realize a gain in the $00^\circ 1 \rightarrow 10^\circ 0$ band approaching $\gamma = \alpha[\exp(\Delta\varepsilon/kT) - 1]$, where α is the absorption coefficient of the $00^\circ 1 \leftarrow 10^\circ 0$ band in the absence of optical pumping [6.33]. Experimentally, a maximum gain of 23% per meter was observed at 100 Torr with a peak pump-power density of about 1 MW/cm^2. (The pump-power requirement scales roughly as p^2.) Laser oscillation was obtained for CO_2 pressures in the range of 80–600 Torr at gas temperatures of 40–240 °C. A maximum energy conversion efficiency of 14% was achieved [6.34].

A few relatively weak HBr laser lines in the 4.5 μm region are found to be absorbed by N_2O gas. This spectral coincidence enabled *Chang* and *Wood* [6.35] to achieve laser action in pure N_2O gas at pressures up to 7.5 atm in a 4 mm long resonator by optical pumping [6.3]. Laser action cannot be achieved at higher pressures mainly because the center of the ν_3 absorption band of N_2O is located too far from most of the HBr laser lines which remain unabsorbed even at high N_2O pressures.

In order to absorb the HBr laser output more effectively, a small amount of CO_2 gas was mixed into the high-pressure N_2O gas. It is known that the ν_3 excitation can be resonantly transferred from CO_2 to N_2O by collisions and that the rate of this $V-V$ transfer is two orders of magnitude faster than the rate of collisional deactivation of the ν_3 state. With CO_2 "seeding", the maximum attainable operating pressure of the N_2O laser increased dramatically to 42 atmospheres [6.3]. Oscillation occurs primarily in the *R*-branch of the N_2O emission band near 10.5 μm. The relevant energy levels and transitions are shown in Fig. 6.13. The optimum CO_2 concentration depended somewhat on the resonator length and the operating pressure. However,

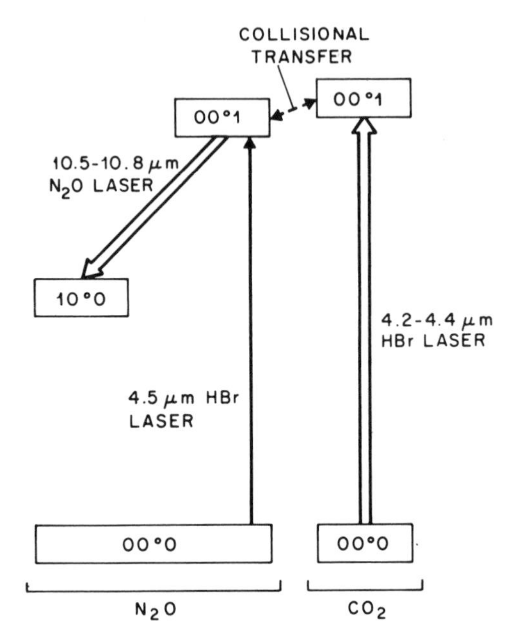

Fig. 6.13. Energy-level diagram for the N_2O:CO_2 optical-transfer laser

6 to 8% CO_2 was found to be a good compromise. The penetration distance for the pump beam should be an order of magnitude longer in N_2O:CO_2 mixture than in pure CO_2, and the transition from the pumped to the unpumped region should be less abrupt. Presumably this factor contributes to the ability of the N_2O:CO_2 laser to attain a higher operating pressure.

Shown in Fig. 6.14 are curves relating the maximum operating pressure of an N_2O:CO_2 laser to the absorbed pump energy from an HBr laser for four different resonator lengths from 1 mm to 8 mm. The optimum resonator length is seen to be dependent on the operating pressure, (e.g., 4 mm at 10 atm and 2 mm at 20 atm) and qualitatively reflects the dependence of pump penetration depth on the total pressure.

N_2O is a linear three-atomic molecule with vibrational energy levels completely analogous to the CO_2 molecule. However, unlike the symmetric CO_2 molecule for which every other rotational level is missing, N_2O is asymmetric with all rotational levels present. Consequently, for a given total population in a vibrational state, the population per rotational level is approximately a factor of two lower in N_2O than in CO_2. This is one reason why an electrically excited N_2O laser is much inferior to a similar CO_2 laser in terms of gain and power output. This disadvantage for N_2O disappears at pressures above 10 atm, since the emission band no longer consists of discrete vibrational-rotational lines. Consequently, at high pressures, N_2O is no longer inferior to CO_2 as a laser medium. In fact, N_2O is the preferred medium since its rotational lines merge into a smooth continuum at a lower pressure ($\sim$5 atm) than in CO_2.

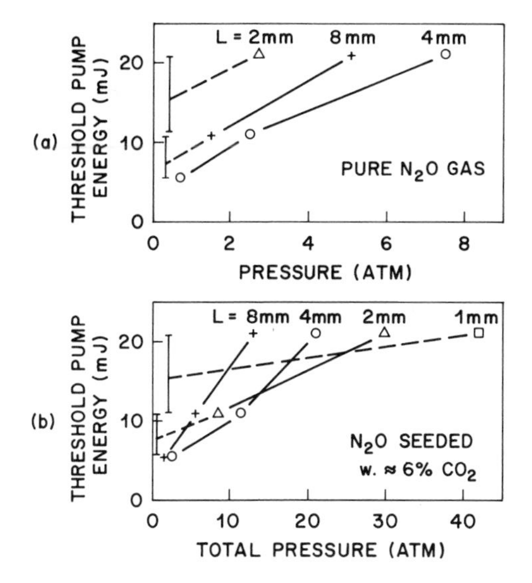

Fig. 6.14a and b. Dependence of maximum operating pressure on the pump energy in 1, 2, 4, and 8 mm long N_2O optical-transfer lasers (from [6.3])

The broad continuous gain profile of high-pressure CO_2 and N_2O lasers can in principle be used to generate either mode-locked picosecond pulses or continuously tunable infrared output. The very short resonator used in the above optically pumped high-pressure N_2O and CO_2 lasers is particularly convenient for jump-free tuning of the oscillation frequency over a large free spectral range by simple length tuning of the resonator. Experimentally, *Chang* and *Wood* [6.36] have obtained continuous tuning of a single laser mode over $5\,cm^{-1}$ near $10.5\,\mu m$ with a resolution of $0.014\,cm^{-1}$. The high-pressure laser uses a mirror spacing of 1.8 mm fine tuned by varying the gas pressure in the cell. All but one transverse mode is suppressed by a small (0.4 mm) iris in the resonator. The two to three longitudinal modes that oscillate simultaneously in the laser have a spacing of $2.75\,cm^{-1}$ and are easily separated by a low-resolution spectrometer.

In analogy to the term "chemical-transfer laser" the above-described $N_2O{:}CO_2$ laser can be termed an "optical-transfer laser". Another form of optical-transfer laser has been reported by *Lin* [6.24] in connection with the optically pumped NO laser discussed in Subsection 6.2.2. Here NO molecules are optically pumped by a Xe/Ar flash lamp and are excited into the $A^2\Sigma^+$, $C^2\Pi$, and $D^2\Sigma^+$ electronic states. The NO vibrational frequencies in these states are higher than the vibrational frequency in the ground electronic state and fall within $\pm 30\,cm^{-1}$ of the ν_3 frequency of CO_2 centered at $2349\,cm^{-1}$. Consequently, the vibrational energy in excited NO molecules can be efficiently transferred to CO_2 molecules. Experimentally, CO_2 laser output at $10\,\mu m$ was observed when typically 2 Torr of CO_2 was added to a mixture of 10% NO in Ar (or He) at 20 Torr pressure and excited by a Xe/Ar flash lamp with 1.8 kJ input energy.

One other form of optical-transfer laser was proposed earlier in 1971 by *Eletskii* et al. [6.37]. These authors proposed to "seed" a mixture of CO_2:N_2:He with an alkali-metal vapor such as Cs and use the resonance radiation from a lamp containing the same alkali-metal vapor to optically pump the seed atoms. Based on the known cross sections for the conversion of electronic excitation of alkali-metal atoms to vibrational excitation of N_2 molecules, the authors concluded that a continuous-wave CO_2 laser based on such a scheme is feasible. However, to the knowledge of this author, no experimental realization of the scheme has yet been reported.

The versatility of the optical-transfer scheme has been demonstrated recently by *Kildal* and *Deutsch* [6.38]. Here the second harmonic of the 9.6 μm $P(24)$ line output from a TEA CO_2 laser was generated in a $CdGeAs_2$ crystal and was used to excite CO molecules into the $J=13$ level of the $n=1$ vibrational state. The vibrational energy of CO matches the energies of the ν_3 state of CO_2, N_2O, OCS, the ν_2 state of C_2H_2 (acetylene), and the $\nu_1+\nu_3$ state of CS_2 within 206 cm^{-1} or less. This makes vibrational energy transfers between CO and these other molecules quite efficient. Laser action was observed at 10.6 μm in CO:CO_2, at 10.8 μm in CO:N_2O, near 9.3 μm in CO:OCS [6.39], near 11.5 μm in CO:CS_2:H_2, and near 8 μm ($\nu_2-\nu_5$ band) in CO:C_2H_2:H_2 [6.40]. The resonator length varied from 0.4 to 30 cm. The available pump energy per pulse at 4.8 μm was 15 mJ and a maximum of 0.48 mJ was obtained at 10.6 μm. Direct optical pumping of OCS near 4.8 μm also produced laser action near 8.3 μm.

An analogous scheme involving photolysis and electronic-to-vibrational energy transfer has recently produced laser action in CO_2, N_2O, C_2H_2, and HCN [6.41]. In this scheme, Br_2 molecules are flash photolyzed to produce excited Br atoms in the $^2P_{1/2}$ state at 3685 cm^{-1}. This energy is transferred to the 10^01 and 02^01 states of CO_2 followed by relaxation to the 00^01 state giving rise to laser action at 10.6 μm. Similar mechanisms are involved in the other molecules investigated. This method has produced new laser bands in HCN near 3.85 μm ($00^01 \rightarrow 01^00$), 8.48 μm ($00^01 \rightarrow 10^00$) and 7.25 μm (perhaps $10^00 \rightarrow 01^00$).

A BCl_3 laser tentatively interpreted as being optically pumped by a CO_2 laser was reported in 1968 by *Karlov* et al. [6.42]. To a conventional discharge-excited CO_2:N_2:He laser operating at a pressure of 7 Torr they added less than 1 Torr of BCl_3 to the gas mixture and observed simultaneous outputs at 10.6 μm and several wave-lengths near 20 μm. They postulated that most of the 20-μm lines represented laser action on the $\nu_3 \rightarrow \nu_1$ and $\nu_3 \rightarrow \nu_2$ transitions of BCl_3 molecules which had been optically pumped into the ν_3 level by the 10.6-μm radiation inside the laser resonator.

Continued work on transfer lasers using optically pumped CO as the transfer medium [6.38], has produced laser action in SiH_4 for the first time. *Deutsch* and *Kildal* [6.140] observed 6 laser lines in the 7.90–7.99 μm region corresponding to the $\nu_3-\nu_4$ band in SiH_4 at pressures up to 35 Torr. The same authors also extended the operating pressure of their CO–CO_2–He transfer laser (10 μm) to 16 atm by using a 1:9:57 mixture.

6.3.2 Lasers Based on Hot Bands in OCS, SF_6, CO_2, NO, and H_2O

An optically pumped OCS laser oscillating at 18.98 μm and 19.06 μm has been reported by *Schlossberg* and *Fetterman* [6.43]. The pump source is a 3 MW, 250 ns, TEA CO_2 laser operating on the $P(22)$ 9.57-μm line. The absorbing transition is identified as the $P(5)$ line of the $(00^\circ0)\to(02^\circ0)$ overtone band of OCS. The two emission lines are identified as the $Q(4)$ and the $P(5)$ lines of the $(02^\circ0)\to(01^10)$ vibrational band. The relevant energy levels and transitions are shown in Fig. 6.15. An output power of approximately 1 kW can be obtained at an OCS pressure of 0.5 Torr. The resonator is a 1 m, internal-mirror device similar to the ones used in optically pumped FIR lasers (to be described in Subsection 6.4.4).

Similar overtone-pumped laser systems have been proposed and analyzed in [6.33], but laser action has not been experimentally demonstrated.

Recently, *Barch* et al. [6.44] reported an optically pumped SF_6 laser operating at 15.900 μm. The experimental setup is similar to the one used in [6.43], but simultaneous pumping by $P(12)$ and $P(14)$ lines (each 300 kW) near 10.5 μm from a TEA CO_2 laser is required. The optimum SF_6 pressure is 55 mTorr. The authors postulated that the SF_6 molecules were either two-photon or two-step pumped from the ground state via ν_3 to $2\nu_3$, which appears to be in Fermi resonance with $\nu_2+2\nu_4$ or $2\nu_6+2\nu_4$. Laser action takes place from $2\nu_3$ to either $\nu_2+\nu_4$ or $2\nu_6+\nu_4$. The transition is essentially a hot band of ν_4 which is shifted from the ν_4 fundamental.

The need for a high power laser near 15.9 μm for laser isotope separation of UF_6 [6.141] has given great impetus to the search for such a laser through optical pumping. The first laser to appear in this wavelength range was the 15.9 μm SF_6 laser [6.44] which we have just discussed. More recently, *Osgood* [6.142] reported an optically pumped HBr–CO_2 transfer laser oscillating in either the 16 μm or the 14 μm region. This laser uses a 1:1 mixture of HBr and CO_2 gases at a total pressure of 0.5–7 Torr. The HBr molecules are optically pumped by the 4.1 μm (1→0 band) output from an HBr laser. The resulting excitation is then transferred by collisions to the $00^\circ1$ level of CO_2 which is at near resonance with the $v=1$ state of HBr. After sufficient energy transfer has taken place, a strong stimulating pulse at 9.6 μm from a separate CO_2 TEA laser is used to drive the CO_2 molecules from $00^\circ1$ to $02^\circ0$, thereby establishing population inversion between $02^\circ0$ and 01^10. This leads to an output pulse at 16 μm after a buildup time of 100–200 ns. Cooling the CO_2 gas reduces the thermal population of the 01^10 state and improves the degree of population inversion. At lower operating pressures the 16 μm pulse lasts as long as the pump pulse, and usually only the rotational level in the $02^\circ0$ state which is populated by the 9.6 μm stimulating pulse gives rise to laser oscillation. At higher operating pressures, the 16 μm pulse becomes shorter than the pump pulse, and the laser tends to oscillate on the lines near the peak of the emission band. If the laser cavity is also made to be resonant at 10.6 μm, but with sufficient loss such that the buildup of the 10.6 μm pulse is somewhat delayed, then a 14 μm output corresponding to the $10^\circ0\to01^10$ transition is

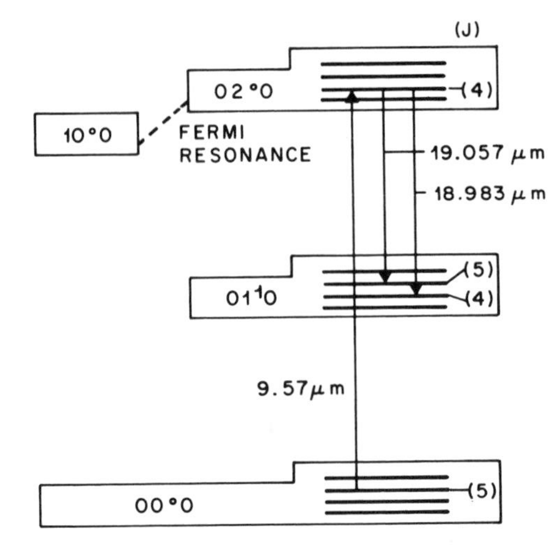

Fig. 6.15. Energy-level diagram for the optically pumped OCS laser

also observed. Notice that the 10°0 state is actually a component of the [10°0, 02°0] Fermi dyad.

A gas dynamic approach to the excitation of the above laser has also been demonstrated by *Stregack* et al. [6.143]. In this approach, a mixture of N_2–He or D_2–Ar is first excited in an electrical discharge. The resulting vibrationally excited gas is then accelerated through an array of supersonic nozzles and mixed with injected CO_2 molecules (typically 0.3 Torr CO_2 in 6 Torr total pressure). A very high vibrational temperature in the ν_3 manifold of CO_2 and a low vibrational temperature in the ν_2 manifold are achieved in this way. Outputs in the 16 μm or 14 μm band can then be obtained either by using an external stimulating pulse at 9.6 μm or 10.6 μm or by generating *Q*-switched stimulating pulses internally. The use of gas dynamic approach eliminates the need for an HBr laser and permits the operation of the laser at very high repetition rates.

Buchwald et al. [6.144] have demonstrated another optically pumped CO_2 laser which has some similarities to the above schemes. They use the $P_2(6)$ or $P_2(5)$ line of an HF laser to pump $^{12}C^{18}O_2$ or $^{12}C^{16}O^{18}O$ molecules to the 10°1 state. This leads to laser action on the $10^\circ1 \rightarrow 01^11$ (17.5 μm or 16.8 μm) band and the $10^\circ1 \rightarrow 10^\circ0$ (4.3 μm) band of the pumped molecule. The latter transition also gives rise to cascade transitions on many lines in the $10^\circ0 \rightarrow 01^10$ band between 16.6 μm and 18.1 μm. These systems also oscillate in the usual 10 μm difference band.

Prior to the above work, a very similar set of laser bands has been made to oscillate in the normal CO_2 molecules by electronic-to-vibrational energy transfer using photodissociated Br [6.145], also cf. [6.41]. The same system has also produced laser action in the 2→1 vibrational band (~5 μm) of NO and the $2\nu_2-\nu_2$ band (7–8 μm) of H_2O as well as an unidentified line at 16.9 μm in H_2O [6.146].

7.3.3 Lasers Based on Fundamental Bands in HF and NH_3

An optically pumped laser based on the scheme illustrated in Fig. 6.10 was demonstrated in 1972 by *Skribanowitz* et al. [6.147]. They used the $R(J'-1)$ lines, where $J'=2$ to 5, of the $1-0$ band from a pulse HF laser to pump the corresponding absorption lines of HF gas and obtained laser action on the $P(J'+1)$ lines of the same band. Incidentally, higher output powers at these P-branch lines are readily available from the pump laser itself. Directional asymmetry in the laser gain was observed as in [6.45].

Recently *Chang* and *McGee* [6.148] developed an optically pumped NH_3 laser based on the above mentioned scheme. Here, a megawatt, pulsed CO_2 laser tuned to the $R(16)$ line at 9.294 μm was used to pump NH_3 from the $a(6,0)$ level in the ground vibrational state to the $s(7,0)$ level in the v_2 vibrational state. Laser action was observed at 12.81 μm corresponding to the $aP(8,0)$ line of the v_2 fundamental band. The peak output power was ~ 5 kW at 2–8 Torr pressure. At room temperature, the Boltzmann factor between the starting level $a(6,0)$ and the terminal level $a(8,0)$ is ~ 4. This means that a population inversion for the laser transition can be achieved when the pump intensity is greater than the saturation intensity I_s. The identification of the energy levels was aided by detailed absorption measurements at the pump frequency which revealed that the $aR(6,0)$ absorption line of NH_3 is about $0.02\,cm^{-1}$ higher in frequency than the $R(16)$ line of the CO_2 laser. Previously, the same pump line has been used to obtain a far infrared laser line at 90.5 μm in NH_3 [6.57] and the upper level was thought to be $s(7,1)$.

The 12.81 μm NH_3 laser discussed above has been used to extend the tuning range of an InSb spin-flip Raman laser to 16.8 μm [6.149]. Since the tuning range covers the v_3 band region of UF_6, and the efficiency of the NH_3 laser can potentially be improved to 20% or higher, this tunable infrared source may become useful for isotope separation experiments.

6.4 Lasers Based on Pure-Rotational or Inversion Transitions

Rotational transitions are associated with the permanent dipole moment of molecules and generally occur in the far-infrared (FIR) region from about 30 μm to 2 mm and in the microwave region. Before 1970, the FIR portion of the electromagnetic spectrum was only very sparsely covered by coherent monochromatic sources. The situation has changed dramatically since then with the advent of optically pumped FIR lasers. Multitudes of cw laser lines providing stable output at the mW level and a few pulsed laser lines providing high peak power up to 1 MW are now available. It is expected that the optical pumping technique will eventually lead to thousands of laser lines in the FIR region.

Also included for discussion in this section are inversion transitions in NH_3 which have been made to exhibit laser action in the FIR by the same excitation technique used for rotational transitions. In principle, however, inversion transitions are more akin to vibrational-rotational transitions.

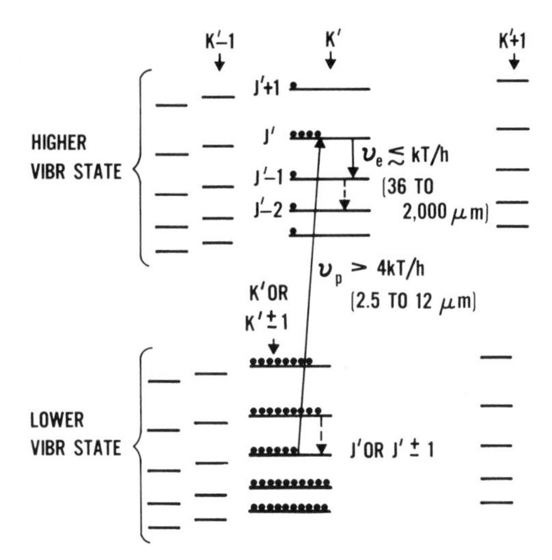

Fig. 6.16. Partial energy-level diagram for a symmetric-top molecule showing possible FIR laser transitions and the corresponding pump transition (from [6.1])

6.4.1 Basic Principles

The basic principles of optically pumped FIR lasers are quite simple and can be understood by referring to Fig. 6.16 which is a partial energy-level diagram for a symmetric-top molecule. Two vibrational states of the molecule are shown in Fig. 6.16, the lower of which is either the ground vibrational state or a low-lying vibrational state. Each vibrational state is seen to consist of a large number of rotational levels with different quantum numbers J and K. In optical pumping, molecules are excited by an infrared laser beam from a certain rotational level in the lower vibrational state to a rotational level (J', K') in the upper vibrational state. Since the thermal population in the upper vibrational state is very small, an inversion of population between levels (J', K') and $(J'-1, K')$ is readily established. Laser action can then be obtained on the rotational transition connecting these levels provided that the molecule has permanent dipole moment. Sometimes, laser action on the cascade transition from $(J'-1, K')$ to $(J'-2, K')$ as well as laser action in the lower vibrational state due to the depletion of population from the starting level (a reversed three-level scheme, see Fig. 6.1b) are also observed. These descriptions are also valid for linear molecules, except that in linear molecules only $K=0$ levels are present.

In symmetric-top molecules the laser transitions must obey the selection rules $\Delta J = -1$, and $\Delta K = 0$. In the case of an asymmetric-top molecule, these selection rules are relaxed and several laser transitions with different wavelengths and varying strengths may exist for a given upper level. Similarly, many more absorption lines are present in a given pump band if the molecule is asymmetric. In molecules containing at least one OH, NH_2, or CX_2Y group, laser action could also take place on transitions corresponding to the internal rotation of such a molecular subgroup with respect to the rest of the molecule.

In ammonia, the average dipole moment is zero due to the tunneling motion of the N atom through the H_3 triangle. However, FIR transitions are allowed between inversion states of different parities. These transitions follow essentially the selection rules for parallel vibrational bands in symmetric-top molecules: i.e., $\Delta J=0, \pm 1$, and $\Delta K=0$.

Since the photon energy of FIR radiation is $\lesssim kT$, the energy separation between the rotational levels is quite small. The close spacing of energy levels makes it very difficult to achieve population inversion by other means, e.g., by electrical discharge. This kind of difficulty does not arise in optical pumping due to the high degree of monochromaticity of the pump radiation. However, a close coincidence between the pump wavelength and the absorption wavelength is required. Fortunately, there are many infrared laser lines and there are also many absorption lines for each molecule. Consequently, the probability of spectral coincidence is reasonably high.

The operating pressure of optically pumped FIR lasers is usually between 0.01 and 1 Torr. Two kinds of problems are encountered in the operation of these lasers at higher pressures. One is collisional equilibration among rotational levels, which is particularly rapid for polar molecules. The other is self-absorption of the emission frequency by rotational transitions in lower vibrational states, which may be negligible at lower pressures due to the slight relative shifts of rotational spectra associated with different vibrational states.

A representative experimental system is shown schematically in Fig. 6.17. The principal device in the system is a meter-long, internal-mirror, hole-coupled, FIR resonator which is filled to a suitable pressure with the molecules to be excited. The FIR output emanating from the horn-shaped coupling hole is first deflected by a beam-splitter flange and then exits from the vacuum vessel as a collimated beam through a polyethylene lens/window. The infrared pump beam is focused into the resonator through a small window on the beam-splitter flange and the coupling hole on the output mirror. The pump beam is then reflected many times between the resonator mirrors until it is completely absorbed. The infrared pump laser, whether of the cw variety or of the high-power pulsed TEA variety, is usually equipped with a diffraction-grating end-mirror to restrict operation to only one of the many transitions available from the laser. Further details on experimental techniques will be discussed in Subsection 6.4.4.

The laser gain that can be achieved can be estimated from (6.3) and (6.4). For rotational transitions in a symmetric-top molecule, the transition dipole moment is given by

$$\mu_{jl}^2=\mu_v^2\,\frac{J'^2-K'^2}{J'(2J'+1)}\,, \tag{6.5}$$

where μ_v is the permanent dipole moment of the molecule in the excited vibrational state. The degeneracy factors are given by

$$g_j=g_{J'}=2J'+1\,, \tag{6.6a}$$

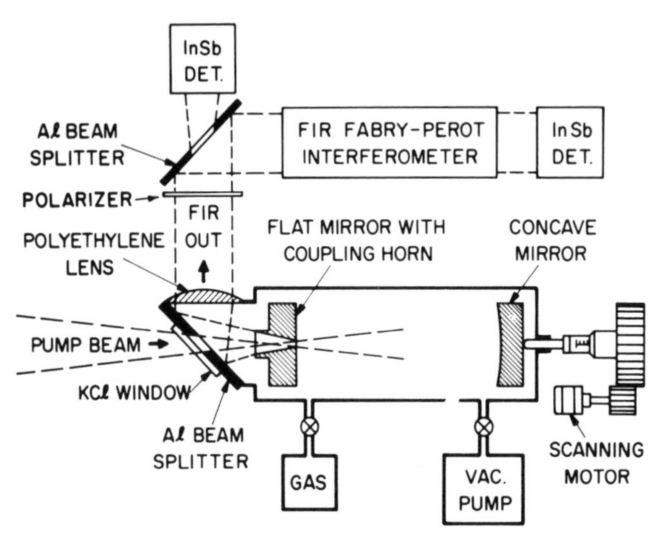

Fig. 6.17. Schematic diagram of a representative optically pumped FIR laser system (from [6.1])

and

$$g_l = g_{J'-1} = 2J' - 1\,. \tag{6.6b}$$

At typical operating pressures the FIR laser line is collision broadened and the line-shape function is given by [6.13]

$$S_L(\nu, \nu_{jl}) = \frac{\Delta\nu_L/2\pi}{(\nu-\nu_{jl})^2 + (\Delta\nu_L/2)^2} \tag{6.7}$$

where $\Delta\nu_L$ is the Lorentz linewidth (FWHM). By combining (6.3–7) we obtain an expression for the laser gain at the line center:

$$\gamma(\mathrm{cm}^{-1}) = 3.972 \times 10^{-5} \frac{\mu_v^2}{\lambda_{jl}\Delta\nu_L} \frac{(J'^2 - K'^2)}{(J'^2 + 0.5J')} \left(n_{J'} - \frac{2J'+1}{2J'-1} n_{J'-1} \right). \tag{6.8}$$

The corresponding expression for the $\Delta J = 0$ inversion transition in ammonia is

$$\gamma(\mathrm{cm}^{-1}) = 7.944 \times 10^{-5} \frac{\mu_v^2}{\lambda_{jl}\Delta\nu_L} \frac{K'^2}{J'(J'+1)} (n_{J'a} - n_{J's})\,. \tag{6.9}$$

The value of μ_v is usually on the order of 1 Debye (3.3346×10^{-30} Coulomb-meter), which is very large compared to the dipole moment of, say, the CO_2 laser transition (0.0356 Debye). This large value of dipole moment more than offsets the large value of λ_{jl} (in μm) in the denominator of (6.8). The ratio of inversion density to linewidth that can be achieved is on the order of 10^5 to $10^6\,\mathrm{cm}^{-3}\,\mathrm{Hz}^{-1}$. When these typical values are inserted into (6.8), we find that the range of small signal gain that can be expected for a laser of this kind is from 10^{-4} to $10^{-1}\,\mathrm{cm}^{-1}$, i.e., 1% per meter to 1% per mm, which is more than adequate for achieving laser action.

One interesting feature of this type of optically pumped FIR lasers is that when the pump beam is linearly polarized, the FIR output is also almost always linearly polarized. The plane of polarization has a definite orientation for each FIR transition and is either parallel or perpendicular to the plane of polarization of the pump beam. This phenomenon is due to the fact that the dipole moment of any transition, whether vibrational-rotational or pure-rotational, is oriented either perpendicular or parallel to the space-fixed total-angular-momentum vector of the molecule. Molecules that happen to have their total-angular-momentum vector oriented relative to the pump field in the direction for maximum pump absorption therefore absorb and emit more strongly than molecules oriented along other directions. As a result, the excited medium exhibits a polarized gain pattern and the laser oscillates with its electric field polarized in the preferred direction. In fact, polarization of excited molecules by optical pumping is a well-known phenomenon in classical optical-pumping experiments (see Sect. 6.1) as well as in photo-luminescence.

The polarization-dependent laser gain can be expressed as

$$\gamma_\theta = \Pi_\theta \gamma\,, \tag{6.10}$$

where γ is given by (6.8) and Π_θ represents the angular gain pattern $(1/2 < \Pi_\theta < 2)$. The gain pattern depends on the angular distribution of the total-angular-momentum vectors of the molecules excited into the upper laser level. For the derivation of expressions for Π_θ, we introduce an angular distribution function Ψ (in steradian^{-1}) of excited molecules, which is always normalized such that

$$\int \Psi(\Omega)\, d\Omega = 1\,,$$

where Ω is a two-dimensional (spherical) angular variable. The pump radiation will be taken to be linearly polarized along the x-direction, and its intensity will be assumed to be below the saturation level. The molecules in the source level i are assumed to be completely randomly oriented. For Q-branch $(\Delta J = 0)$ pump transitions, the transition dipole moment is parallel to the total-angular-momentum vector whose direction is given by a unit vector $\hat{\boldsymbol{r}}$. From the fact that the transition probability is proportional to the square of the directional cosine $\hat{\boldsymbol{r}} \cdot \hat{\boldsymbol{x}}$, and from the normalization condition for Ψ, we obtain

$$\Psi = \frac{3}{4\pi} (\hat{\boldsymbol{r}} \cdot \hat{\boldsymbol{x}})^2\,.$$

For R- and P-branch ($\Delta J = \pm 1$) pump transitions, the transition dipole moment is perpendicular to and rotates about $\hat{r}$. It can be shown, by resolving this rotating dipole moment into two orthogonal out-of-phase components, that the directional cosine in this case is given by $|\hat{r} \times \hat{x}|$. From this and the normalization condition, we obtain

$$\Psi = \frac{3}{8\pi} |\hat{r} \times \hat{x}|^2 .$$

To evaluate the angular gain pattern Π_θ, let us assume that the laser gain is probed by a weak FIR probe signal which is linearly polarized along a direction $\hat{\theta}$. For $|\Delta J| = 1$ rotational transitions, the transition dipole moment is perpendicular to and rotates about the total-angular-momentum vector of any given molecule in the upper laser level, and the FIR transition probability is proportional to the square of the directional cosine $|\hat{r} \times \hat{\theta}|$. By integrating the transition probability over all molecular orientations and by observing the normalization condition that $\Pi_\theta = 1$ when the excited molecules are completely randomly oriented ($\Psi = 1/4\pi$), we obtain

$$\Pi_\theta = \int \Psi(\Omega) \tfrac{3}{2} |\hat{r} \times \hat{\theta}|^2 \, d\Omega . \tag{6.11a}$$

For $\Delta J = 0$ FIR transitions (such as the inversion transition in ammonia) the transition dipole moment is parallel to $\hat{r}$ and we obtain

$$\Pi_\theta = \int \Psi(\Omega) \, 3(\hat{r} \cdot \hat{\theta})^2 \, d\Omega . \tag{6.11b}$$

From these expressions, Π_x and Π_y can be evaluated for various combinations of pump- and emission-transition selection rules. The results are summarized in Fig. 6.18.

The degree of polarization of the laser gain is of course reduced by collisional reorientation of molecules. However, for excited states with $K \ll J$, the characteristic time for molecular reorientation by collisions is much longer than the lifetime of the upper laser level. The $K \ll J$ transitions are, therefore, not expected to be significantly depolarized by collisions. The few transitions that are experimentally observed to be depolarized are most likely rotational transitions associated with $K \approx J$ states which are pumped via an R- or a P-branch transition. For these transitions, the degree of polarization is low (4:3), and the reorientation time can be comparable to the upper-state lifetime.

Another fact to be borne in mind in the calculation of laser gain is that under typical operating conditions the pump transition is usually Doppler broadened. This means if a single-frequency pump source is used then a "hole" will be burnt in the Doppler-broadened line and only molecules in a narrow velocity class will be excited into the upper state. The effect this has on the performance of the laser was discussed in Section 6.1. Finally, an accurate estimate of the inversion density can be obtained only by considering the

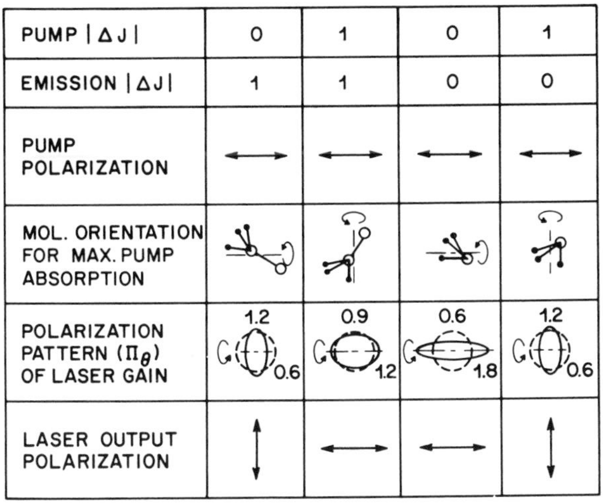

PUMP |ΔJ|	0	1	0	1
EMISSION |ΔJ|	1	1	0	0
PUMP POLARIZATION	⟷	⟷	⟷	⟷
MOL. ORIENTATION FOR MAX. PUMP ABSORPTION				
POLARIZATION PATTERN (Π_θ) OF LASER GAIN	1.2 / 0.6	0.9 / 1.2	0.6 / 1.8	1.2 / 0.6
LASER OUTPUT POLARIZATION	↕	⟷	⟷	↕

Fig. 6.18. Polarization dependence of laser gain for various combinations of selection rules followed by pump and emission transitions

effects of many relaxation and transfer processes and the saturation of both the pump and emission transitions. Further discussions of the theoretical aspects of optically pumped FIR lasers will be given in Subsection 6.4.5.

6.4.2 Summary of Observed Laser Lines

So far, 27 different molecules have exhibited FIR laser action by optical pumping and more than 500 laser lines have been reported. Their output wavelengths span the six-octave range from approximately 30 μm to 2 mm. As shown in Fig. 6.19, the number of laser lines per octave is a maximum in the 20–40 cm^{-1} range and decreases montonically toward both ends of the spectrum. The shape of this distribution curve is influenced by several factors. For instance, almost all FIR resonators used by various investigators had relatively high loss for $\lambda > 2$ mm, and in some cases, the FIR resonators were highly lossy even for $\lambda > 700$ μm. Furthermore, the detectors used in some studies did not respond at either one or the other end of the FIR region. However, the shape of this distribution function is not likely to be altered drastically by the removal of these factors for the following reasons: The FIR laser gain is proportional to $\nu_{jl}\Delta n_{jl}$, where $\nu_{jl} = 2BJ'$, B being the rotational constant of the molecule, and the inversion density Δn_{jl} that can be achieved in optical pumping depends to a large extent on the population available at the source level i in the lower vibrational state. On the basis of these facts, it can be shown that a symmetric-top molecule will exhibit the highest gain when $J' \approx (kT/hB)^{1/2}$ and the emission frequency is about equal to $(4BkT/h)^{1/2}$. For the gain to be within an order of magnitude of this maximum value, the emission frequency should fall within the following bounds:

$$(0.2BkT/h)^{1/2} < \nu_e < (20BkT/h)^{1/2} .$$

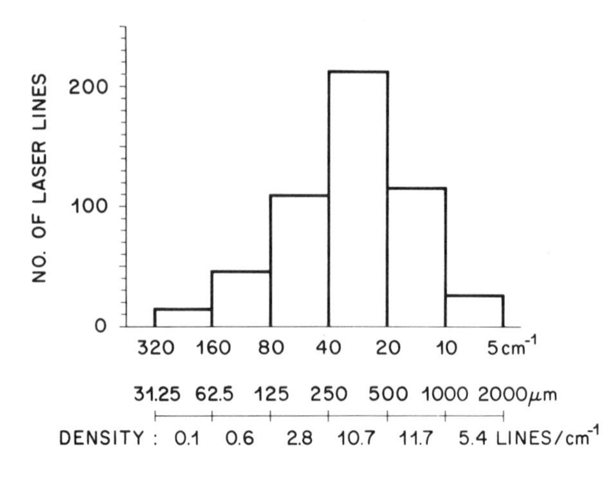

Fig. 6.19. Spectral distribution of optically pumped FIR laser lines

For the molecule CH_3F ($B=0.84\,cm^{-1}$), this frequency range turns out to be $6\,cm^{-1}$ to $60\,cm^{-1}$. Other molecules would cover a similar range, falling either somewhat higher or somewhat lower in frequency. An asymmetric-top molecule usually covers two or three overlapping frequency ranges each associated with a permanent dipole moment along one of the molecular axes. Furthermore, a molecule with an asymmetric internal-rotation group such as OH or NH_2 can cover another wavelength range associated with internal rotation or libration of these molecular subgroups. The rotational transitions of gaseous molecules are, therefore, expected to cover the entire FIR region and have a distribution function peaked in the submillimeter-wave region.

A list of molecules that have been investigated and the number of FIR laser lines observed in each molecule is given in Table 6.1. Except for the HF molecule which was pumped by an HF laser, and two lines in NH_3 produced by pumping with an N_2O laser, all of the FIR laser lines were produced by using a CO_2 laser as the pump source. The molecules listed have not all been investigated with the same degree of completeness. Nevertheless, the general trend is clear. By using a CO_2 laser as the pump source, an asymmetric-top molecule with OH or NH_2 group can yield 30 to 60 laser lines. Other asymmetric-top and symmetric-top molecules typically yield 15 to 30 laser lines per molecule. These numbers would approximately double if N_2O and isotopic-CO_2 lasers were also utilized as the pump source. Since there are more than 100 other volatile polar molecules with absorption bands in the 10 μm region, and since other IR lasers such as CO, HF, DF, HCl, and HBr can all be used as pumping sources, this method of pumping can potentially yield on the order of 10^4 FIR laser lines. If tunable IR sources such as spin-flip Raman lasers and parametric oscillators were also utilized as the pump source, then the potential number of available FIR laser lines would be well in excess of 10^5. The density of laser lines is presently already in excess of 10 lines per cm^{-1} in the 10–40 cm^{-1} region (as is indicated in the lower part of Fig. 6.19).

Table 6.1. Classification of molecules studied according to the molecular type and the number of optically pumped FIR laser lines observed in each molecule

Type of molecule	Mol. formula	No. of lines	References
Diatomic molecule	HF	7	[6.45, 46]
Symmetric top	CH_3F	16	[6.47–50]
w/o internal	CH_3Cl	24	[6.50–52]
rotation	CH_3Br	35	[6.52]
	CH_3I	20	[6.52, 53]
	CH_3CN	31	[6.49, 52, 54]
	CH_3CCH	15	[6.49]
Symm. top w. inversion	NH_3	30	[6.50, 55–58]
Symm. top w. internal rot.	CH_3CF_3	1	[6.50]
Asymmetric top	CH_2Cl_2	1	[6.51]
w/o internal	CH_2CHCl	18	[6.48, 54]
rotation	CH_2CHCN	19	[6.54, 59]
	CH_2CF_2	14	[6.54, 59, 60]
	O_3	3	[6.56]
	D_2O	10	[6.61, 62]
Asymmetric top w. OH group	CH_3OH	82	[6.48, 50, 54, 56, 60, 63–67, 134]
	CH_3OD	16	[6.63, 134]
	CD_3OD	8	[6.68]
	CH_3CH_2OH	1	[6.51]
	HCOOH	51	[6.54, 56, 69, 70]
	CH_2OHCH_2OH	42	[6.69]
Asymm. top w. NH_2	NH_2NH_2	26	[6.71]
group	CH_3NH_2	35	[6.54, 59, 69]
Asymmetric top	CH_3CH_2F	23	[6.54, 56]
w. heavier	CH_3CH_2Cl	4	[6.51]
internal-rotation	CH_3CHF_2	4	[6.50, 60]
group	CH_3OCH_3	6	[6.69]
Total	27	542	

A complete listing of published laser lines is beyond the scope of this review. It is also difficult to consolidate various published data due to greatly varying experimental conditions under which the data were obtained and due to very poor accuracy with which the wavelength and the output power were determined in some experiments. We have compiled instead a list of 120 "strong" laser lines pumped by a CO_2 laser, which is presented as Table 6.2. These lines represent about 22% of all reported FIR laser lines and are considered to be strong. The definition of the term "strong" is necessarily ambiguous and somewhat arbitrary, and some errors are unavoidable in classifying the strengths of laser lines. Roughly speaking, though, all these lines can be expected to produce 10^{-4} W or more cw or quasi-cw output in a typical laboratory system. Those marked with an asterisk are considered very-strong and should produce at least a few milliwatts of output power.

Recently, *Dyubko* et al. have reported numerous new optically pumped FIR laser lines including: 19 lines from 224 to 1990.75 μm in CD_3Cl, 9 lines

Table 6.2. Strong FIR laser lines pumped by a CO_2 laser
[* = very strong; 9 P, 9 R = 9.4 μm band; 10 P, 10 R = 10.4 μm band; 11 P = (01^11)–(11^10) band]

λ (μm)	Molecule	Pump	λ (μm)	Molecule	Pump
37.5	CH_3OH	9 P 32	286.	CH_3OH	10 R 48
40.2	CH_3OH	9 P 34	294.3	CH_3Br	10 R 28
*41.7	CH_3OH	9 P 32	*295.	CH_3OD	9 R 8
57.	CH_3OD	9 R 8	299.	CD_3OD	10 R 24
70.1	$(CH_2OH)_2$	9 P 34	*307.	CH_3OD	9 R 8
*70.6	CH_3OH	9 P 34	307.7	CH_3Cl	11 P 19
95.8	$(CH_2OH)_2$	9 R 10	311.1	CH_3Br	10 P 20
*96.5	CH_3OH	9 R 10	311.2	CH_3Br	10 P 40
103.0	CH_3OD	9 P 30	311.2	CH_3Br	10 R 50
117.1	$(CH_2OH)_2$	9 P 14	312.	CH_3OD	10 R 10
*117.4	CH_3OD	9 P 26	332.9	CH_3Br	10 R 6
*118.8	CH_3OH	9 P 36	*334.0	CH_3Cl	9 P 42
118.9	$(CH_2OH)_2$	9 P 34	346.3	CH_3CN	9 P 16
148.5	CH_3NH_2	9 P 24	*349.3	CH_3Cl	10 R 18
170.6	CH_3OH	9 P 36	352.8	CH_3Br	9 P 18
184.	CD_3OD	10 R 24	369.1	CH_3OH	9 P 16
192.8	CH_3F	10 R 32	*372.7	CH_3F	9 P 50
193.5	N_2H_4	10 P 24	372.9	CH_3CN	10 P 20
197.	$(CH_2OH)_2$	9 P 38	375.0	CH_2CF_2	10 P 12
202.4	CH_3OH	9 P 36	377.5	CH_3I	9 R 16
233.	CH_3OH	9 R 10	*380.0	CH_3Br	10 R 18
241.0	CH_3Cl	10 P 10	387.3	CH_3CN	9 R 12
245.0	CH_3Br	9 P 28	388.	HCOOH	9 R 16
246.	CH_3OH	10 R 38	390.5	CH_3I	10 P 42
258.	CH_2Cl_2	10 P 26	392.3	CH_3OH	9 P 36
264.0	CH_3Br	10 R 10	*393.6	HCOOH	9 R 18
271.3	CH_3Cl	10 P 20	394.2	HCOOH	?9 R 16
275.0	CH_3Cl	9 R 14	406.	CD_3OD	10 R 12
277.	$(CH_2OH)_2$	9 P 38	406.0	HCOOH	?9 R 16
282.0	CH_3CN	9 P 50	414.	HCOOH	9 R 22

from 378 to 1253.73 μm in CH_3I, and 32 lines from 272 to 1550 μm in CD_3I [6.150]; 70 lines from 184 to 1290 μm in CD_3OH [6.151]; 34 lines from 254 to 1230 μm in HCOOH, 65 lines from 240 to 1730 μm in HCOOD, 20 lines from 265 to 1240 μm in DCOOH, and 57 lines from 218 to 1280 μm in DCOOD [6.152]. This brings the total number of optically pumped FIR laser lines to the order of 800 and the total number of gases and vapors studied to 33.

6.4.3 Specific Examples and Special Topics

The first optically pumped FIR laser was reported by *Chang* and *Bridges* in 1970 [6.47]. They observed laser action in CH_3F by pumping the gas with a Q-switched CO_2 laser tuned to the 9.55-μm $P(20)$ line. Based on a previous study on CH_3F [6.72] the molecule was expected to be excited from the $J=12$ level in the ground vibrational state to the $J=12$ level in the ν_3

Table 6.2 (continued)

λ (μm)	Molecule	Pump	λ (μm)	Molecule	Pump
415.	CH_2CF_2	10 P 14	585.7	CH_3Br	9 P 40
*415.0	CH_3Br	10 R 2	647.9	CH_3CCH	10 P 14
418.3	CH_3Br	10 P 26	652.7	CH_3CN	9 P 30
420.0	HCOOH	?9 R 18	658.5	CH_3Br	9 P 56
*428.	HCOOH	9 R 20	*660.7	CH_3Br	10 R 20
432.1	HCOOH	?9 R 18	671.0	CH_3I	10 P 28
*447.2	CH_3I	10 P 18	675.3	CH_3CCH	9 P 40
453.4	CH_3CN	9 R 16	713.7	CH_3CN	10 P 32
458.	CH_3CHF_2	10 P 20	715.4	CH_3Br	10 R 14
466.3	CH_3CN	9 R 16	719.3	CH_3I	10 P 22
492.	$(CH_3)_2O$	10 P 34	749.3	CH_3Br	10 P 14
494.7	CH_3CN	9 P 6	831.1	CH_3Br	10 P 28
*496.1	CH_3F	9 P 20	870.8	CH_3Cl	9 P 52
*508.4	CH_3I	9 P 34	884.	CH_2CF_2	10 P 12
508.5	CH_3Br	10 R 42	900.	CH_3CH_2Cl	10 R 30
512.	HCOOH	9 R 28	*925.5	CH_3Br	10 R 46
513.2	HCOOH	9 R 24	*944.0	CH_3Cl	9 R 12
517.3	CH_3I	10 P 14	958.3	CH_3Cl	9 P 38
520.	$(CH_3)_2O$	10 P 12	990.5	CH_3Br	10 P 10
529.3	CH_3I	10 P 36	*1063.9	CH_3I	10 P 38
531.1	CH_3Br	10 P 24	1097.1	CH_3CCH	9 P 8
533.	CH_3CHF_2	10 P 20	1174.9	CH_3CCH	10 P 44
*545.2	CH_3Br	10 P 38	1221.8	$C^{13}H_3F$	9 P 32
*545.4	CH_3Br	10 R 32	*1253.7	CH_3I	10 P 32
554.4	CH_2CF_2	10 P 14	1310.4	CH_3Br	10 R 4
566.4	CH_3CCH	9 P 18	1572.6	CH_3Br	10 P 4
568.	CH_2CF_2	10 P 24	1720.	CH_3CH_2Cl	10 R 28
570.5	CH_3OH	9 P 16	1814.4	CH_3CN	10 P 46
576.2	CH_3I	10 P 16	1886.9	CH_3Cl	9 P 26
578.9	CH_3I	10 R 34	1965.3	CH_3Br	10 P 28

vibrational state yielding an emission wavelength of 496 μm. Shown in Fig. 6.20 is the experimentally observed dependence of the FIR output from this CH_3F laser on resonator-length tuning for two different pump frequencies separated by about 50 MHz. The spacing of the main peaks corresponds to a resonator-length change of 248 μm and confirms the laser wavelength to be 496 μm. A detailed examination of Fig. 6.20 coupled with accurate frequency measurements using microwave techniques reveals three pairs of laser lines near 496 μm (B, b), 452 μm (A, a) and 541 μm (C, c). Each pair consists of two lines separated by 30–40 MHz corresponding to $K=1$ and $K=2$ components. The relevant energy levels for CH_3F are shown in Fig. 6.21. These laser transitions were identified by comparing the measured frequencies of A, a transitions with the known rotational constants in the ground vibrational state and from the polarization of the FIR output relative to the polarization of the pump radiation. Measured accurate frequencies of B, b and C, c

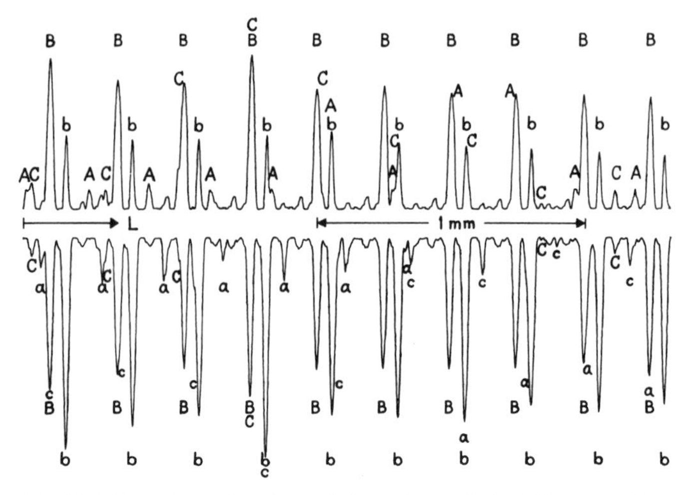

Fig. 6.20. Experimentally observed dependence of the FIR output from a CH_3F laser on the resonator-length tuning. The upper and lower traces are taken at somewhat different pump frequencies 50 MHz apart (from [6.47])

transitions have yielded new accurate spectroscopic constants for the ν_3 vibrational state of CH_3F.

Subsequently, ten additional laser lines have been reported for CH_3F [6.49, 50]. But the 496 μm laser lines remain the most extensively studied optically pumped FIR laser lines. In cw operation, a CH_3F laser typically provides 10^{-2} W of output power at 20 to 50 mTorr. With a *Q*-switched CO_2 laser as the pump source, the output is typically 10^{-1} W at 0.1 to 0.2 Torr. And with a high-power TEA CO_2 laser as the pump source [6.66, 73–81], an output spike as high as 0.5 MW at 3 Torr has been reported. In the case of TEA-laser pumping, a large nonresonant CH_3F gas cell is usually employed. This configuration leads to the maximum energy extraction. However, the resulting "superradiant" output pulse contains many irregular temporal spikes and its spectral width is relatively wide (>100 MHz) [6.76, 81]. The spikes can be made regular by mode locking the TEA laser, [6.61] and the spectral width can be made narrower than 30 MHz (at the expense of the output power) by using an off-axis or transversely pumped FIR resonator [6.76, 78, 80]. Thus the best approach to both a very high FIR output power and good mode properties appears to be the use of an FIR oscillator-amplifier chain [6.61, 76].

A spectral width of about 700 MHz was observed for a superradiant 496 μm CH_3F laser by *Brown* et al. [6.76]. Apparently $K=0$–7 transitions (instead of just $K=1, 2$) are all excited. This may result from the wide bandwidth of the TEA pump laser used and/or off-resonance pumping (see following discussion of NH_3 lasers) at high pump intensities. The high pump intensity may also lead to a secondary gain peak (corresponding to the Stokes

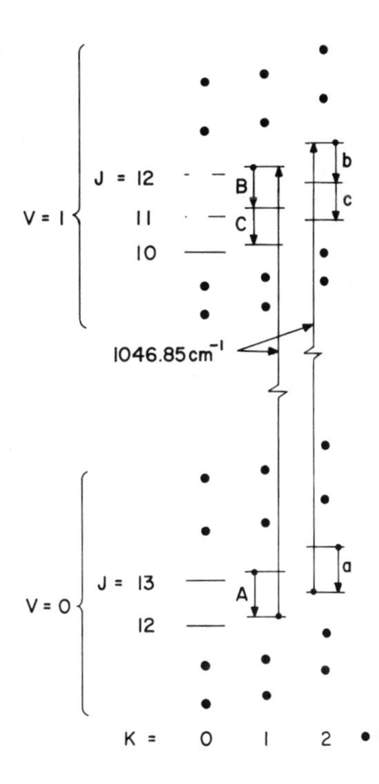

Fig. 6.21. Partial energy-level diagram showing the pump and laser transitions in CH_3F (from [6.47])

frequency) for each of the K-component due to Raman-like process discussed in Section 6.1.

Several groups have reported progress toward a high-power narrow-line pulsed 496 μm laser system intended primarily for performing Thomson scattering on Tokamak plasmas [6.123]. By adding a 1.8 m long, zigzag pumped amplifier section to their zigzag pumped oscillator [6.80], *Drozdowicz* et al. [6.153] obtained 6 kW output at 496 μm with a linewidth of 30 MHz. The maximum photon conversion efficiency was 15% at 0.1 Torr. *Brown* et al. [6.154] studied a collinearly pumped oscillator-amplifier system which used special Ge etalon mirrors having maximum transmission at 9.6 μm and suitable reflectivities at 496 μm. The 1.4 m long oscillator delivered 1.7 kW with a linewidth of <60 MHz at 1.5 Torr. This was amplified to 30 kW in a 1.8 m amplifier, giving a photon conversion efficiency of 35% at 2.7 Torr. *Semet* and *Luhmann* [6.155] also studied a collinearly pumped system, but they used a Ge plate at Brewster's angle to transmit 9.6 μm and reflect (orthogonally polarized) 496 μm. Their 87 cm-long dielectric waveguide laser produced 11 kW of output with a linewidth of less than 40 MHz at 2 Torr. This was amplified to a maximum of 156 kW. To obtain this output power, the pump pulse to the amplifier section was delayed by 45 ns. Without the delay, the efficiency of the amplifier was about a factor of two lower. *Evans* et al. [6.156] chose to use slightly off-axis pumping. Their 2.4 m oscillator was made to oscillate on a single longitudinal mode by using two intracavity aperture stops and a Fox-Smith interferometer with a copper-mesh output mirror. The peak output from the oscillator was 1 kW and had a linewidth of less than 25 MHz. They were able to amplify this output to 250 kW with a linewidth of 50–60 MHz. In still another study, *Plant* and *Detemple* [6.168] found that for a 3.2 m CH_3F cell the photon conversion efficiency was 18% in the "superradiant" mode, 12% in the amplifier mode, and ~1.5% in the cavity oscillator mode.

Ammonia is spectroscopically one of the most extensively studied symmetric-top molecules. The positions of its spectral lines relative to the CO_2 and N_2O laser lines were studied in detail by *Shimizu* [6.82] using Stark spectroscopy. His spectroscopic data enabled *Chang* et al. [6.55] to obtain cw laser action at 81.5 and 263.4 μm in NH_3 by using a 1.5 W N_2O laser tuned to the $P(13)$ line at 10.78 μm as the pump source. These laser lines in the ν_2 vibrational state of NH_3 differ from most other FIR laser lines in that the 81.5 μm line is a rotation-inversion transition ($\Delta J=1$) and the 263.4 μm line is a pure-inversion transition ($\Delta J=0$). The relative output polarizations of these laser lines also support their identifications.

Although the ν_2 absorption band of NH_3 is closely overlapped by the CO_2 laser bands, only one CO_2 laser line has so far been found to coincide well enough with an NH_3 absorption line to permit FIR laser operation with less than 100 W of pump power [6.56]. The lack of additional coincidences has, however, been overcome to some extent by two approaches. In the first approach, *Fetterman* et al. [6.50] demonstrated that by using sufficiently intense pump radiation, e.g., from a TEA CO_2 laser, FIR laser action can be achieved even when the pump frequency is as much as 950 GHz (40 collisional halfwidths) off the absorption line. This is because a homogeneously broadened transition can be saturated even when the pump frequency is located very far off in the wing of the absorption line [see (6.27)]. Basically, the requirement for off-resonance pumping is $\mu E_p/h \gtrsim \delta\nu$, where μ is the transition dipole moment for the absorption line, E_p is the field strength of the pump radiation and $\delta\nu$ is the frequency mismatch. *Gullberg* et al. [6.57] have further extended the off-resonance pumping scheme to many other NH_3 lines. In the second approach, *Fetterman* et al. [6.58] obtained new cw FIR laser lines in NH_3 by means of Stark tuning. The Stark field not only tunes certain components of an NH_3 absorption line that is normally off-resonance from a CO_2 laser line into exact resonance but also modifies the parity selection rule for the absorption line such that even a normally forbidden transition becomes allowed.

In the above application of Stark tuning, the Stark shift occurs mainly on the pump transition, while the shift of the FIR emission frequency is quite small. The opposite situation arises in other symmetric-top molecules when the pump transition is a Q-branch transition and the value of K is large. *Inguscio* et al. [6.83] have estimated that a tuning range of 6 GHz might be realizable for the 496 μm CH_3F laser by utilizing several K transitions.

The molecule for which the largest number of FIR laser lines has been reported so far is methanol (CH_3OH). A total of 82 FIR laser lines all pumped by a cw or quasi-cw CO_2 laser has been reported and many of them qualify as strong lines. This and the ready availability of the material make CH_3OH a very convenient general-purpose laboratory source of FIR laser lines. Such an abundance of FIR laser lines, however, is not without a drawback. Very often a single CO_2 laser line simultaneously excites several FIR laser lines, making it difficult to tune the FIR resonator to the desired wavelength and oscillation mode and at the same time suppress all the unwanted output frequencies. The abundance of FIR laser lines in CH_3OH is due to its asymmetric molecular structure and the presence of an FIR-active OH-torsional mode. The torsional mode is probably responsible for some of its laser lines with wavelengths shorter than 100 μm.

The simplest FIR laser molecule studied is HF. *Skribanowitz* et al. [6.45, 46] observed laser action on seven rotational transitions (36.5 to 252.7 μm) in the $v=1$ vibrational state of HF by using $v=1\rightarrow 0$ vibrational-rotational laser lines (2.7 μm) from a 3–4 kW pulsed HF laser as the pump source. The HF pressure in the 12 cm long gain cell ranged from 50 mTorr to 6 Torr. By placing the gain cell in a ring resonator and by passing the pump beam through the gain cell only in one direction, they observed that at approximately 15 times above threshold the forward FIR circulating power inside the ring resonator is about two orders of magnitude greater than the backward circulating power. This is in qualitative agreement with a previous theory [6.84] which predicts asymmetric gain in such a system when the transitions are Doppler broadened.

As can be expected from the large permanent dipole moment (1.82 Debye) of HF and a small rotational partition function, the gain was found to be exceptionally high ($>1\,cm^{-1}$). This high gain permitted *Skribanowitz* et al. [6.85] to study the emission behavior of the excited molecular system in the absence of optical feedback. At lower (1 to 20 mTorr) HF pressures and in longer (30 to 100 cm) gain cells, the FIR emission from both ends of the cell was found to be in the form of a delayed pulse with a ringing tail. The delay time ranged from 0.5 to 2 μs and the width of the main pulse was roughly 1/4 to 1/2 the delay time. Decreasing the pressure or the pump power increased the pulse delay and width and decreased its amplitude.

The observed behavior of the output pulse is consistent with the picture of a Bloch-vector pendulum falling from the fully inverted position, but is not consistent with the rate-equation description of amplified fluorescence (a monotonic radiative decay). Numerical calculations based on a semiclassical treatment of the response of an extended, inhomogeneously broadened amplifying medium to a short input pulse yielded pulse shapes and delay times that were in good agreement with the experimental observations. Hence, it is evident that the process of FIR emission involves macroscopic molecular coherence arising from the mutual interaction of the excited molecules through the radiation field. The system therefore exhibits some characteristics of superradiance in the less rigorous semiclassical sense; e.g., the peak output intensity is proportional to N^2 (N is the total number of inverted molecules), and the output pulse has a ringing tail and tends toward a π-pulse in the limit that $\gamma L \gg 1$ (γ is the gain coefficient). However, it is not known whether the initial FIR photon originated thermally from outside the molecular system, or from inelastic scattering of the pump photons, or from a spontaneous process within the system. Consequently, the experiment does not constitute a rigorous proof of the observation of superradiance in the more subtle quantum electrodynamic sense [6.86].

In addition to HF and CH_3F, intense emission pulses from a mirrorless gain cell have also been observed in NH_3, D_2O, CH_3CN, CH_3Cl, and CH_3Br, all of which used a TEA CO_2 laser as the pump source [6.61].

6.4.4 Experimental Techniques

The basic experimental setup for an optically pumped FIR laser has already been described in Subsection 6.4.1. In this subsection we shall discuss those factors that are important for a good performance of such a laser system.

As discussed previously, the operating pressure of an optically pumped FIR laser is usually well below 1 Torr. At such low pressures, the absorption coefficient of the gas being pumped is often too low to produce a significant absorption of the pump power in a single pass. It is, therefore, very important to design the FIR resonator in such a way that it traps the pump beam for as many round trips as possible. This is usually accomplished by using gold-coated mirrors that are highly reflecting at both the FIR wavelengths and the pump wavelength. The pump beam is usually introduced into the resonator by focusing it through a small hole at the center of a resonator mirror. Mirror curvatures are chosen to produce maximum trapping of the pump beam. Shown in Fig. 6.22 is an experimental curve for the resonator-length dependence of the reflectance (i.e., the relative amount of power reflected from the coupling hole with the input beam focused at the same hole) of a slightly lossy hole-coupled resonator comprising one concave mirror and one flat mirror with a coupling hole. It is seen that the lowest reflectance, corresponding to the maximum trapping of the pump beam, is obtained for $L/R = 0.3$ to 0.4 or 0.6 to 0.7, where R is the radius of curvature of the curved mirror. To obtain the minimum diffraction loss at FIR wavelengths, one would have chosen $L = 0.5R$, corresponding to a semiconfocal resonator, but such a resonator is a very poor radiation trap because the pump beam would refocus on the coupling hole only after two round trips.

In principle, many normal modes of the resonator are simultaneously excited by the focused pump beam at the entrance hole. Fine tuning of the resonator would reveal resonances due to these resonator modes. The trapping efficiency of the resonator is also modulated by these fine resonances. This is particularly noticeable when the transverse modes are degenerate (a confocal

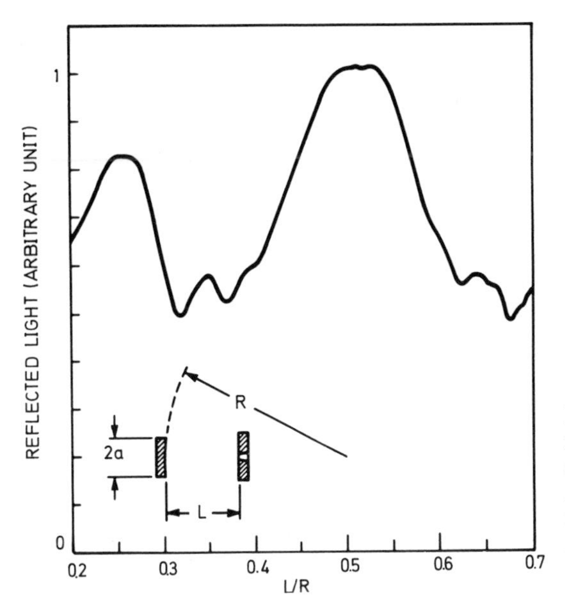

Fig. 6.22. Reflection of pump power from a weakly absorbing FIR resonator as a function of L/R, where L is the mirror spacing and R is the radius of curvature of the concave mirror (from [6.1])

resonator or resonator with two flat mirrors) or when only a few transverse modes have low loss.

The same coupling hole for the pump input is often used to couple out the FIR signal. To reduce the diffraction angle of the FIR output, the coupling hole is usually shaped as a radiation horn of 12–15° total cone angle. The far-field radiation pattern from such a radiation horn should be satisfactory for most applications. However, the use of a single coupling hole makes it impossible to optimize the input and output couplings separately. The use of mesh mirrors [6.65, 87], on the other hand, is incompatible with the requirement of efficient pump trapping. This dilemma has been solved by the recent development of a new type of hybrid output mirror [6.88]. The hybrid mirror consists of a fine metal mesh deposited on an Si substrate and overcoated with a multilayer dielectric film for high reflectance at the pump wavelength. The mirror is partially transmitting at FIR wavelengths but is essentially totally reflecting at the pump wavelength. While the pump beam still uses a small hole for coupling, the FIR output coupling can be independently optimized by adjusting the mesh pattern. Furthermore, a high beam quality can be obtained for the FIR output.

Further design data on mesh-dielectric hybid mirrors [6.88] have been published by *Danielewicz* and *Coleman* [6.157].

Another obstacle to efficient cw operation of an optically pumped FIR laser in a polar molecule is the slowness of vibrational deactivation of the lower laser level relative to rotational relaxation between the upper and lower laser levels. Under typical operating conditions, collisions with the resonator walls are relied on to remove population from the lower laser level. The deactivation time is determined by the molecular diffusion time across the tube radius. Hence, for efficient operation of the laser, the cell diameter should not be made larger than necessary. This suggests the use of a waveguide resonator.

Hodges and *Hartwick* [6.89] found that FIR resonators with a quartz or brass waveguide can operate at higher pressures and produce higher FIR output power than a conventional open resonator. Using a similar metal-waveguide resonator, *Tanaka* et al. [6.65] achieved approximately 10% photon conversion efficiency at 81.5 μm in NH_3. For a dielectric waveguide, the guide radius must be 100λ or greater to avoid excessive radiative waveguide losses [6.90]. In a metallic waveguide, the low loss modes excited by a linearly polarized pump beam are mainly TE_{1l} and TE_{0l} modes. With hole coupling, the TE_{1l} modes are observed to dominate [6.89, 91]. On the other hand, the TE_{0l} modes, which have a null at the center, have been found to dominate in a mesh-coupled waveguide resonator [6.87]. As long as the waveguide loss is not excessive, the smaller the waveguide diameter the lower the threshold pump power and the higher the optimum operating pressure. However, since the mode volume varies as the square of the waveguide diameter, and because the waveguide loss does increase with the decreasing waveguide diameter, the output power will at some point begin to decrease with decreasing waveguide diameter [6.91]. The optimum waveguide size depends on the waveguide mode used, but in general, a 12 mm ID tube appears to be satisfactory [6.51, 54, 65, 67, 89, 91]. However, for operation near 1 mm and beyond, it appears advantageous to have a waveguide of 25 mm ID or larger [6.65, 91].

Further details of mode properties of waveguide lasers with mesh mirrors [6.87] have been published by *Wood* et al. [6.158]. The properties of a FIR laser using a rectangular waveguide was studied by *Yamanaka* et al. [6.159] who found that the best result was obtained when the wide dimension of the waveguide was oriented along the preferred direction of polarization for the FIR signal.

Shown in Fig. 6.23 is one possible design for a waveguide resonator. Plunger-type end-mirrors have been found to be preferable over disc-type end-mirrors since a gap between the mirror and the tube end introduces large losses in the resonator. Waveguide resonators are particularly attractive for their compactness. The use of a waveguide resonator in conjunction with a waveguide CO_2 laser as the pump source results in a very compact, practical FIR laser system [6.90].

Another method of increasing the lower laser level deactivation rate is to introduce a buffer gas into the gain cell. *Chang* and *Lin* [6.92] have found that, in comparison with a cw laser with pure CH_3F, a laser using a 1:1 mixture of CH_3F and C_6H_{14} (*n*-hexane) optimizes at about 100% higher CH_3F partial pressure and produces 55% higher output power at 496 μm. Based on theoretical considerations, the effectiveness of C_6H_{14} as a buffer gas is attributed to its large vibrational heat content relative to its molecular weight. The excited CH_3F molecules are deactivated by rapid vibrational energy

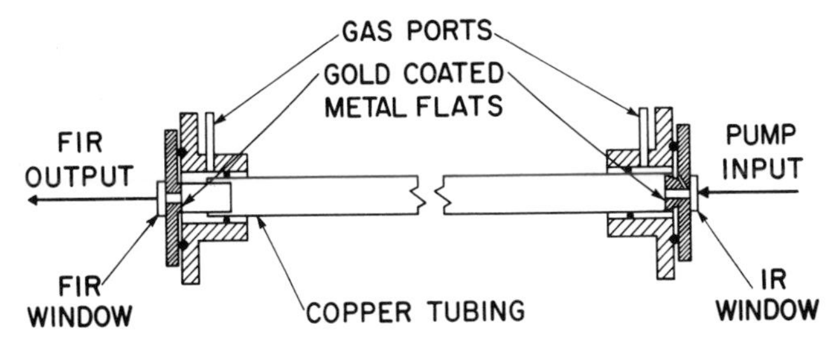

Fig. 6.23. Illustrative design of a waveguide FIR laser derived from the work of *Hodges* et al. [6.89, 90]

transfer to C_6H_{14} molecules which then transport the excess vibrational energy to the resonator wall. Other hydrocarbon molecules including C_2H_6, C_5H_{10}, C_7H_{16}, and C_8H_{18} all showed similar but somewhat smaller effect. Simple molecules He and H_2 are somewhat effective in increasing the operating pressure but do not significantly improve the output power capability of the laser, while nitrogen shows no beneficial effect at all. In general, a good buffer gas should have little or no absorption at both the pump and the FIR wavelengths, and should be a good absorber and carrier of vibrational energy. Both conventional and waveguide type FIR lasers could benefit from the addition of a good buffer gas.

For the detection of FIR signals, the following liquid-helium-cooled detectors with submicrosecond response are often used: Ge:Ga photoconductor for $\lambda < 170\,\mu m$, InSb hot-electron bolometer for $\lambda > 150\,\mu m$, and GaAs photoconductor for $80\,\mu m < \lambda < 500\,\mu m$. For measurements of cw or average power, it may be more convenient to use one of the slower room-temperature thermal detectors such as a Golay cell, vacuum thermocouple, flake thermistor, or pyroelectric detector. These room-temperature detectors are less sensitive but still have NEPs lower than $10^{-9}\,W/Hz^{1/2}$. For detection of high-power ns FIR pulses, absolute frequency measurements, or GHz-bandwidth heterodyne detection, point-contact diodes [6.47, 61, 75, 77, 93], Schottky barrier diodes [6.94, 95], and Josephson junction diodes [6.96] can be used. By comparing with the harmonics of an accurate microwave frequency, an accuracy of one part in 10^6 or better can be achieved in absolute frequency measurements. This has been done for 19 FIR laser lines in CH_3F and CH_3OH [6.47, 93, 94].

For routine measurements of FIR wavelengths, the use of an FIR Fabry-Perot interferometer equipped with metal-mesh mirrors can provide an accuracy on the order of 0.01% [6.48]. For the routine identification of laser lines, the laser resonator itself can be used as a scanning interferometer with an accuracy of $\approx 1\%$. For the determination of the polarization of FIR output, FIR grid polarizers are available.

The preferred mode of operation for the CO_2 pump laser is pulse-modulated operation [6.48]. In this mode of operation, current pulses of 0.1

to 0.5 ms duration and several hundred mA amplitude are superimposed on a keep-alive discharge current of a few mA. Peak output powers one to two orders of magnitude higher than the cw level can be obtained. By adjusting the gas mixture and the pulse repetition rate, an average power approaching the cw output level can also be obtained. In comparison with true cw operation at about the same average power, the higher peak power available in the pulse-modulated mode makes it easier to exceed the threshold condition for weak laser lines. Also, by using a fast detector in conjunction with electronic gating, a higher signal-to-noise ratio can be obtained in signal detection. In comparison with Q-switched operation, pulse-modulated operation delivers much more pump energy per pulse and, therefore, produces a much higher inversion density in the gas. The output frequency is also much more stable in pulse-modulated operation than in Q-switched operation.

Experimentally, it is found that in order to obtain a stable FIR output the pump frequency must be stable within a few MHz. This means the CO_2 laser should operate in a single transverse mode and should be made tunable over as much of the gain linewidth as possible. The latter requirement can be met if the CO_2 laser resonator is kept shorter than 2 m. A different approach to the pump laser design is to use a long (>5 m) resonator and operate the laser at low (<5 Torr) gas pressure such that several longitudinal modes will oscillate simultaneously within the gain line. Fine tuning of the laser is much less important in this case. Furthermore, if the spectral coincidence with the absorption line happens to be almost exact, then the multimode pump source will be able to saturate the entire Doppler-broadened absorption line (by burning several holes in the inhomogeneous absorption line) and thus achieve a higher conversion efficiency [6.97]. Recently, *Fetterman* et al. [6.161] reported that the saturated output power from a cw FIR NH_3 laser could be increased by as much as ten-fold by sweeping the pump frequency across the inhomogeneous absorption line by means of a combined DC and AC (~ 100 kHz) Stark field. The improvement arised from the pumping of molecules in all velocity classes as was discussed above.

6.4.5 Theoretical Analyses of Laser Performance

The basic expressions for FIR laser gain have already been given in Subsection 6.4.1. However, the dependence of inversion density and output power on pump power, gas pressure, various relaxation rates, and resonator parameters remains to be shown. Since the problem involves a large number of energy levels, we shall develop our theory somewhat heuristically by considering the detailed balance of energy and population flow in steady state at each stage of the entire excitation-relaxation process without writing down a system of rate equations. In order that we can be constantly in touch with real situations, the theory will be developed specifically for the 496 μm CH_3F laser. However, the formalism is completely general, and the resulting expressions are applicable to many systems.

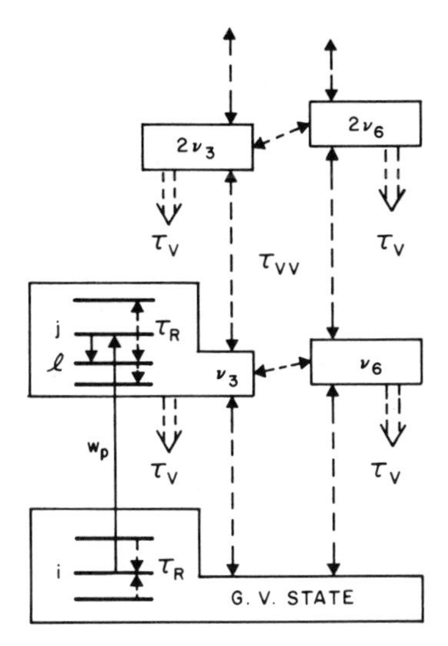

Fig. 6.24. Relaxation model used in the analysis of a CH_3F laser

The model we adopt for the relaxation process is illustrated in Fig. 6.24. The upper laser level j is optically pumped at a rate of w_p. The resulting population is rotationally thermalized in the v_3 state at a rate τ_R^{-1}. For simplicity we shall assume that the same relaxation rate τ_R^{-1} applies to both $\Delta K=0$ processes and (normally slower) $\Delta K \neq 0$ processes [6.98]. At the next stage of the relaxation process the excited vibrational energy is rapidly thermalized among the nv_3 levels and is also transferred to the nearly resonant v_6 manifold all at the $V-V$ transfer rate, τ_{VV}^{-1}. Resonant $V-V$ transfer results only in a redistribution of population and not in any net removal of vibrational energy from the molecular system. Consequently, the vibrational temperature of the v_3 and v_6 manifolds rises to T_3 which is significantly higher than the ambient temperature T_0. For simplicity, we assume that all other vibrational manifolds are unaffected and remain at T_0. The excess vibrational energy accumulated in the molecular system relaxes slowly either by collisions with the resonator walls or by $V-T/R$ transfer at a rate τ_V^{-1}. At higher gas pressures and in large-bore resonators, the $V-T/R$ process dominates and $\tau_V p \approx 1.7 \times 10^{-3}$ s·Torr [6.99]. At lower gas pressures and in small-bore resonators, wall collisions dominate and τ_V is determined by the diffusion time. For this case $\tau_V/pr \approx 5.6 \times 10^{-3}$ s/Torr·cm, where r is the resonator tube radius [6.99]. For the resonant $V-V$ processes, $\tau_{VV}p \approx 2$ to 9×10^{-6} s·Torr [6.100]. For rotational relaxation, $\tau_R p \approx 10^{-8}$ s·Torr [6.101, 102].

On account of its analytical simplicity, we shall first consider the high-pressure case for which both the pump and the laser transitions are homogeneously broadened. The results will be generalized to medium- and low-pressure cases by simple reinterpretation of some of the variables appearing in the equations.

The pump beam, after entering the resonator through a small coupling hole, is assumed to be reflected between the end-mirrors for a large number of round trips. The equivalent resonator loss coefficient will be denoted by a_p and should have a value between 10^{-4} and $10^{-3}\,cm^{-1}$ in a properly designed resonator. The effective pumping rate per unit volume is given by

$$w_p = \frac{P_p}{V_p} \cdot \frac{\alpha}{a_p + \alpha} \tag{6.12}$$

where P_p is the input pump power, V_p is the effective volume pumped by the infrared radiation and α is the absorption coefficient of the gas given by

$$\begin{aligned} \alpha &= F_{ij}\left(n_i^* - \frac{g_I}{g_3 g_J} n_j^*\right) S_h(\nu_p, \nu_{ij}) \\ &= 3.979 \times 10^{-10}\, \lambda_p^2 A_{ij}\left(n_i^* - \frac{g_i}{g_j} n_j^*\right) S_h(\nu_p, \nu_{ij}) \end{aligned} \tag{6.13}$$

where the pump wavelength λ_p is in μm, the Einstein coefficient A_{ij} is in s^{-1}, n in cm^{-3}, the homogeneous line-shape function S_h in Hz^{-1}, g_I, g_J are rotational degeneracy factors, and g_3 is the vibrational degeneracy factor. The asterisks are reserved for the discussion of medium- and low-pressure cases. They have no significance at the moment and can be ignored. Equation (6.12) indicates that when $\alpha > a_p$, substantially all the pump power is absorbed by the gas. If, on the other hand, $\alpha \ll a_p$, then the pumping rate is directly proportional to α. In the limit of low pump power, $n_j \approx 0$, and we have

$$\alpha_0 = F_{ij} n_{i0}^* S_h(\nu_p, \nu_{ij}) \,. \tag{6.14}$$

The saturation behavior of α will be discussed after we evaluate the steady-state population distribution under cw pumping.

Under steady-state pumping at a rate of w_p an excess vibrational energy density of $w_p \tau_V$ will accumulate in the molecular gas. This energy will be thermalized in the ν_3 and ν_6 vibrational manifolds. However, due to the finite $V-V$ transfer rate, the actual population of the ground vibrational state will be deficient by an amount of $w_p \tau_{VV}/\varepsilon_p$ (where $\varepsilon_p = h\nu_p$ is the energy of the pump photon) while the actual population of the ν_3 vibrational state will have an excess amount of $w_p \tau_{VV}/\varepsilon_p$ over the thermal value. The total density of molecules is given by [6.100]

$$\begin{aligned} n^* = (n_g^* + w_p \tau_{VV}/\varepsilon_p)\,[1 - \exp(-\varepsilon_1/kT_1)]^{-g_1} \\ \cdot [1 - \exp(-\varepsilon_2/kT_2)]^{-g_2} \ldots \,, \end{aligned} \tag{6.15}$$

where n_g^* is the population density of the ground vibrational state, $\varepsilon_m = h\nu_m$ is the vibrational energy of the m-th vibrational mode, g_m is its degeneracy number, and T_m is the vibrational temperature of the corresponding vibrational manifold. Equation (6.15) can be inverted to give

$$n_g^* = n^*[1-\exp(-\varepsilon_1/kT_1)]^{g_1}[1-\exp(-\varepsilon_2/kT_2)]^{g_2}\ldots - w_p\tau_{VV}/\varepsilon_p. \tag{6.16}$$

The population density in the v_3 state is given by

$$n_3^* = g_3(n_g^* + w_p\tau_{VV}/\varepsilon_p)\exp(-\varepsilon_3/kT_3) + w_p\tau_{VV}/\varepsilon_p. \tag{6.17}$$

The vibrational heat content under optical pumping is given by [6.100]

$$H_p^* = H_0^* + w_p\tau_V = n^* \sum_m \frac{g_m\varepsilon_m \exp(-\varepsilon_m/kT_m)}{1-\exp(-\varepsilon_m/kT_m)} + w_p\tau_{VV}, \tag{6.18}$$

where H_0 is the vibrational heat content of the gas in the absence of optical pumping and is given by

$$H_0^* = n^* \sum_m \frac{g_m\varepsilon_m \exp(-\varepsilon_m/kT_0)}{1-\exp(-\varepsilon_m/kT_0)}. \tag{6.19}$$

Under steady-state pumping $T_6 = T_3 \neq T_0$, and $T_m = T_0$ for all other vibrational manifolds. Equations (6.18) and (6.19) then yield

$$w_p(\tau_V - \tau_{VV}) + n^*(g_3+g_6)\,\varepsilon_p[\exp(\varepsilon_p/kT_0)-1]^{-1} = n^*(g_3+g_6)\,\varepsilon_p[\exp(\varepsilon_p/kT_3)-1]^{-1}, \tag{6.20}$$

where we have used the approximation $\varepsilon_6 \approx \varepsilon_3 \approx \varepsilon_p$ to simplify the expression. From (6.20) we obtain

$$\exp(\varepsilon_p/kT_3) = 1 + n^*(g_3+g_6)\,\varepsilon_p[w_p(\tau_V-\tau_{VV}) + n^*(g_3+g_6)\,\varepsilon_p \exp(-\varepsilon_p/kT_0)]^{-1}, \tag{6.21}$$

where we have made use of the fact that $\exp(\varepsilon_p/kT_0) \gg 1$.

From (6.16), (6.17), and (6.21) we obtain

$$n_g^* = n_{g0}^*[1-\exp(-\varepsilon_p/kT_0)]^{-(g_3+g_6)} \cdot [1 + w_p(\tau_V-\tau_{VV})/(g_3+g_6)\,\varepsilon_p n^* + \exp(-\varepsilon_p/kT_0)]^{-(g_3+g_6)} - w_p\tau_{VV}/\varepsilon_p, \tag{6.22}$$

and

$$\begin{aligned} n_3^* &= g_3 n_{g0}^* [1 - \exp(-\varepsilon_p/kT_0)]^{-(g_3+g_6)} \\ &\quad \cdot [w_p(\tau_V - \tau_{VV})/(g_3+g_6)\,\varepsilon_p n^* + \exp(-\varepsilon_p/kT_0)] \\ &\quad \cdot [1 + w_p(\tau_V - \tau_{VV})/(g_3+g_6)\,\varepsilon_p n^* + \exp(-\varepsilon_p/kT_0)]^{-(g_3+g_6+1)} \\ &\quad + w_p \tau_{VV}/\varepsilon_p\,, \end{aligned} \tag{6.23}$$

where n_{g0}^* is the population of the ground vibrational state in the absence of optical pumping.

The population densities on i and j rotational levels are related to n_g^* and n_3^* by

$$n_i^* = (n_g^* + w_p\tau_R/\varepsilon_p)\, Q_R^{-1} g_I \exp(-\varepsilon_{Ri}/kT_R) - w_p\tau_R/\varepsilon_p \tag{6.24}$$

and

$$n_j^* = (n_3^* - w_p\tau_R/\varepsilon_p)\, Q_R^{-1} g_J \exp(-\varepsilon_{Rj}/kT_R) + w_p\tau_R/\varepsilon_p \tag{6.25}$$

where Q_R is the rotational partition function [6.99], ε_R is the rotational energy (the term value), and $w_p\tau_R/\varepsilon_p$ represents an unthermalized hump (or dip) in population distribution among the rotational levels due to optical pumping. In the absence of optical pumping we have

$$\begin{aligned} n_{i0}^* &= n_{g0}^* Q_R^{-1} g_I \exp(-\varepsilon_{Ri}/kT_R) \\ &= n_{g0}^* f_i\,, \end{aligned} \tag{6.26}$$

where f_i is the fraction of molecules in the rotational level i. Now by using (6.22–26) in (6.13, 14) we obtain

$$\alpha = \frac{\alpha_0}{1 + (P_p/P_{ps})\,[(a_p+\alpha_0)/(a_p+\alpha)]}\,, \tag{6.27}$$

where the saturation pump power is given by

$$P_{ps} = \frac{(a_p+\alpha_0)\, h\nu_p V / F_{ij} S_h(\nu_p, \nu_{ij})}{f_i(n_{g0}^*/n^*)\,[\tau_V + (\tau_V - \tau_{VV})\,(1 + \Delta\varepsilon_{Rij}/kT_R)/(g_3+g_6)] + \tau_R(1 + g_I/g_J g_3)} \tag{6.28}$$

where $\Delta\varepsilon_{Rij} = \varepsilon_{Ri} - \varepsilon_{Rj}$ is the difference of rotational energy between levels i and j. Equation (6.28) can also be written in the form

$$\begin{aligned} \frac{P_{ps}}{h\nu_p V}\,\frac{\alpha_0}{a_p+\alpha_0} &= \frac{w_{ps}}{\varepsilon_p} \approx \frac{n_{i0}^*}{f_i\tau_V(1+g_{eff}^{-1}) + 2\tau_R} \\ &= \frac{n_{g0}^*}{\tau_V(1+g_{eff}^{-1}) + 2\tau_R f_i^{-1}} \end{aligned} \tag{6.29}$$

where $g_{\rm eff}=g_3+g_6$ (in CH_3F $g_3=1$, $g_6=2$). The presence of the parameter $f_i(\approx 10^{-3})$ in the expression arises from the rapid rotational relaxation which makes it impossible to saturate a single rotational level without saturating the entire vibrational state. This feature is not obvious from rate equation analysis using a simpler energy level model [6.103]. The presence of $g_{\rm eff}$ in the expression arises from rapid $V-V$ transfer between ν_3 and ν_6 modes. In practice, all the vibrational manifolds are populated by $V-V$ transfers to some degree. Consequently, $g_{\rm eff}$ is greater than g_3+g_6 in a real system.

At typical operating pressures, the pump transition is mainly Doppler broadened. This means, single-frequency pump radiation would burn a hole into the Doppler profile. The width of this hole is given by [6.104]

$$\Delta\nu_{\rm h}=\Delta\nu_{\rm c}(1+P_{\rm p}/P_{\rm ps})^{1/2}\,, \tag{6.30}$$

where $\Delta\nu_{\rm c}$ is the collisional linewidth. The saturation power $P_{\rm ps}$ is still given by (6.28) except that $S_{\rm h}(\nu_{\rm p},\nu_{ij})$ is now replaced by $2/\zeta\pi\Delta\nu_{\rm h}$, where $\zeta=1$ if $\nu_{\rm p}=\nu_{ij}$, but $\zeta=2$ when $\nu_{\rm p}$ is sufficiently detuned from the center of the Doppler line. This is because off-line-center pumping leads to two holes symmetrically located about the center of the Doppler line with each hole "seeing" only one-half of the circulating pump power.

Normally, the characteristic time for velocity cross relaxation is very much longer than $\tau_{\rm R}$ [6.105, 106]. Consequently, only molecules belonging to a small velocity class (determined by the hole width) participate in the entire excitation-relaxation process. Since the hole is predominantly homogeneously broadened, (6.12–29) are still applicable except that we have to keep in mind that $S_{\rm h}=2/\zeta\pi\Delta\nu_{\rm h}$ and that all n_x^*'s imply population densities of a restricted velocity class. Explicitly, n^*, $n_{\rm g0}^*$, and n_{i0}^* are, respectively, equal to ξn, $\xi n_{\rm g0}$, and ξn_{i0}, where

$$\xi=S_{\rm D}(\nu_{\rm p},\nu_{ij})\cdot\zeta\pi\Delta\nu_{\rm h}/2\,, \tag{6.31}$$

and $S_{\rm D}$ is a Doppler lineshape function. All other n_x^*'s are implicit functions of ξ due to their dependence on n^*, $n_{\rm g0}^*$, or n_{i0}^*.

The saturation behavior of α for a Doppler-broadened absorption line is given by [6.104].

$$\alpha(\nu_{\rm p})=\alpha_0(\nu_{\rm p})\,(1+P_{\rm p}/P_{\rm ps})^{1/2}\,. \tag{6.32}$$

A more general result for mixed broadening with velocity cross relaxation (but without $V-V$ transfer) has been obtained by *Tucker* [6.107].

With off-line-center pumping, the FIR emission lineshape should exhibit two peaks corresponding to the two holes burnt in the pump transition. However, the Doppler effect will now be reduced by a factor of $\nu_{\rm p}/\nu_{\rm FIR}$. The FIR lineshape is therefore a superposition of two homogeneous lines. If the pump power is below saturation, then the two lines will collapse into one as the pump frequency is tuned to the center of the absorption line. If the pump power is sufficient to drive the system well into saturation, the FIR emission may again exhibit two peaks due to resonant-modulation splitting (cf. Sect. 6.1), and the FIR power output may show a dip (as ν_p is tuned through the absorption line center) corresponding to the Lamb dip of the pump transition.

The inversion density for the principal FIR laser transition in the limit of vanishing FIR intensity is given by

$$\begin{aligned}\Delta n_{jl}^* &= n_j^* - \frac{g_j}{g_l} n_l^* \\ &= w_p \tau_R/\varepsilon_p + (n_3^* - w_p \tau_R/\varepsilon_p) f_j [1 - \exp(h\nu_{jl}/kT_R)] \\ &\approx w_p \tau_R / h\nu_p - n_3^* f_j h\nu_{jl}/kT_R \end{aligned} \tag{6.33}$$

where n_3^* is given by (6.23). The last term in (6.33) represents the adverse (absorption-causing) effect of rotational thermalization in the ν_3 state. The above derivation assumes that the rotational energy is completely thermalized after one collision. This is not a completely realistic model since rotational relaxation of polar molecules tends to follow the $\Delta J = \pm 1$ selection rule [6.98]. A more accurate expression for the population distribution among the rotational levels can be obtained by solving a Fokker-Planck equation [6.108].

A complete expression for FIR small signal gain can now be obtained by substituting (6.23) and (6.33) in (6.8) or (6.9). For finite FIR intensity (I), the FIR laser gain is given by

$$\gamma(\nu_{FIR}) = F_{jl}[\Delta n_{jl}^* - (1 + g_j/g_l)(I\gamma\tau_R/h\nu_{FIR})] \cdot S_L(\nu_{FIR}, \nu_{jl}), \tag{6.34}$$

where Δn_{jl}^* is given by (6.33), and $S_L(\nu_{FIR}, \nu_{jl})$ is given by (6.7) with

$$\Delta\nu_L = \Delta\nu_c[1 + (P_p/P_{ps})(\nu_{FIR}/\nu_p)^2]^{1/2}. \tag{6.35}$$

From (6.34) we obtain

$$\gamma(\nu_{FIR}) = \gamma_0/(1 + I/I_s), \tag{6.36}$$

where

$$I_s = \frac{h\nu_{FIR}}{\tau_R(1 + g_j/g_l) F_{jl} S_L(\nu_{FIR}, \nu_{jl})} \tag{6.37}$$

is the FIR saturation intensity. The small signal gain γ_0 is affected by pump saturation through its dependence on w_p. In a more general analysis, *Tucker* [6.107] has concluded that the effect of FIR intensity on pump saturation is negligibly small.

The FIR output power is given by [6.109]

$$P_{\mathrm{FIR}}=A_{\mathrm{FIR}}I_{\mathrm{s}}T[(\gamma_0-a_{\mathrm{FIR}})L+\ln(1-T)^{1/2}][1+(1-T)^{1/2}]^{-1}\cdot[1-(1-T)^{1/2}\exp(-a_{\mathrm{FIR}}L)]^{-1} \quad (6.38)$$

where A_{FIR} is the effective cross-sectional area of the FIR mode, T is the transmittance of the output mirror, a_{FIR} is the effective FIR loss coefficient (including mirror absorption but not T), and L is the resonator length.

A semiquantitative study of the 570.5 µm CH_3OH laser has been carried out by *Henningsen* and *Jensen* [6.103] both theoretically and experimentally. The theoretical treatment was based on a four-level model with no allowance made for inhomogeneous broadening of the pump transition at lower pressures. The resulting expressions were made to fit the experimental data with the help of several adjustable parameters. The study provided some useful illustrations of the trends in the pressure dependences of threshold pump power and differential conversion efficiency and the pressure and pump power dependences of the FIR output. Their study of the FIR gain linewidth also clearly demonstrated the reduction of the FIR linewidth below the Doppler linewidth due to hole burning.

The gain and the saturation power of optically pumped FIR lasers can be improved by the addition of a suitable buffer gas. *Chang* and *Lin* [6.92] have studied the problem both experimentally and theoretically. The effect of a polyatomic buffer gas such as *n*-hexane on the laser gain is shown to be

$$\begin{aligned}\gamma/\gamma'=1&-[2+(n_3'\Delta f_{jl}/\Delta n_{jl}')](\beta_{\mathrm{b}}/\beta_{\mathrm{a}})(n_{\mathrm{b}}/n_{\mathrm{a}})\\&+(n_3'-w_{\mathrm{p}}\tau_{VV}'/h\nu_{\mathrm{p}})(\Delta f_{jl}/\Delta n_{jl}')(h\nu_{\mathrm{p}}-\bar{H}_{\mathrm{a}}')\\&\cdot[\bar{H}_{\mathrm{b}}'(\alpha_{\mathrm{aa}}/\alpha_{\mathrm{ba}})-\bar{H}_{\mathrm{a}}'(\alpha_{\mathrm{ab}}/\alpha_{\mathrm{aa}})](kT_V'^2\bar{C}_{\mathrm{a}}')^{-1}(n_{\mathrm{b}}/n_{\mathrm{a}})\\&+(w_{\mathrm{p}}\tau_{VV}'/h\nu_{\mathrm{p}})(\Delta f_{jl}/\Delta n_{jl}')(\psi_{\mathrm{b}}/\psi_{\mathrm{a}})(n_{\mathrm{b}}/n_{\mathrm{a}})\\&+\text{higher order terms in }(n_{\mathrm{b}}/n_{\mathrm{a}})^n,\end{aligned} \quad (6.39)$$

where a "primed" quantity is its value in the absence of the buffer gas (but in the presence of optical pumping), subscripts a and b refer to active gas and buffer gas, respectively, and $\Delta f_{jl}=f_j-f_l$. The β_x's are broadening coefficients defined by

$$\Delta\nu_{\mathrm{c}}=\beta_{\mathrm{a}}n_{\mathrm{a}}+\beta_{\mathrm{b}}n_{\mathrm{b}}.$$

The $\bar{H}_x$'s are vibrational heat content per molecule, and $\bar{C}_{\mathrm{a}}=d\bar{H}_{\mathrm{a}}/dT_V$ is the vibrational heat capacity per molecule. The α's are related to τ_V's (assumed to be diffusion dominated) by

$$\tau_{Vx}\approx\tau_{Dx}=\alpha_{xx}n_x+\alpha_{xy}n_y.$$

It can be shown from the kinetic theory of gases that

$$\frac{\alpha_{xx}}{\alpha_{yx}} = \frac{\alpha_{xx}}{\alpha_{xy}} = \frac{D_{xy}}{D_{xx}} = \frac{\sqrt{2}(1+m_x/m_y)^{1/2}}{(1+n_y/n_x)}\left(\frac{\sigma_{xx}}{\sigma_{xy}}\right)^2$$

where D's are binary diffusion coefficients, m's are molecular weights, and σ's are gas kinetic cross sections. The ψ_x's are defined through the relation

$$\tau_{VV}^{-1} = \psi_a n_a + \psi_b n_b = \tau_{VV}'^{-1} + \psi_v n_b .$$

The second term on the right-hand side of (6.39) represents the adverse effect of buffer gas addition on the laser gain due to increased linewidth and rotational relaxation rate. The third term represents the beneficial effect of buffer gas addition on the laser gain due to the absorption and transportation of vibrational energy by buffer gas molecules. The fourth term represents another beneficial effect of buffer gas caused by the enhancement of $V-V$ transfer rate. The third term indicates that a good buffer gas should have a large $\bar{H}_b/\alpha_{ba}$ ratio, which translates roughly into a large number of vibrational modes relative to the molecular weight. This requirement is well satisfied by the hydrocarbons. The same term also suggests that buffer gases are particularly useful for lasing gases with small vibrational heat content $\bar{H}_a$ and small vibrational heat capacity $\bar{C}_a$. The fourth term suggests the advantage of having vibrational states in the buffer molecule which are in near resonance with the pumped vibrational state of the active molecule. Buffer gases also increase the saturation pump power P_{pa} not only through the reduction of τ_V and τ_{VV} but also by the enhancement of g_{eff} in (6.28) and (6.29).

In their original work on the CH_3F laser, *Chang* and *Bridges* [6.47] observed microsecond delays of the 496 μm pulse from the 0.3 μs pump pulse. The delay was found to decrease with increasing CH_3F pressure. Further experimental and theoretical studies of this phenomenon were made by *Bluyssen* et al. [6.110]. From numerical calculations based on a set of rate equations for a three-level system, they concluded that the delayed pulsing and its dependence on the pressure is entirely consistent with the rate equation description of oscillation build-up in a resonator provided that hole burning in the pump transition is properly taken into account.

DeTemple and *Danielewicz* [6.162] have presented a rate equation analysis of the performance of a cw waveguide laser. Theoretical predictions on the dependences of the FIR output on the pump power, the gas pressure, and the mirror reflectivity were found to be in good agreement with experimental observations. They concluded that a power conversion efficiency of 0.2–0.6% could be realistically achieved for the 496 μm CH_3F laser. Some relevant parameters for the pump transition of this laser have been determined experimentally by *Hodges* and *Tucker* [6.163].

The performance of a pulsed FIR laser with strong pumping has been considered theoretically by *Tucker* [6.163] and *Temkin* and *Cohn* [6.165], using rate equation models. The essential conclusion of both papers is that while the FIR gain may be very high at the beginning of the pump pulse, it diminishes rapidly with time due to the filling up of rotational levels in the excited vibrational state. The time constant for this to happen is given by $\tau_R f_j^{-1}$, where τ_R is the time constant for rotational relaxation and f_j is the fraction of the population of the excited vibrational state that would be in the level j in rotational thermal equilibrium (cf. [6.26]). For a given pump pulse length, the maximum FIR output would be obtained near the gas pressure at which $\tau_R f_j^{-1}$ is equal to the pulse length. If a 100 ns pulse is used to pump a CH_3F laser, this pressure would occur at 2.5 Torr if only one K component is pumped, and at 1.8 Torr if $K=1$ to $K=6$ components are all excited to saturation.

6.4.6 Applications

The fact that optically pumped FIR lasers can provide mW level cw output at hundreds of wavelengths throughout the FIR region, and kW to MW level pulsed output at some FIR wavelengths provides excellent opportunities both to improve and extend existing applications that have been depending on other types of FIR sources and to develop entirely new applications in the FIR region made possible by these new sources. Some scientific applications have already produced valuable results and others are being actively developed. Industrial and military applications are expected to emerge as soon as the devices become more highly developed.

Optically pumped FIR lasers are by themselves a very useful spectroscopic tool for studying excited vibrational states of polar molecules. The multiple passage of the pump beam and the conversion of high-infrared power into a highly detectable FIR signal provide a method of detecting coupled vibrational-rotational and pure-rotational transitions that is much more sensitive than other conventional spectroscopic means. By careful measurement of the emission frequency and output polarization, and from the accurately known pump frequency, the possible values of J and $|\Delta J|$ involved in the coupled transitions can be determined within very narrow bounds. Correlations among all the emission frequencies observed in a given molecule and their corresponding pump frequencies yield further information that is particularly powerful in revealing the presence of overlapping hot bands which are often too weak to be seen by conventional methods [6.49]. With the help of other data available from conventional infrared and microwave spectroscopy, the above-described information often leads to a complete identification of the observed transitions [6.47, 57, 62, 111–113]. Spectroscopic constants of excited vibrational states can then be deduced with much higher accuracy than is now possible with conventional methods [6.49].

The value of FIR sources in magnetospectroscopy has long been recognized [6.114, 115]. Now, many previously inaccessible parts of the FIR spectrum have been opened up to magnetospectroscopy by the advent of optically pumped FIR lasers. As a result, a sensitive new method of determining the identity and the concentration of impurities in high-purity semiconductors has become available [6.116]. New laser wavelengths have also contributed to an extensive study of the field-dependent cyclotron-resonance linewidth in InSb [6.117], as well as to a detailed investigation of low-lying crystal-field states of Tb^{3+} in $TbPO_4$ [6.118]. The high stability of optically pumped FIR lasers, arising from the availability of ultrastable CO_2 pump lasers and the absence of an electrical discharge in the FIR resonator, also makes possible very high resolution magnetospectroscopy [6.119, 120]. The ability of optically pumped FIR lasers to deliver multikilowatt output power, on the other hand, has made possible the observation of saturation and power broadening in magnetospectroscopy [6.120].

Megawatt-level narrow-linewidth CH_3F lasers are being actively developed [6.77, 121] for a number of important diagnostic experiments in Tokamak plasmas. First of all, the spatial distribution and the effective charge of impurity ions in the plasma can be measured by the scattering of such a high-power FIR laser beam [6.122]. Secondly, when the deuterium-tritium plasma reaches the state of high purity, Thomson scattering of a high-power FIR laser beam will become the only remaining method of determining the important ion temperature in an experimental controlled-fusion device [6.123]. Such a high-power source is also suitable for transient and localized heating of a Tokamak plasma at the harmonics of the electron cyclotron frequency. The transverse thermal conductivity of the plasma can then be determined by monitoring the resulting temporal and spatial variations of the electron temperature with a ruby laser [6.123]. It may also be possible to determine the lifetime of trapped electrons in a Tokamak device by the temporal-echo technique using two 496 μm pulses [6.123]. High-power FIR sources may also be used to excite parametric decay instabilities in plasmas for experimental studies [6.124, 125]. A CH_3F laser with a focused intensity of 40 kW/cm^2 has been used to study the strong enhancement of avalanche ionization and plasma heating due to cyclotron resonance in gases [6.125, 126, 166]. Finally, the usefulness of stable sources at many FIR wavelengths for interferometric measurement of plasma densities seems quite apparent.

High-power FIR lasers may be useful as the pump source for millimeter- or submillimeter-wave parametric oscillators [6.127]. FIR lasers may also serve as local oscillators for FIR radiometers for use in astronomy.

Recently, an optically pumped CH_3OH laser has been used to study higher-order mixing processes in a Josephson junction [6.167].

The high stability, the very narrow linewidth [6.103], and the easy accessibility of their frequencies to the harmonics of microwave frequencies, make FIR lasers good secondary frequency standards in the FIR region.

Optically pumped FIR lasers are eminently useful for any industrial application requiring an FIR source, such as medium-resolution length measurement in machine shops [6.128], ellipsometry for thickness measurement of semiconductor epitaxial layers [6.129], imaging through dielectric materials [6.169], etc. With further developments in detectors and modulators, they may also become viable for space communication [6.130] and waveguide transmission [6.131].

In the military domain, FIR lasers are of interest in imaging and surveillance through cloud, dense fog, and camouflage [6.132]. The narrow beam angle and some atmospheric attenuation can be used to advantage for secure communication and high-resolution, all-weather aircraft landing systems [6.133]. The short wavelength is convenient for scaled-down measurements on radar targets [6.133].

6.5 Conclusions

To date, optical pumping has been most successful in generating new laser lines in the FIR. It is replacing electrical discharge as the primary method of laser excitation in that spectral region. Some of the advantages of optical pumping over electrical discharge are as follows:

1) A multitude of stable discrete wavelengths can be generated over the entire FIR region, giving mW level cw output or multi-kW level pulsed output.

2) Discharge-associated fluctuations, thermal drift, and molecular dissociation are almost completely absent in the FIR resonator.

3) Pump sources, such as the CO_2 laser, are well developed and highly stable.

4) The output is linearly polarized without the need for any FIR polarizing element.

5) A reasonably good overall efficiency can be achieved because the pump source can be highly efficient and the pump energy is deposited only in the intended energy level.

6) A very long resonator is not required because the gain is reasonably high. In some cases, the gain is sufficiently high to permit generation of high-power superradiant pulses in a traveling-wave configuration.

7) It is feasible to construct a compact and perhaps permanently sealed laser system.

With the rapid development of optically pumped lasers in the FIR region, the frontier for the search of new laser wavelengths has now shifted back to the middle-infrared region. Work in this region is only beginning, but some encouraging progress has already been made. The combination of optical pumping and $V-V$ transfer has proven to be a versatile and powerful excitation technique. This technique is the basis of a continuously tunable high-pressure N_2O:CO_2 laser. At lower pressures, oscillation in know laser bands of CO_2, N_2O, OCS, CS_2, C_2H_2 as well as new laser bands in HCN has been demonstrated by this technique. New laser lines in OCS and SF_6 have also been obtained by direct optical pumping. In the middle-infrared region, optical pumping is somewhat handicapped by the lack of a pump laser with a wavelength shorter than 5 μm that is comparable to the CO_2 laser in terms of power, efficiency and convenience. In principle, however, there is no reason why optical pumping cannot provide a large number of laser lines in the middle-infrared region.

In the near-infrared region, laser action by optical pumping has been demonstrated in I_2 and Na_2. Eventually, they could provide millions of discrete laser lines throughout the visible and near-IR regions. Although part of this region is also covered by liquid-phase dye lasers, the better beam quality and more stable single-frequency operation possible in a gas-phase system may be of advantage in certain applications.

After a decade of nearly dormant existence, optical pumping, the oldest excitation technique for gas lasers, has shown a remarkable resurgence. Already, it has gained some practical importance, and there are ample reasons to believe that a much greater potential is yet to be tapped. Continued vigorous research in this field should bring forth rapid advances in the coming years.

Acknowledgment. The author thanks *O. R. Wood, II*, for a critical reading of the manuscript.

References

6.1 T.Y.Chang: IEEE Trans. MTT-**22**, 983 (1974)
6.2 T.Y.Chang, O.R.Wood: Appl. Phys. Lett. **23**, 370 (1973)
6.3 T.Y.Chang, O.R.Wood: Appl. Phys. Lett. **24**, 182 (1974)
6.4 W.Happer: Rev. Mod. Phys. **44**, 169 (1972)
6.5 A.L.Schawlow, C.H.Townes: Phys. Rev. **112**, 1940 (1958)
6.6 N.Bloembergen: Phys. Rev. **104**, 324 (1956)
6.7 A.E.Siegman: *Microwave Solid-State Masers* (McGraw-Hill 1964) pp. 292—294
6.8 A.Javan: Phys. Rev. **107**, 1579 (1957)
6.9 P.W.Anderson: J. Appl. Phys. **28**, 1049 (1957)
6.10 K.Shimoda, T.Shimizu: *Progress in Quantum Electronics*, Vol. 2, pt. 2 (Pergamon Press, 1972) pp. 61—76
6.11 A.M.Clogston: J. Phys. Chem. Solids **4**, 271 (1958)
6.12 N.Bloembergen, Y.R.Shen: Phys. Rev. **133**, A37 (1964)
6.13 J.P.Gordon: In *Laser Technology and Applications*, ed. by *S.L.Marshall* (McGraw-Hill 1968) pp. 31—34
6.14 S.Jacobs, G.Gould, P.Rabinowitz: Phys. Rev. Lett. **7**, 415 (1961)
6.15 P.Rabinowitz, S.Jacobs, G.Gould: Appl. Opt. **1**, 513 (1962)
6.16 P.Rabinowitz, S.Jacobs: In *Quantum Electronics III*, ed. by *Grivet* and *Bloembergen* (Columbia University Press, 1964) pp. 489—498
6.17 P.P.Sorokin, J.R. Lankard: Phys. Rev. **186**, 342 (1969)
6.18 P.P.Sorokin, J.R.Lankard: J. Chem. Phys. **51**, 2929 (1969)
6.19 P.P.Sorokin, J.R.Lankard: J. Chem. Phys. **54**, 2184 (1971)
6.20 J.J.Wynne, P.P.Sorokin: J. Phys. B. Atom. Mol. Phys. **8**, L37 (1975)
6.21 R.L.Byer, R.L.Herbst, H.Kildal, M.D.Levenson: Appl. Phys. Lett. **20**, 463 (1972)
6.22 G.Herzberg: *Spectra of Diatomic Molecules*, 2nd ed. (Van Nostrand-Reinhold, 1950) pp. 240—280
6.23 M.Henesian, R.L.Herbst, R.L.Byer: IEEE J. QE-**11**, 26D (1975)
6.24 M.C.Lin: IEEE J. QE-**10**, 516 (1974)
6.25 C.K.N.Patel: Phys. Rev. Lett. **12**, 588 (1964)
6.26 A.L.Golger, V.S.Letokhov: Kvantovaya Elektron. **13**, 30 (1973); Soviet J. Quant. Electron. **3**, 15 (1973)

6.27 C.K.N.Patel: Phys. Rev. **136 A**, 1187 (1964)
6.28 C.K.N.Patel: Appl. Phys. Lett. **6**, 12 (1965)
6.29 I.Wieder: Phys. Lett. **24 A**, 759 (1967)
6.30 P.A.Bokhan: Opt. i Spektroskopiya **32**, 826 (1972); Opt. Spectr. **32**, 435 (1972)
6.31 T.Y.Chang, O.R.Wood: Appl. Phys. Lett. **21**, 19 (1972)
6.32 V.I.Balykin, A.L.Golger, Yu.R.Kolomiiskii, V.S.Letokhov, O.A.Tumanov: Zh. Eksperim. i Teor. Fiz. Pis. Red. **19**, 482 (1974); JETP Lett. **19**, 256 (1974)
6.33 A.L.Golger, V.S.Letokhov: Kvantovaya Elektron. (Moscow) **17**, 106 (1973); Soviet J. Quant. Electron. **3**, 428 (1974)
6.34 V.I.Balykin, A.L.Golger, Yu.R.Kolomiiskii, V.S.Letokhov, O.A.Tumanov: Kvantovaya Elektron. (Moscow) **1**, 2386 (1974); Soviet J. Quant. Electron. **4**, 1325 (1975)
6.35 T.Y.Chang, O.R.Wood: Appl. Phys. Lett. **22**, 93 (1973)
6.36 T.Y.Chang, J.D.McGee, O.R.Wood: Opt. Commun. **18**, 57 (1976)
6.37 A.V.Eletskii, L.Ya.Efremenkova, B.M.Smirnov: Doklady Akademii Nauk USSR **194**, 298 (1970); Soviet Phys.-Doklady **15**, 843 (1971)
6.38 H.Kildal, T.F.Deutsch: Appl. Phys. Lett. **27**, 500 (1975)
6.39 T.F.Deutsch: Appl. Phys. Lett. **8**, 334 (1966)
6.40 C.F.Shelton, F.T.Byrne: Appl. Phys. Lett. **17**, 436 (1970)
6.41 A.B.Petersen, C.Wittig, S.R.Leone: Appl. Phys. Lett. **27**, 305 (1975)
6.42 N.V.Karlov, Yu.B.Konev, Yu.N.Petrov, A.M.Prokhorov, O.M.Stel'makh: Zh. Eksperim. i Teor. Fiz. Pis. Red. **8**, 22 (1968); JETP Lett. **8**, 12 (1968)
6.43 H.R.Schlossberg, H.R.Fetterman: Appl. Phys. Lett. **26**, 316 (1975)
6.44 W.E.Barch, H.R.Fetterman, H.R.Schlossberg: Opt. Commun. **15**, 358 (1975)
6.45 N.Skribanowitz, I.P.Herman, R.M.Osgood,Jr., M.S.Feld, A.Javan: Appl. Phys. Lett. **20**, 428 (1972)
6.46 I.P.Herman, J.C.MacGillivray, N.Skribanowitz, M.S.Feld: In *Laser Spectroscopy*, ed. by *R.G.Brewer, A.Mooradian* (Plenum Press, New York 1974) pp. 379—412
6.47 T.Y.Chang, T.J.Bridges: In *Proc. Symp. Submillimeter Waves*, ed. by *J.Fox* (Polytechnic Press, Brooklyn, New York 1970) pp. 93—98; Opt. Commun. **1**, 423 (1970)
6.48 T.Y.Chang, T.J.Bridges, E.G.Burkhardt: Appl. Phys. Lett. **17**, 249 (1970)
6.49 T.Y.Chang, J.D.McGee: Appl. Phys. Lett. **19**, 103 (1971)
6.50 H.R.Fetterman, H.R.Schlossberg, J.Waldman: Opt. Commun. **6**, 156 (1972)
6.51 D.A.Jennings, K.M.Evenson, J.J.Jimenez: IEEE J. QE-**11**, 637 (1975)
6.52 T.Y.Chang, J.D.McGee: IEEE J. QE-**12**, 62 (1976)
6.53 S.F.Dyubko, V.A.Svich, L.D.Fesenko: Opt. i Spektroskopiya **37**, 208 (1974); Opt. Spectr. **37**, 118 (1974)
6.54 H.E.Radford: IEEE J. QE-**11**, 213 (1975)
6.55 T.Y.Chang, T.J.Bridges, E.G.Burkhardt: Appl. Phys. Lett. **17**, 357 (1970)
6.56 R.J.Wagner, A.J.Zelano, L.H.Ngai: Opt. Commun. **8**, 46 (1973)
6.57 K.Gullberg, B.Hartmann, B.Kleman: Physica Scripta **18**, 177 (1973)
6.58 H.R.Fetterman, H.R.Schlossberg, C.D.Parker: Appl. Phys. Lett. **23**, 684 (1973)
6.59 S.F.Dyubko, V.A.Svich, L.D.Fesenko: Zh. Eksperim. i Teor. Fiz. Pis. Red. **16**, 592 (1972); JETP Lett. **16**, 418 (1972)
6.60 D.T.Hodges, R.D.Reel, D.H.Barker: IEEE J. QE-**9**, 1159 (1973)
6.61 T.K.Plant, L.A.Newman, E.J.Danielewicz, T.A.DeTemple, P.D.Coleman: IEEE Trans. MTT-**22**, 988 (1974)
6.62 F.Keilmann, R.L.Sheffield, J.R.R.Leite, M.S.Feld, A.Javan: Appl. Phys. Lett. **26**, 19 (1975)
6.63 S.F.Dyubko, V.A.Svich, L.D.Fesenko: Zh. Tekhn. Fiz. **43**, 1772 (1973); Soviet Phys.-Tech. Phys. **18**, 1121 (1974)
6.64 Yu.S.Domnin, V.M.Tatarenkov, P.S.Shumyatskii: Kvantovaya Elektron. (Moscow) **1**, 603 (1974); Soviet J. Quant. Electron. **4**, 401 (1974)
6.65 A.Tanaka, A.Tanimoto, N.Murata, M.Yamanaka, H.Yoshinaga: Japan. J. Appl. Phys. **13**, 1491 (1974)
6.66 J.R.Izatt, B.L.Bean, G.F.Caudle: Opt. Commun. **14**, 385 (1975)
6.67 A.Tanaka, M.Yamanaka, H.Yoshinaga: IEEE J. QE-**11**, 853 (1975)

6.68 S.Kon, E.Hagiwara, T.Yano, H.Hirose: Japan. J. Appl. Phys. **14**, 731 (1975)
6.69 T.K.Plant, P.D.Coleman, T.A.DeTemple: IEEE J. QE-**9**, 962 (1973)
6.70 S.F.Dyubko, V.A.Svich, L.D.Fesenko: Kvantovaya Elektron. (Moscow) **17**, 128 (1973); Soviet J. Quant. Electron. **3**, 446 (1974)
6.71 S.F.Dyubko, V.A.Svich, L.D.Fesenko: Zh. Prikl. Spektroskopiya **20**, 718 (1974); J. Appl. Spectr. **20**, 545 (1974)
6.72 T.Y.Chang, C.H.Wang, P.K.Cheo: Appl. Phys. Lett **15**, 157 (1969)
6.73 F.Brown, E.Silver, C.E.Chase, K.J.Button, B.Lax: IEEE J. QE-**8**, 499 (1972)
6.74 T.A.DeTemple, T.K.Plant, P.D.Coleman: Appl. Phys. Lett. **22**, 644 (1973)
6.75 F.Brown, S.R.Horman, A.Palevsky: Opt. Commun. **9**, 28 (1973)
6.76 F.Brown, S.Kronheim, E.Silver: Appl. Phys. Lett. **25**, 394 (1974)
6.77 D.E.Evans, L.E.Sharp, B.W.James, W.A.Peebles: Appl. Phys. Lett. **26**, 630 (1975)
6.78 L.E.Sharp, W.A.Peebles, B.W.James, D.E.Evans: Opt. Commun. **14**, 215 (1975)
6.79 R.J.Temkin, D.R.Cohn, Z.Drozdowica: Opt. Commun. **14**, 314 (1975)
6.80 D.R.Cohn, T.Fuse, K.J.Button, B.Lax, Z.Drozdowica: Appl. Phys. Lett. **27**, 280 (1975)
6.81 D.E.Evans, B.W.James, W.A.Peebles, L.E.Sharp: Intern. Conf. Infrared Physics, ETH Zurich, Switzerland (1975), Conference Digest, p. C224; Infrared Phys. **16**, 193 (1976)
6.82 F.Shimizu: J. Chem. Phys. **52**, 3572 (1970)
6.83 M.Inguscio, P.Minguzzi, F.Strumia: Intern. Conf. Infrared Physics, ETH Zurich, Switzerland (1975), Conference Digest, p. C229
6.84 N.Skribanowitz, M.S.Feld, R.E.Francke, M.J.Kelly, A.Javan: Appl. Phys. Lett. **19**, 161 (1971)
6.85 N.Skribanowitz, I.P.Herman, J.C.MacGillivray, M.S.Feld: Phys. Rev. Lett. **30**, 309 (1973)
6.86 L.Allen, J.H.Eberly: *Optical Resonance and Two-Level Atoms* (John Wiley and Sons, New York 1975) pp. 181—194
6.87 R.A.Wood, N.Brignall, C.R.Pidgeon, F.Al-Berkdar: Opt. Commun. **14**, 301 (1975)
6.88 E.J.Danielewicz, T.K.Plant, T.A.DeTemple: Opt. Commun. **13**, 366 (1975)
6.89 D.T.Hodges, T.S.Hartwick: Appl. Phys. Lett. **23**, 252 (1973)
6.90 D.T.Hodges, T.S.Hartwick: Intern. Conf. Submillimeter Waves and Their Applications, Georgia Institute of Technology, Atlanta, GA (1974), Conference Digest, p. 59
6.91 M.Yamanaka, H.Yoshinaga: Intern. Conf. Submillimeter Waves and Their Applications, Georgia Institute of Technology, Atlanta, GA (1974), Conference Digest, p. 26
6.92 T.Y.Chang, C.Lin: J. Opt. Soc. Am. **66**, 362 (1976)
6.93 F.R.Petersen, K.M.Evenson, D.A.Jennings, J.S.Wells, K.Goto, J.J.Jimenez: IEEE J. QE-**11**, 838 (1975)
6.94 H.R.Fetterman, B.J.Clifton, P.E.Tannenwald, C.D.Parker: Appl. Phys. Lett. **24**, 70 (1974)
6.95 H.R.Fetterman, B.J.Clifton, P.E.Tannenwald, C.D.Parker, H.Penfield: IEEE Trans. MTT-**22**, 1013 (1974)
6.96 Y.Taur, J.H.Classen, P.L.Richards: IEEE Trans. MTT-**22**, 1005 (1974)
6.97 M.Yamanaka, Y.Homma, A.Tanaka, M.Takada, A.Tanimoto, H.Yoshinaga: Japan. J. Appl. Phys. **13**, 843 (1974)
6.98 T.Oka: In *Advances in Atomic and Molecular Physics*, Vol. **9**, ed. by *D.R.Bates, I.Estermann* (Academic Press, New York, London 1973) pp. 124—206
6.99 E.Weitz, G.Flynn, A.M.Ronn: J. Chem. Phys. **56**, 6060 (1972)
6.100 G.Herzberg: *Infrared and Raman Spectra of Polyatomic Molecules* (Van Nostrand, 1945) pp. 501—530
6.101 G.Birnbaum: J. Chem. Phys. **46**, 2455 (1967)
6.102 J.Schmidt, P.R.Berman, R.G.Brewer: Phys. Rev. Lett. **31**, 1103 (1973)
6.103 J.O.Henningsen, H.G.Jensen: IEEE J. QE-**11**, 248 (1975)
6.104 A.Yariv: *Quantum Electronics* (John Wiley and Sons, New York 1967) pp. 263—268
6.105 R.L.Shoemaker, S.Stenholm, R.G.Brewer: Phys. Rev. A **10**, 2037 (1974)
6.106 J.W.C.Johns, A.R.W.McKellar, T.Oka, M.Romheld: J. Chem. Phys. **62**, 1488 (1975)
6.107 J.R.Tucker: Intern. Conf. Submillimeter Waves and Their Applications, Georgia Institute of Technology, Atlanta, GA (1974), Conference Digest, p. 17

6.108 B.F. Gordiets, S.A. Reshetnyak, L.A. Shelepin: Kvantovaya Elektron. (Moscow) **1**, 591 (1974); Soviet J. Quant. Electron. **4**, 329 (1974)
6.109 W.W. Rigrod: J. Appl. Phys. **36**, 2487 (1965)
6.110 H.J.A. Bluyssen, R.E. McIntosh, A.F. VanEtteger, P. Wyder: IEEE J. QE-**11**, 341 (1975)
6.111 L.O. Hocker: Opt. Commun. **10**, 157 (1974)
6.112 G. Duxbury, T.J. Gamble, H. Herman: IEEE Trans. MTT-**22**, 1108 (1974)
6.113 G. Graner: Opt. Commun. **14**, 67 (1975)
6.114 K.J. Button, B. Lax: In *Proc. Symp. Submillimeter Waves*, ed. by *J. Fox* (Polytechnic Press, Brooklyn, New York 1970) p. 401
6.115 K.J. Button: Laser Focus, p. 29 (Aug. 1971)
6.116 H.R. Fetterman, J. Waldman, C.M. Wolfe, G.E. Stillman, C.D. Parker: Appl. Phys. Lett. **21**, 434 (1972)
6.117 R. Kaplan, B.D. McCombe, R.J. Wagner: Solid State Commun. **12**, 967 (1973)
6.118 J.F.L. Lewis, G.A. Prinz: Phys. Rev. B **10**, 2892 (1974)
6.119 H.R. Fetterman, D.M. Larsen, G.E. Stillman, P.E. Tannenwald, J. Waldman: Phys. Rev. Lett. **26**, 975 (1971)
6.120 H.R. Fetterman, H.R. Schlossberg, J. Waldman: Laser Focus, p. 42 (Sept. 1972)
6.121 F. Brown, D.R. Cohn: IEEE Trans. MTT-**22**, 1112 (1974)
6.122 D.E. Evans, M.L. Yeoman: Phys. Rev. Lett. **33**, 76 (1974)y
6.123 D.L. Jassby, D.R. Cohn, B. Lax, W. Halverson: Nucl. Fusion **14**, 745 (1974)
6.124 D.L. Jassby: J. Appl. Phys. **44**, 919 (1973)
6.125 B. Lax, D.R. Cohn: IEEE Trans. MTT-**22**, 1049 (1974)
6.126 B. Lax, D.R. Cohn: Appl. Phys. Lett. **23**, 363 (1973)
6.127 F. Brown: Intern. Conf. Infrared Physics, ETH Zurich, Switzerland (1975), Conference Digest, p. I—50; Infrared Phys. **16**, 171 (1976)
6.128 H.A. Gebbie: In *Proc. Symp. Submillimeter Waves*, ed. by *J. Fox* (Polytechnic Press, Brooklyn, New York 1970) p. 1
6.129 R.O. DeNicola: In *Lasers in Industry*, ed. by *S.S. Charschan* (Van Nostrand-Reinhold, 1972) pp. 326—334
6.130 L.A. Hoffman, T.S. Hartwick, H.J. Wintroub: In *Proc. Symp. Submillimeter Waves*, ed. by *J. Fox* (Polytechnic Press, Brooklyn, New York 1970) p. 519
6.131 S.E. Miller: Bell System Tech. J. **33**, 1209 (1954)
6.132 R.C. Hofer, H. Jacobs, J. Schumacher: In *Proc. Symp. Submillimeter Waves*, ed. by *J. Fox* (Polytechnic Press, Brooklyn, New York 1970) p. 553
6.133 H.R. Fetterman, H.R. Schlossberg: Microwave J., p. 35 (Nov. 1974)
6.134 S. Kon, T. Yano, E. Hagiwara, H. Hirose: Japan. J. Appl. Phys. **14**, 1861 (1975)
6.135 M. Rokni, S. Yatsiv: Phys. Lett. **24 A**, 277 (1967)
6.136 D. Grischkowsky, P.P. Sorokin, J.R. Lankard: IXth International Conference on Quantum Electronics, Amsterdam (1976), postdeadline paper S. 4; Opt. Commun. **18**, 205 (1976)
6.137 P.P. Sorokin, J.R. Lankard: IEEE J. QE-**9**, 227 (1973)
6.138 J.L. Carlsten, P.C. Dunn: Opt. Commun. **14**, 8 (1975)
6.139 D. Cotter, D.C. Hanna, R. Wyatt: Opt. Commun. **16**, 256 (1976)
6.140 T.F. Deutsch, H. Kildal: IXth International Conference on Quantum Electronics, Amsterdam (1976), paper N3; Opt. Commun. **18**, 124 (1976)
6.141 R.J. Jensen, J.G. Marinuzzi, C.P. Robinson, S.D. Rockwood: Laser Focus, May 1976, p. 51
6.142 R.M. Osgood, Jr.: Appl. Phys. Lett. **28**, 342 (1976)
6.143 J.A. Stregack, T.J. Manuccia, N. Harris, B. Wexler: IXth International Conference on Quantum Electronics, Amsterdam (1976), paper N1; Opt. Commun. **18**, 123 (1976)
6.144 M.I. Buchwald, C.R. Jones, H.R. Fetterman, H.R. Schlossberg: IXth International Conference on Quantum Electronics, Amsterdam (1976), postdeadline paper S. 9
6.145 A.B. Petersen, C. Wittig, S.R. Leone: J. Appl. Phys. **47**, 1051 (1976)
6.146 A.B. Petersen, C. Wittig, S.R. Leone: IXth International Conference on Quantum Electronics, Amsterdam (1976), paper N4; Opt. Commun. **18**, 125 (1976)
6.147 N. Skribanowitz, I.P. Herman, M.S. Feld: Appl. Phys. Lett. **21**, 466 (1972)

6.148 T.Y.Chang, J.D.McGee: Appl. Phys. Lett. **28**, 526 (1976)
6.149 C.K.N.Patel, T.Y.Chang, V.T.Nguyen: Appl. Phys. Lett. **28**, 603 (1976)
6.150 S.F.Dyubko, L.D.Fesenko, O.I.Baskakov, V.A.Svich: Zh. Prikla. Spectroskopii **23**, 317 (1975)
6.151 S.F.Dyubko, V.A.Svich, L.D.Fesenko: Radiofizika **18**, 1434 (1975)
6.152 S.F.Dyubko, V.A.Svich, L.D.Fesenko: Zh. Tekh. Fiz. **45**, 2458 (1975)
6.153 Z.Drozdowicz, R.J.Temkin, K.J.Button, D.R.Cohn: Appl. Phys. Lett. **28**, 328 (1976)
6.154 F.Brown, P.D.Hislop, S.R.Kronheim: Appl. Phys. Lett. **28**, 654 (1976)
6.155 A.Semet, N.C.Luhmann,Jr.: Appl. Phys. Lett. **28**, 659 (1976)
6.156 D.E.Evans, L.E.Sharp, W.A.Peebles, G.Talor: IXth International Conference on Quantum Electronics, Amsterdam (1976), paper B2; Opt. Commun. **18**, 9 (1976)
6.157 E.J.Danielewicz, P.D.Coleman: Appl. Opt. **15**, 761 (1976)
6.158 R.A.Wood, N.Brignall, C.R.Pidgeon, F.Al-Berkdar: Infrared Phys. **16**, 201 (1976)
6.159 M.Yamanaka, H.Tsuda, S.Mitani: Opt. Commun. **15**, 426 (1975)
6.160 G.Kramer, C.O.Weiss: Appl. Phys. **10**, 187 (1976)
6.161 H.R.Fetterman, C.D.Parker, P.E.Tannenwald: IXth International Conference on Quantum Electronics, Amsterdam (1976), paper B3; Opt. Commun. **18**, 10 (1976)
6.162 T.A.DeTemple, E.J.Danielewicz: IEEE J. QE-**12**, 40 (1976)
6.163 D.T.Hodges, J.R.Tucker: Appl. Phys. Lett. **27**, 667 (1975)
6.164 J.R.Tucker: Opt. Commun. **16**, 209 (1976)
6.165 R.J.Temkin, D.R.Cohn: Opt. Commun. **16**, 213 (1976)
6.166 M.P.Hacker, R.J.Temkin, B.Lax: IXth International Conference on Quantum Electronics, Amsterdam (1976), post-deadline paper U5; Opt. Commun. **18**, 226 (1976)
6.167 J.O.Henningsen, P.E.Lindelof, B.S.Steingrimsson: Appl. Phys. Lett. **27**, 702 (1975)
6.168 T.K.Plant, T.A.DeTemple: J. Appl. Phys. **47**, 3042 (1976)
6.169 T.S.Hartwick, D.T.Hodges, D.H.Barker, F.B.Foote: Appl. Opt. **15**, 1919 (1976)

Additional References with Titles

G. B. Abdullaev, L. A. Kulevskii, P. V. Nikles, A. M. Prokhorov, A. D. Savel'ev, E. Yu. Salaev, V. V. Smirnov: Difference-frequency generation in a GaSe crystal with continuous tuning in the 560–1050 cm^{-1} range. Sov. J. Quant. Electron. **6**, 88 (1976)

M. F. Becker, M. H. Kang, K. M. Chung: Third-harmonic generation in SF_6 at 10.6 μm. Opt. Commun. **18**, 147 (1976)

M. I. Buchwald, C. R. Jones, H. R. Fetterman, H. R. Schlossberg: Direct optically pumped multiwavelength CO_2 laser. Opt. Commun. **18**, 209 (1976)

R. L. Byer: A 16-μm source for laser isotope enrichment. IEEE J. Quant. Electron. QE-**12**, 732 (1976)

T. Y. Chang, J. D. McGee: Off-resonant infrared laser action in NH_3 and C_2H_4 without population inversion. Appl. Phys. Lett. **29**, 725 (1976)

T. Y. Chang, J. D. McGee, V. T. Nguyen, C. K. N. Patel: A 13.9 to 16.6-μm spin-flip raman laser pumped by a new optically pumped NH_3 laser at 12.81 μm. Opt. Commun. **18**, 208 (1976)

T. Y. Chang, J. D. McGee, O. R. Wood: Continuous tuning of a single laser mode over 5 cm^{-1} in a high-pressure N_2O/CO_2 transfer laser. Opt. Commun. **18**, 279 (1976)

D. Cotter, D. C. Hanna, R. Wyatt: Infrared stimulated raman generation: Effects of gain focusing on threshold tuning behavior. Appl. Phys. **8**, 333 (1975)

D. Cotter, D. C. Hanna, R. Wyatt: A high-power widely tunable infrared source based on stimulated electronic raman scattering in Cs vapor. Opt. Commun. **16**, 256 (1976)

R. Frey, F. Pardire, J. Ducuing: Tunable infrared generation by resonant 4-wave interaction in H_2. Opt. Commun. **18**, 204 (1976)

A. Z. Grasiuk, I. G. Zubarev, A. V. Kotov, S. I. Mikhailov, V. G. Smirnov: High-power tunable IR raman laser on compressed H_2 and its application. Opt. Commun. **18**, 211 (1976)

A. Z. Grasiuk, I. G. Zubarev, A. V. Kotov, S. I. Mikhailov, V. G. Smirnov: Tunable compressed hydrogen infrared raman laser. Sov. J. Quant. Electron. **6**, 568 (1976)

D. Grischkowsky, P. P. Sorokin, J. R. Lankard: An atomic 16 micron laser. Opt. Commun. **18**, 205 (1976)

D. C. Hanna, D. Cotter, R. Wyatt: Generation of tunable infrared radiation by stimulated Raman scattering in alkali vapors. Opt. Commun. **18**, 112 (1976)

L. O. Hocker, C. F. Dewey: Difference-frequency generation in pronstite from 11 to 23 μm. Appl. Phys. **11**, 137 (1976)

H. Kildel, T. F. Deutsch: Infrared third-harmonic generation in molecular gases. IEEE J. Quant.Electron.QE-**12**,429(1976)

H. Kildel, T. F. Deutsch: Infrared third-harmonic generation in molecular gases. Opt. Commun. **18**, 146 (1976)

G. Kramer, C. O. Weiss: Frequencies of sum optically pumped submillimeter laser lines. Appl. Phys. **11**, 187 (1976)

N. Lee, R. L. Aggarwal, B. Lax: Generation of 12 μm radiation by difference-frequency mixing of CO_2 laser radiation in GaAs. Appl. Phys. Lett. **29**, 45 (1976)

N. Lee, B. Lax, R. L. Aggarwal: High-power far-infrared generation in GaAs. Opt. Commun. **18**, 50 (1976)

A. Mooradian: Recent advances in tunable lasers. Sov. J. Quant. Electron. **6**, 420 (1976)

C. A. Moore, L. S. Goldberg: Tunable uv and IR picosecond pulse generation by nonlinear mixing using a synchronous mode-locked dye laser. Opt. Commun. **16**, 21 (1976)

V.T. Nguyen, E.G. Burkhardt: CW tunable laser sideband generation from 5.5 to 6.5 μm by light scattering from spin motion in a spin-flip Raman laser. Appl. Phys. Lett. **28**, 187 (1976)

C.K.N. Patel, T.Y. Chang, V.T. Nguyen: Spin-flip Raman laser at wavelengths up to 16.8 μm. Appl. Phys. Lett. **28**, 603 (1976)

M.J. Pellin, J.T. Yardley: Resonance enhancement of infrared four wave mixing by infrared active molecules. Appl. Phys. Lett. **29**, 304 (1976)

M. Rosenbluth, R.J. Temkin, K.J. Button: Submillimeter laser wavelength tables. Appl. Opt. **15**, 2635 (1976)

E.D. Shaw: Far IR generation from a spin-flip laser. Appl. Phys. Lett. **29**, 28 (1976)

A.A. Vedenov, G.D. Myl'nikov, D.N. Sobolenko: Conversion of CO_2 laser radiation to far infrared. Sov. J. Quant. Electron. **6**, 424 (1976)

B. Walker, G.W. Chantry, D.G. Moss, C.C. Bradley: The behavior of a 10.6 μm pulsed spin-flip Raman laser as a spectroscopic source. Brit. J. Appl. Phys. (J. Phys. D.) **9**, 1501 (1976)

R.B. Weisman, S.A. Rice: Tunable infrared ultrashort pulses from a mode-locked parametric oscillator. Opt. Commun. **19**, 28 (1976)

Subject Index

Springer Series in Optical Sciences

Editor: D. L. MacAdam

Vol. 1 W. Koechner

Solid-State Laser Engineering

36 figures, XI, 620 pages. 1976. ISBN 3-540-90167-1

This book written from an industrial vatange point provides a description of the engineering aspects of solid-state laser construction and operation. In contrast to most books on lasers, which are mainly concerned with an introduction of physical or quantum mechanical principles of laser action or with the application of laser systems, *Solid-State Laser Engineering* deals principally with the design and fabrication of lasers and laser systems. After introducing the basic concepts of optical amplification, the author discusses laser oscillators and amplifiers with particular reference to analytical models and measurable quantities. — In the subsequent chapters a detailed account of the various components of solid-state lasers is presented and materials, systems and performance parameters of the most important solid-state lasers are discussed.

Vol. 2 R. Beck, W. Englisch, K. Gürs

Table of Laser Lines in Gases and Vapors

IV, 130 pages. 1976. ISBN 3-540-07808-8

Vol. 3 **Tunable Lasers and Applications**

Proceedings of the Loen Conference, Norway, 1976.
Editors: A. Mooradian, T. Jaeger, P. Stokseth
238 figures, VIII, 404 pages. 1976. ISBN 3-540-07968-8

Contents: Tunable and High Energy UV-Visible Lasers. — Tunable IR Laser Systems. — Isotope Separation and Laser Driven Chemical Reactions. — Nonlinear Excitation of Molecules. — Laser Photokinetics. — Atmospheric Photochemistry and Diagnostics Photobiology. — Spectroscopic Applications of Tunable Lasers.

Vol. 4 V. S. Letokhov, V. P. Chebotayev

Nonlinear Laser Spectroscopy

193 figures, 22 tables, approx. 500 pages. 1977. ISBN 3-540-08044-9. In preparation

Contents: Introduction. — Elements of the Theory of Resonant Interaction of a Laser Field and Gas. — Narrow Saturation Resonances on Doppler-Broadened Transition. — Narrow Resonance of Two-Photon Transitions Without Doppler Broadening. — Nonlinear Resonances of Coupled Doppler-Broadened Transitions. — Narrow Nonlinear Resonances in Spectroscopy. — Nonlinear Atomic-Laser Spectroscopy. — Nonlinear Molecular-Laser Spectroscopy. — Nonlinear Narrow Resonances in Quantum Electronics. — Narrow Nonlinear Resonances in Experimental Physics. — References. — Subject Index.

Springer-Verlag Berlin Heidelberg New York